D0883647

Mathematics of Optimization: How to do Things Faster

Pure and Applied
UNDERGRADUATE TEXTS · 30

The Sally
SERIES

Mathematics of Optimization: How to do Things Faster

Steven J. Miller

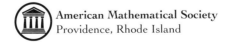

American Mathematical Society
Providence, Rhode Island

EDITORIAL COMMITTEE

Gerald B. Folland (Chair)   Steven J. Miller
Jamie Pommersheim   Serge Tabachnikov

2010 *Mathematics Subject Classification.* Primary 46N10, 65K10, 90C05, 97M40, 58C30, 11Y16, 68Q25.

For additional information and updates on this book, visit
www.ams.org/bookpages/amstext-30

Library of Congress Cataloging-in-Publication Data

Names: Miller, Steven J., 1974- author.
Title: Mathematics of optimization : how to do things faster / Steven J. Miller.
Description: Providence, Rhode Island : American Mathematical Society, [2017] — Series: Pure
    and applied undergraduate texts ; volume 30 | Includes bibliographical references and index.
Identifiers: LCCN 2017029521 | ISBN 9781470441142 (alk. paper)
Subjects: LCSH: Mathematical optimization–Problems, exercises, etc. | Operations research–
    Problems, exercises, etc. | Management science–Problems, exercises, etc. | AMS: Functional
    analysis – Miscellaneous applications of functional analysis – Applications in optimization, con-
    vex analysis, mathematical programming, economics. msc | Numerical analysis – Mathematical
    programming, optimization and variational techniques – Optimization and variational tech-
    niques. msc | Operations research, mathematical programming – Mathematical programming
    – Linear programming. msc | Mathematics education – Mathematical modeling, applications
    of mathematics – Operations research, economics. msc | Global analysis, analysis on manifolds
    – Calculus on manifolds; nonlinear operators – Fixed point theorems on manifolds. msc —
    Number theory – Computational number theory – Algorithms; complexity. msc | Computer
    science – Theory of computing – Analysis of algorithms and problem complexity. msc
Classification: LCC QA402.5 .M5534 2017 | DDC 519.6–dc23 LC record available at https://
lccn.loc.gov/2017029521

Copying and reprinting. Individual readers of this publication, and nonprofit libraries acting
for them, are permitted to make fair use of the material, such as to copy select pages for use
in teaching or research. Permission is granted to quote brief passages from this publication in
reviews, provided the customary acknowledgment of the source is given.

Republication, systematic copying, or multiple reproduction of any material in this publication
is permitted only under license from the American Mathematical Society. Permissions to reuse
portions of AMS publication content are handled by Copyright Clearance Center's RightsLink®
service. For more information, please visit: http://www.ams.org/rightslink.

Send requests for translation rights and licensed reprints to reprint-permission@ams.org.

Excluded from these provisions is material for which the author holds copyright. In such cases,
requests for permission to reuse or reprint material should be addressed directly to the author(s).
Copyright ownership is indicated on the copyright page, or on the lower right-hand corner of the
first page of each article within proceedings volumes.

© 2017 by the American Mathematical Society. All rights reserved.
The American Mathematical Society retains all rights
except those granted to the United States Government.
Printed in the United States of America.

♾ The paper used in this book is acid-free and falls within the guidelines
established to ensure permanence and durability.
Visit the AMS home page at http://www.ams.org/

10 9 8 7 6 5 4 3 2 1   22 21 20 19 18 17

To my three J's
(my brother Jeff, his wife Jackie, and their daughter Justine),
especially to my brother the engineer
for many conversations over the years on efficiency,
as well as a willingness to never grow up
and still play with Lego bricks with me!

Contents

Acknowledgements

This book is the outgrowth of operations research classes taught at Williams College and Mount Holyoke College, where in addition to working through the content of this book, most of the students did applied projects; these classes ranged from senior capstone classes to transition courses in applying mathematics to real world problems. It is a pleasure to thank the students for many helpful discussions related to the course and this book, and for work editing and helping to write many of the chapters, especially (but not limited to) Nathan Anderson, Kushatha Fanikiso, Ross Flieger-Allison, Alex Gonye, Kirby Gordon, Alexander Greaves-Tunnell, Intekhab Hossain, Julio Luquin, Eliza Matt, Connor Mulhall, Ashwin Narayan, Chetan Patel, Seth Perlman, Faraz Rahman, Reidar Riveland, Eric Robinson (who also carefully read large portions of an earlier draft), Thomas Rosal, Jirapat Samranvedhya, Aaditya Sharma, Benjamin Solis-Cohen, Michael Stone, Gregory Szumel, Minh Tran, Anthony Tsao, Rhys Watkins, and Hallee Wong.

I would like to especially thank Gabriel Staton, who transcribed notes from recorded lectures of an earlier version of the class, the team at the AMS (especially Sergei Gelfand and Christine M. Thivierge) for all their help in seeing this project to completion, Wyatt Millstone, who carefully read the entire manuscript in less than a month, catching numerous small mistakes and greatly clarifying the exposition, and my brother Jeff Miller, for many helpful conversations on efficiency and coding. Much of the inspiration for choice of topics and presentation comes from Joel Franklin's excellent book, *Mathematical Methods of Economics: Linear and Nonlinear Programming, Fixed-Point Theorems* [**Fr**]. I picked it up at a used book sale as a college freshman and over the years learned much from it, and I highly recommend it for those wishing a more advanced treatment of many of the topics in this book (especially Linear and Quadratic Programming and Fixed Point Theorems).

Some of this book was inspired by work done with the CCICADA Center at Rutgers, a Department of Homeland Security University Center. The author was partially supported by NSF grants DMS0600848, DMS0970067 and DMS1561945 during the writing of this book, and gratefully acknowledges their support.

Preface

In the first N math classes of your career, you can be misled as to what the world is truly like. How? You're given exact problems and you find exact solutions. The real world is far more complicated, though this leads to interesting mathematics to handle these issues. Often, we cannot solve problems exactly; moreover, the problems we try to solve are themselves merely approximations to the world! We are forced to develop techniques to approximate not just the solution, but even the statement of the problem. Additionally, we often need the solutions quickly; it does us little to no good to have a method that requires years to reach a solution, as businesses cannot afford to wait that long.

This book is meant to serve as an introduction to Optimization Theory, which has a large overlap with Operations Research (OR). It has been used for a senior capstone class, tying together much of the math seen in the undergraduate curriculum, as well as for a transition course on the issues in applying mathematics to the real world. The choice depends on the interests of instructor and students, and leads to different topics being emphasized; we highlight some possibilities for various semester classes in the next section.

While this book touches on some standard problems in OR, our subject matter is deliberately wider, as one of our goals is to showcase commonalities between very different areas of mathematics. Since many problems in OR are related to finding optimal choices, there is a natural connection between these themes, which gives us the freedom to revisit much of the mathematics you may have seen with a different perspective. Thus, large parts of this book are concerned with developing mathematical theory, while other parts deal with implementing it. These two themes build naturally on each other and show how each influences the development of the other.

Operations Research was born (or perhaps it's better to say came of age) as a discipline during the tumultuous events of World War II as a way of efficiently finding optimal or close to optimal solutions. Throughout the chapters below, we constantly re-examine problems you have seen, but now keep in mind the computational cost of our approach. For example, primes play a critical role in modern encryption and e-commerce. While it's trivial to determine if n is prime (simply check to see if any number from 2 to $\sqrt{n}$ divides it), for industrial applications we often deal with numbers as large as or larger than 10^{300}, making such an approach unusable. If we're slightly more clever, we can significantly reduce the run-time from n to on the order of $\sqrt{n}$, since if n factors as xy, then at least one of the two factors is at most $\sqrt{n}$. Sadly, this is still far too slow to be of use, as it requires checking at least 10^{150} numbers. To put that in perspective, the universe is a little less than $5 \cdot 10^{17}$ seconds old, and there are less than 10^{100} particles and photons in it. Thus, if every such object were a supercomputer capable of checking 10^{20} numbers per second (far better than our current computers can do!) and if they had been running since the dawn of time, we would have examined fewer than 10^{140} numbers!

The simple example above highlights the predicament we face. The approach we take is to discuss these issues and develop a good amount of the theory to resolve them. As this is a first course, we are not able to fully prove the efficiency of many modern algorithms; however, we are frequently able to motivate and justify the material by looking at related problems which are easier to analyze and which showcase many of the techniques. We refer the reader to the literature for fuller explanations.

While much of this book concerns linear programming, a powerful method to solve or approximately solve numerous optimization problems, our purpose is *not* to write a linear programming textbook. Those exist and do the job better than we can as they are devoted primarily to that topic. See for example [**Ar, BC, BP, Fr, GKT, Me, Ra, Va**] for a subset of excellent books covering linear programming and its generalizations (among other applied topics in operations research and optimization theory). Our goal is to showcase connections between lots of mathematical areas, while at the same time introducing the issues and difficulties of applying math to problems in the real world. Thus this book is a mix of theory and applications, with numerous options for the instructor to design the course appropriate for their class. In order not to get carried away in digressions, a lot of good, related subjects are relegated to exercises or supplemental material, though many topics that would not typically be mentioned in an operations course will be discussed, as they serve as excellent springboards.

As mentioned above, this book has been used to teach both senior capstone classes as well as introductory and advanced applied mathematics courses. To assist the instructor and students, all the lectures, homeworks and additional comments from the different versions of this course taught at Williams are available online at

    http://web.williams.edu/Mathematics/sjmiller/public_html/377Fa16/
   https://web.williams.edu/Mathematics/sjmiller/public_html/416/index.htm
   https://web.williams.edu/Mathematics/sjmiller/public_html/317/index.htm

(exam questions are also available from the author). Our purposes are to introduce the reader to a variety of important problems and techniques, while at the same time showcasing how material from earlier courses is in fact important! Thus our approach is a bit non-standard, as normally one would not find number theory mixed with linear programming. The advantage of such an approach is that instead of exploring in great detail many algorithms whose analysis requires significant technical machinery, we can gain insights from subjects that are more readily accessible to an undergraduate; for instructors or students with less familiarity in these areas, the online lectures should provide a detailed supplement.

We hope that after reading this book you'll have a new appreciation for the challenges in applying mathematics, as well as the power of theory and computers. The following quote, from Jamie Lesser (one of my shared thesis students at Williams, where she was a Math/CS double major), illustrates this nicely. A key part of her thesis (Optimal Precision Monte Carlo Integration and Application in Computer Graphics) was figuring out how to precisely formulate the problem, which we emphasize consistently in the pages below.

> One of the most important parts of research is learning how to frame a question in a good way, a way that highlights the key features of the problem and allows us to then apply our tools to solve it. This is not easy, and this is often the step that requires the most creativity.

Opportunities for efficiency surround us, and frequently there are easy approaches. In the very first chapter we'll see faster ways to multiply numbers and matrices then you've probably seen before, and their implementation is not significantly harder than what you're used to. Figure 1 gives a wonderful example of how frequently a very simple idea can solve a complicated problem well. The picture is of my two kids, right before we went on the Peter Pan ride at Disney World. People visit to go on rides, not to wait in line. Guests want to plan accordingly but need the right data; in particular, they want to know how long a wait will be so they can determine if it's worth staying in line for a ride or if it's better to move on to something else. Disney has a terrific, low-budget, low-tech solution to constantly check and see if their wait time estimates are accurate. Visitors are periodically given a red card as they enter a line, which is scanned as it's distributed. When they get to the end of the line, the red card is collected and scanned again. This allows the company to collect and compare waiting times, providing simple, almost real-time updates to themselves and visitors.

We end with one final warning, returning to a point we mentioned above. There are two aspects to efficiency: the first is how fast the algorithm is to run or implement, while the second is how close it is to the solution. Sometimes it's better to have a quick algorithm that gets us closer to the solution instead of a slow algorithm that is more accurate but takes too much time to run. Linear programming has great examples of this – frequently the Simplex Method runs fast and we can bound how close we are to the optimal solution. Since the parameters of the problem are often approximations or guesses, the danger in only reaching an approximate optimal solution might be minimal. Thus, one of the goals of this book is to change your perspective; it's usually enough to get close to the answer quickly. I'll leave you with one of my favorite television quotes from my childhood. It's from Disney's *Zorro and Son* television series from 1983. Zorro is old and not

Figure 1. Cameron and Kayla helping management assess wait times at Disney World.

able to defend the people as he did in his youth, and sends for his son to help. He arrives, and the eager father wants to see what the son has learned. The son is happy to oblige, and takes out his sword. His father always drew a $\mathcal{Z}$ in his battles (on enemy soldiers or items on the scene). Holding the blade carefully, and steadying it by using his other hand for support, he slowly and beautifully carves a $\mathcal{Z}$, which he proudly shows to his father. Zorro shakes his head in disgust:

> In this business, you usually don't have time for anything fancy. You've got to get in, make your $\mathcal{Z}$, and get out!

Course Outlines

The prerequisites for this book are calculus and linear algebra. While real analysis is useful, we review most of the terminology and results we need from analysis, using our problems as a motivation for the theory. For classes assuming strong backgrounds in analysis, one can go a bit deeper in the rigor of the proofs. In a one-semester course it should be possible to get through most, but not all, of the material. At Williams all of the students did projects and presentations (including scheduling help for the local schools to working with regional health care providers and writing up theoretical applications such as the proofs of the Four Color Theorem and Kepler's conjecture); if these are not done, then one has additional time to cover a bit more.

Below is a brief summary of the various parts, followed by how they can fit together into different courses.

- Part I reviews many classical problems, discussing approaches that allow us to find answers in significantly less time than the standard methods. Topics in the first chapter include how the way Babylonians multiplied has applications to cell phones and streaming video, how the ability to multiply numbers (and matrices) significantly faster than expected leads to secure crypto-systems and e-commerce, and how results in linear algebra can allow us to efficiently compute any desired Fibonacci number without going through earlier terms in the sequence. In the second chapter, we review some basic combinatorics and discuss connections between Pascal's triangle and chaos. We end with the Euclidean algorithm to quickly compute the greatest common divisor (gcd) of two integers. This is a beautiful example for many reasons. First, there is an obvious, straightforward brute force approach that is trivial to implement but too slow to be practical. The Euclidean algorithm gives a fast, constructive method to find the gcd. Interestingly, a worse-case analysis of the run-time gives an upper bound that is comparable to the horribly inefficient brute force method; however, a slightly more clever counting shows that it runs *significantly* faster than we might have expected. We use this to suggest that many of the algorithms we'll encounter later might also run in far less time than a worst-case analysis indicates; while this is not a proof and there is no direct connection, the success here at least gives us hope.

- Parts II and III form the core of the book. We develop some of the general theory of linear programming, including the Simplex Method to efficiently find solutions, and binary integer programming. We spend a lot of time describing how symmetries can greatly reduce computational complexity and how they may be exploited. Frequently a related problem is easier to study than the original, and we can pass bounds from that problem back to the original. We describe how the Simplex Method works in complete detail. The method of proof is quite easy: we show that there are two steps (called Phase I and Phase II). Assuming we can do Phase II we then show how to do Phase I, and then we use Phase I to prove we can do Phase II; while this looks like a circular argument, there is a nice twist (not mentioned here, as it's saved as a surprise!) that makes this work. In the course of our analysis we show that it suffices to study solutions of a special form (basic feasible and basic optimal solutions); this is an incredibly important idea, as it allows us to trim the search space to a manageable size. We then delve into applications, especially scheduling problems and integer programming, handling multiple objectives and important uses at once, such as solving (or approximating the solution to) the Traveling Salesman Problem. We conclude with a brief introduction to stochastic programming, where we explore how to handle situations when the values of the parameters are unknown.

- We end with Parts IV and V, which are a collection of special topics related to optimization problems. As there is no dearth of possible material here, we have chosen to include a diverse set that highlights different theoretical tools as well as varied applications. These include Fixed Point Theorems, which play a key role in existence theorems for solving differential equations and Nash Equilibrium, the Gale-Shapley algorithm (which has introduced tremendous efficiencies in assignments from matching residents with hospitals to organ donation), interpolating functions (which allow us to recover complicated behavior from the transmission of a small amount of data), and two problems with a rich history of interactions with theory and computation: the Four Color Theorem (four colors suffice so that every planar graph can be colored such that no adjacent regions are colored the same) and the Kepler conjecture (the best way to stack spheres is the grocer's method, where the centers lie on a hexagonal lattice and the spheres in one row lie above the valleys of the row underneath). These last two problems were solved only through enormous work by computers after mathematicians first trimmed the number of cases to explore down to a reasonable (for computers!) level. While we cannot go into full detail for these two, it's worth seeing the key ideas and learning about the issues and challenges in such work; additionally, we can prove weaker results along the way (such as five colors suffice).

Below we give several suggestions for courses. A good analogy is to view most of the book as a trail through an interesting landscape; you'll encounter many loops as you proceed, and these sections are either self-contained or depend only on core material covered earlier.

Senior Capstone Course: Theory Intensive:

- Chapter 1, Efficient Multiplication I, introduces a lot of the perspectives. The last two sections (on Strassen's algorithm, eigenvalues and Fibonacci numbers) can safely be left as readings for the student.

- The only section that must be covered in Chapter 2, Efficient Multiplication II, is §2.5: The Euclidean Algorithm, as this is used several times throughout the book to give heuristics on the types of problems encountered and the quality of possible resolutions. The other sections can be covered quickly and serve as a quick introduction to fractal geometry and more issues in efficiently computing quantities.

- Chapters 3 through 6 form the core of the book and should be completely done. Linear programming is described in detail, culminating in an analysis of the famous Simplex Method. We could have placed the Stochastic Programming part here, but as one part requires integer valued variables we have chosen to include it at the end of the next part, which deals with advanced programming issues. It is possible to do that chapter right after Chapter 6 and use the one bit in §12.4 which requires binary integer variables as motivation for the next part.

- Chapters 7 through 11 provide a quick introduction to Integer Programming (we have described Chapter 12 above). Section 7.6 on Efficient Sorting can be safely omitted (analyzing the run-time of the Simplex Method is beyond a first course, but one can build some intuition on problems such as these by seeing how run-times are analyzed in other settings, as well as how they can vary among methods designed to do the same task). Chapter 8 on Integer Programming and Chapter 9 on Integer Optimization should be done in their entirety. For Chapter 10 (Multi-Objective and Quadratic Programming) it suffices to cover just the first section where one sees how to handle multiple objectives; interested students can read the rest if they want to know a bit more about the challenges with Quadratic Programming. Similarly, for Chapter 11 on the Traveling Salesman Problem only the first section needs to be done; it's criminal to do Linear Programming and be ignorant of this! The rest of the chapter is a brief description of some techniques; this is a vast subject and it can easily be left to students to read.

- Chapters 13 through 16 deal with Fixed Point Theorems and their applications. For instructors wishing a more rigorous course, these are excellent chapters to cover and should be done fully with two possible exceptions. The first is Chapter 14: Contraction Maps. Contractions are not needed in the later chapters, and thus these can safely be skipped; however, it's worth covering at least the first two sections in order to be aware of the subject. The last four sections are applications to solving differential equations. For those who have taken a first course, this is a great opportunity to revisit material, likely at a higher and more theoretical level, as well as review results from real analysis and see their applications. The other place to omit material is the final section of Chapter 16, where Nash Equilibria are discussed. That topic is also independent of the rest of the book, though it gives one of the most famous applications of the theory of fixed points. It should at the very least be mentioned, as it's one of the most important optimization results in the literature (though sadly it only asserts the existence of an optimal strategy and is not constructive).

- Chapters 17 to 20 are independent of each other; in fact, just the first chapter of the book is needed. These are included to give a sense of issues in optimization. As many of the first three (Gale-Shapley algorithm for stable matchings, Interpolating functions, and the Four Color Problem) as can be done should be done. The final chapter, on the Kepler Conjecture, is meant to be a quick introduction to the incredible technical issues that had to be surmounted; related issues are discussed in the Four Color chapter.

Senior Capstone Course: Theory and Applied:

A natural syllabus for this course is similar to the previous, with the following changes:

- Add in the differential equations sections of Chapter 14: Contraction Maps.
- Add in the Nash Equilibrium section of Chapter 16: Brouwer's Fixed Point Theorem.
- Assign student projects and have presentations to the class and write-ups of work. These are terrific skills to build, but depending on the class size they can take upwards of two weeks for everyone to present.
- Cover advanced topics from Chapters 17 through 20 as time and interest permit.

Transition Course: Emphasizing Applications:

This course is similar to the one above. Depending on the strength, preparation and interests of the students, one can remove some of the more advanced theory (such as the proof of Sperner's lemma and the proof of Brouwer's fixed point theorem). This would free up more time for applications and project work and presentations. To free up additional time for applications, one can cover less of Chapter 2: Efficient Multiplication II.

Part 1

Classical Algorithms

Efficient Multiplication, I

1.1. Introduction

Part of the purpose of an introductory course in applied mathematics is to change our vantage on how we view math. For this reason, we'll revisit some things that you've hopefully seen in previous courses, but our goal will be different. In many pure math courses, the focus is on finding a solution. This is of course a good goal – it's nice to be able to solve the problem! Unfortunately, for many real world applications it's not enough to know that we can do something; we need to be able to do it efficiently.

A great example is factorization. Any integer $N \geq 2$ can be written uniquely as a product of prime powers $p_1^{r_1} p_2^{r_2} \cdots p_k^{r_k}$, where $1 < p_1 < p_2 < \cdots < p_k$ (we need to impose this ordering of the primes to get uniqueness, as we don't want to consider $2^2 \cdot 3$, $2 \cdot 3 \cdot 2$ and $3 \cdot 2^2$ as different factorizations of 12). The proof that such a factorization exists and is unique is fairly straightforward; we leave finding it and analyzing run-times to the reader in Exercises 1.7.1 and 1.7.2. Thus while it's easy to give a procedure to factor a number, *finding* that factorization quickly is a different story. Many modern encryption systems are in fact built around this perceived difficulty (remember that just because no one has published a fast way to factor doesn't mean no fast way exists).

So, being able to do something is not enough; we need to be able to do it well. This means we must be able to solve the problem in a reasonable amount of time with a reasonable amount of resources; it's unlikely to expect to have every subatomic particle in the universe replaced with a supercomputer devoted to performing our computations throughout all eternity (which is less than what we would need to make some of the brute force approaches work for numbers used in practice).

While we can sometimes find better algorithms, often there's a trade-off between efficiency and exactness. Fortunately, for many problems we do not need an exact answer, and frequently a close approximation which can be found quickly suffices. This is terrific, as unlike factorization, some problems are not only computationally difficult but deny a closed form solution. For an advanced example, we know from Galois Theory that there's no closed form expression for the roots of polynomials of degree greater than or equal to 5; however, there are iterative methods (Newton's method, discussed in §13.5, is a great one) that can yield solutions accurate to, say, 32 decimal places, within moments. For many problems, 32 digits of accuracy is more than enough (and if we needed more we could run our algorithm longer).

Of course, there are also problems where we can neither find an exact solution nor a good approximation quickly; not surprisingly, we won't focus too much on those in this book. We'll concentrate on problems where we can either quickly find the answer or at least a very good approximation. As a warm-up, we explore fast ways to multiply numbers and the consequences of these improvements over brute force. In addition to having tremendous applications, these problems provide a nice tour to the ideas and approaches we'll see later.

We begin our tour with a very familiar operation: multiplication. This is a natural choice, as it allows us to start with something you've seen numerous times but with the opportunity of highlighting aspects and applications you've probably never seen.

1.2. Babylonian Multiplication

We start our tour of multiplication with the Babylonians. Much of their mathematics is still with us today. Some is apparent in our everyday life and is due to their working in base 60 (most people use base 10, the **decimal** system, though base 2, **binary**, is very useful in computer science). 60 is a great choice, as it is evenly divisible by many different numbers, which is advantageous when you have to divide quantities into groups. To this day, we can see their influence in 60 seconds in a minute, 60 minutes in an hour, 360 degrees in a circle, and so on.

Other aspects, however, are not as well known. An important one is a result of the difficulty of doing mathematics in base 60. In elementary school we learn the multiplication table; if we can just learn the products $d_1 d_2$ with each $d_i \in \{0, 1, \ldots, 9\}$ we can perform any multiplication. In base 60, however, this would require a mastery of $60 \cdot 60 = 3600$ products!

Of course, it isn't quite as bad as that, as multiplication is commutative and this almost cuts in half the products we must memorize: $\frac{1}{2} 60 \cdot 59 + 60 = 1830$; the first factor is the number of products (where order does not matter) of two distinct numbers, and then of course we must include the squares. While this almost cuts our work in half, it's too much to expect the ancient Babylonian children and merchants to memorize almost 2000 products!

Fortunately, there is a better way. They noticed that

$$ a \cdot b \; = \; \frac{(a+b)^2 - (a-b)^2}{4}. $$

At first this looks like a horrible trade, as we're replacing one multiplication with two products, a subtraction and a division! The gain becomes apparent, however, when we notice that the products are both squares. Thus if we just learn our squares, subtraction, and division by 4, we can do any multiplication! Note that there are only 120 squares to learn (remember $a + b$ can get as large as 120). This is tremendous progress – we've gone from needing to learn 1830 products to just 120, a savings of more than an order of magnitude.

The great idea lurking in the background here is that of a **look-up table**. If there is a computation you will need again and again, do it once, store the value, and then recall it as needed. Additionally, we can store a few key values and use those to interpolate the function at other points (see Chapter 18). These savings are essential for many modern applications. For example, cell phones and other similar devices have had a tremendous infiltration in our society; we seem to be constantly online and connected. While the computing power of these devices has steadily increased, we are putting smaller and smaller devices online, and thus efficiencies in computing are not only welcomed but needed. A terrific example is streaming video; we don't want to have to send every pixel every moment!

1.3. Horner's Algorithm

Our Babylonian example introduced one of the great concepts in modern computing: the look-up table. If there is a problem you need to do again and again, it's often a great strategy to do it once, save the result, and recall as needed. What if, however, we have an entirely new computation (or if memory is expensive)? For a nice example, let's consider a polynomial

$$f(x) = a_n x^n + \cdots + a_1 x + a_0.$$

You should know an easy way to find its value for any given x, say $x = 2$; all we need to do is plug in 2 for x above and evaluate. Thus if

$$f(x) = 3x^4 - 11x^3 + 5x - 6$$

we find

$$f(2) = 3 \cdot 2^4 - 11 \cdot 2^3 + 5 \cdot 2 - 6 = -36.$$

In general, how 'expensive' is this procedure? How many operations does it take? While we'll count both additions and multiplications, for large numbers addition is essentially free relative to the cost of multiplication (think of how many digit operations are required to add two 100 digits numbers versus multiplying them). To evaluate our general f at $x = 2$ we would do

$$a_n \cdot 2^n + a_{n-1} \cdot 2^{n-1} + \cdots + a_1 \cdot 2 + a_0.$$

Notice the kth term requires k multiplications ($k - 1$ to compute 2^k and then one more to multiply that by a_k). So the total number of multiplications needed, if we use this brute force approach, is

$$n + (n - 1) + (n - 2) + \cdots + 1 + 0 = \frac{n(n+1)}{2}$$

(see Exercise 1.7.13). Thus the total cost, in multiplications, of evaluating the polynomial at $x = 2$ is on the order of n^2 (or if you prefer, approximately $n^2/2$). See Exercise 1.7.17 for ways to estimate the sum of the first n integers.

Amazingly, we can do significantly better.

Horner's Algorithm: By grouping the terms and rewriting the polynomial $f(x)$ as
$$f(x) \;=\; (\cdots (((a_n x + a_{n-1})\, x + a_{n-2})\, x + a_{n-3}) \cdots + a_1)\, x + a_0,$$
it costs n multiplications and n additions to evaluate $f(x)$.

Example 1.3.1. *If*
$$f(x) \;=\; 2x^4 + 5x^3 + 7x^2 - 2x + 8,$$
then we re-write it as
$$f(x) \;=\; (((2x + 5)x + 7)\, x - 2)\, x + 8$$
and see there are 4 multiplications and 4 additions, a savings over the 10 multiplications and 4 additions needed for the standard approach.

Thus, Horner's Algorithm reduces the multiplications cost of polynomial evaluation to n, while maintaining the additions cost at n; compared to the brute force approach, there is a square-root savings in the number of multiplications required. To give a sense of how significant this improvement is, for a polynomial of order $n = 1000$, this means we need only do 1000 multiplications instead of a massive $1,000,000$. Efficient algorithms like this one have many useful applications; for instance, Horner's Algorithm is very helpful in the study of fractals, which require thousands of iterations of polynomials. Further, this example illustrates a very important principle: *Just because we have done something one way for a long time does not mean this is the best way to do it.*

It seems like we can't do much better than Horner's Algorithm: given that a polynomial of order n has n terms, it seems that we need at least n multiplications to evaluate the expression. While this is true for a general polynomial, perhaps there are faster methods for special polynomials. *You should get in the habit of always asking questions such as this:* **If we make additional assumptions, are there improvements to be had?** We'll see in the next section that the answer is most definitely 'yes', and the resulting fast multiplication has tremendous applications (such as RSA encryption).

1.4. Fast Multiplication

In the last section we found that polynomial evaluation, which at first appeared to require on the order of n^2 multiplications (if it is of degree n), could be done significantly faster with just n products by cleverly grouping. Unfortunately, for a general polynomial we cannot do better, but perhaps we can for special polynomials. When looking for special cases to investigate, start simple. The simplest degree n polynomial is $f(x) = x^n$. We can write this as
$$f(x) \;=\; (\cdots (((x \cdot x) \cdot x) \cdot x) \cdot x) \cdots x$$
and see that in this case brute force and Horner's Algorithm give the same answer: $n - 1$ multiplications.

Interestingly, we can do significantly better by again cleverly grouping the products. We use the binary expansion of n to save on the required number of multiplications. We'll first illustrate the method with an example. Note that a subscript of 2 after a number whose digits are just 0's and 1's means that it's written in binary; thus 1101_2 is $1 \cdot 2^3 + 1 \cdot 2^2 + 0 \cdot 2^1 + 1 \cdot 2^0$, or 13.

Example 1.4.1. *Let* $f(x) = x^{100}$. *We write 100 in binary:*

$$100 \;=\; 1 \cdot 2^6 + 1 \cdot 2^5 + 0 \cdot 2^4 + 0 \cdot 2^3 + 1 \cdot 2^2 + 0 \cdot 2 + 0 \cdot 1 \;=\; 1100100_2.$$

We now repeatedly square x, *saving the results, and then multiply the powers which correspond to a non-zero digit in the binary expansion:*

$$
\begin{aligned}
x \cdot x &= x^2 \\
x^2 \cdot x^2 &= x^4 \\
x^4 \cdot x^4 &= x^8 \\
x^8 \cdot x^8 &= x^{16} \\
x^{16} \cdot x^{16} &= x^{32} \\
x^{32} \cdot x^{32} &= x^{64}
\end{aligned}
\qquad\qquad
x^{100} \;=\; x^{64} \cdot x^{32} \cdot x^4.
$$

There is of course nothing special about 100.

Fast Multiplication: To compute x^n, write n in binary. Repeatedly square $(x, x^2, x^4, x^8, \ldots)$, save the results, and then multiply the terms corresponding to powers present in the binary expansion of n.

Is this better than brute force? Yes, as far as the number of required multiplications is concerned. For our example with $n = 100$ we now only need to perform 8 multiplications, as opposed to 99 for brute force; in Exercise 1.7.26 you'll show that you need at most $2 \log_2(n)$ products.

Unfortunately, this method does require something that brute force does not: we need to store the products from the repeated squaring. Fortunately, it's easy to save on the storage requirements as well. Let's revisit our example from before. Start with the pair $(1, x)$, and square the second element in the pair. Whenever this yields a power of x that is in the binary expansion of x^{100}, multiply the first element by this power of x. Repeating this process goes as follows:

$$(1, x) \to (1, x^2) \to (1, x^4) \to (x^4, x^4) \to (x^4, x^8) \to (x^4, x^{16})$$
$$\to (x^4, x^{32}) \to (x^{36}, x^{32}) \to (x^{36}, x^{64}) \to (x^{100}, x^{64}).$$

We still get x^{100} in only 8 multiplications (if you don't count the $1 \cdot x^4$ that we used to update our first component), but now we only require storing 1 additional value compared to brute force and the binary expansion of n.

This problem illustrates many of the issues that arise in implementing mathematics: efficiency, storage capacity, and even overflow (see Exercise 1.7.23). Fast multiplication is one of the key ingredients in many modern encryption schemes; without fast and secure methods to encode and decode information, e-commerce as we know it could not exist. We describe one of the most used systems, RSA, in a series of problems beginning with Exercise 2.6.81.

1.5. Strassen's Algorithm

The repeated squaring algorithm from the last section is amazing and probably a bit surprising the first time you see it. It allows us to replace an order N process with one that is order $\log N$ (see Exercise 1.7.29). The real lesson here is that while you may know a way to do something, it can be well worth the time to investigate and see if there is a better way.

A terrific example, similar in spirit to what we've done, is in matrix multiplication. You should hopefully have seen the importance of matrix multiplication (if not, we provide one of many examples in §1.6). How expensive is the brute force approach we've been taught?

While for many applications we're often interested in high powers of a fixed matrix, let's start by looking a bit more generally (similar to Horner's method). Given matrices $\mathbf{A}$ and $\mathbf{B}$, the entry in the ith row and j^{th} column of $\mathbf{AB}$ is the dot product of the ith row of $\mathbf{A}$ and the jth column of $\mathbf{B}$. For two 2×2 matrices we have

$$\mathbf{AB} = \begin{pmatrix} a_{11} & a_{12} \\ a_{21} & a_{22} \end{pmatrix} \begin{pmatrix} b_{11} & b_{12} \\ b_{21} & b_{22} \end{pmatrix} = \begin{pmatrix} a_{11}b_{11} + a_{12}b_{21} & a_{11}b_{21} + a_{12}b_{22} \\ a_{21}b_{11} + a_{22}b_{21} & a_{21}b_{12} + a_{22}b_{22} \end{pmatrix}.$$

Notice this requires 8 multiplications and 4 additions. More generally,

$$\begin{pmatrix} a_{11} & a_{12} & \cdots & a_{1n} \\ \vdots & \vdots & \ddots & \vdots \\ a_{i1} & a_{i2} & \cdots & a_{in} \\ \vdots & \vdots & \ddots & \vdots \\ a_{n1} & a_{n2} & \cdots & a_{nn} \end{pmatrix} \begin{pmatrix} b_{11} & \cdots & b_{1j} & \cdots & b_{1n} \\ b_{21} & \cdots & b_{2j} & \cdots & b_{2n} \\ \vdots & \ddots & \vdots & \ddots & \vdots \\ b_{n1} & \cdots & b_{nj} & \cdots & b_{nn} \end{pmatrix} = \begin{pmatrix} c_{11} & c_{12} & \cdots & \cdots & c_{1n} \\ c_{21} & \ddots & \cdots & \cdots & \vdots \\ \vdots & \vdots & \mathbf{c_{ij}} & \vdots & \vdots \\ \vdots & \vdots & \vdots & \ddots & \vdots \\ c_{n1} & c_{n2} & \cdots & \cdots & c_{nn} \end{pmatrix}$$

with

$$c_{ij} = a_{i1}b_{1j} + a_{i2}b_{2j} + \cdots + a_{in}b_{nj}.$$

Each c_{ij} costs n multiplications and $n-1$ additions; this means the cost of finding C is n^3 multiplications and $n^3 - n^2$ additions.

Thus, if we use brute force, matrix multiplication (of two $n \times n$ matrices) is an order n^3 process; as we can have matrices with more than 10,000 rows, *any* savings on the exponent 3 in the n^3 is highly desirable. The difficulty is that we are finding n^2 dot products, each costing n multiplications. If we wanted a specific c_{ij} we would be out of luck; however, we want *all* of them, and this suggests a way out. Rather than computing each c_{ij} one at a time, perhaps there are other helpful quantities we can compute first, and then use to efficiently find the c_{ij}. This turns out to be possible; the first such approach is due to Strassen, which allows us to do matrix multiplication in order $n^{\log_2 7}$ steps.

Essentially what happens is we save one multiplication. As $\log_2 8 = 3$ and $\log_2 7 \approx 2.8073$, for a matrix with 10,000 rows the brute force would require $1.0 \cdot 10^{12}$ multiplications, while the Strassen algorithm needs about $1.7 \cdot 10^{11}$. Thus we save almost a factor of 6, and the savings only get larger as we increase n (we save a little over a factor of 200 for matrices with a trillion rows)!

Strassen's Algorithm (for $n = 2$**):** Given 2×2 matrices $\mathbf{A} = (a_{ij})$ and $\mathbf{B} = (b_{ij})$, compute the following seven quantities:

$$
\begin{aligned}
m_1 &= (a_{11} + a_{12})(b_{11} + b_{22}) \\
m_2 &= (a_{21} + a_{22})b_{11} \\
m_3 &= a_{11}(b_{12} - b_{22}) \\
m_4 &= a_{22}(b_{21} - b_{11}) \\
m_5 &= (a_{11} + a_{12})b_{22} \\
m_6 &= (a_{21} - a_{11})(b_{11} + b_{12}) \\
m_7 &= (a_{12} - a_{22})(b_{21} + b_{22}).
\end{aligned}
$$

Then $\mathbf{C} = \mathbf{AB}$ can be computed by

$$
\begin{aligned}
c_{11} &= m_1 + m_4 - m_5 + m_7 \\
c_{12} &= m_3 + m_5 \\
c_{21} &= m_2 + m_4 \\
c_{22} &= m_1 - m_2 + m_5 + m_6,
\end{aligned}
$$

which costs 7 multiplications and 18 additions (as opposed to the naive algorithm's 8 multiplications and 4 additions).

While we only discussed the case of two 2×2 matrices, Strassen's algorithm can be extended to arbitrary square matrices. Not surprisingly, it's easier to discuss when n is a power of 2. We split 4×4 blocks into 2×2 blocks, 8×8 blocks into 4×4 blocks, and so on. In these cases, the cost in multiplications of finding the product of two matrices is taken from $n^3 = n^{\log_2 8}$ to $n^{\log_2 7}$, while increasing the required number of additions. While this improvement may seem underwhelming, this was in fact one of the biggest advances in Linear Algebra of the 20th century: when n becomes very large, the savings on computational efficiency are enormous (see Exercise 1.7.31).

Finally, we should note that this is a problem for which it's easy to check that a potential solution works, but very hard to find the solution (the correct set of seven multiplications) in the first place. Sadly, this type of problem is common throughout mathematics.

1.6. Eigenvalues, Eigenvectors and the Fibonacci Numbers

We end our tour of multiplication with a nice example of why we care so much about it. One of the most studied sequences in mathematics is the **Fibonacci numbers** $\{F_n\}$, where $F_0 = 0$, $F_1 = 1$, and for $n > 0$ we set $F_{n+1} = F_n + F_{n-1}$. If this were theoretical mathematics, we might not be very concerned with any difficulties in calculating the values of F_n for given n, for the recursion formula gives us a simple procedure to march down the line and compute any desired term. In practice, however, this can be quite difficult to implement. For example, if we want the trillionth Fibonacci number we would need to compute all the previous ones. Fortunately, it turns out that we can use matrices to jump to the desired term.

Let $\vec{v}_n = \begin{pmatrix} F_n \\ F_{n-1} \end{pmatrix}$, so $\vec{v}_{n+1} = \begin{pmatrix} F_{n+1} \\ F_n \end{pmatrix}$. We can represent this using a matrix equation:

$$\vec{v}_{n+1} \; = \; \begin{pmatrix} 1 & 1 \\ 1 & 0 \end{pmatrix} \vec{v}_n.$$

Indeed, we have

$$\begin{pmatrix} 1 & 1 \\ 1 & 0 \end{pmatrix} \vec{v}_n = \begin{pmatrix} 1 & 1 \\ 1 & 0 \end{pmatrix} \begin{pmatrix} F_n \\ F_{n-1} \end{pmatrix} = \begin{pmatrix} F_n + F_{n-1} \\ F_n \end{pmatrix} = \begin{pmatrix} F_{n+1} \\ F_n \end{pmatrix} = \vec{v}_{n+1}.$$

Therefore, if we now let $\mathbf{A} = \begin{pmatrix} 1 & 1 \\ 1 & 0 \end{pmatrix}$, we find

$$\vec{v}_{n+1} \; = \; \mathbf{A}\vec{v}_n \; = \; \mathbf{A}^2 \vec{v}_{n-1} \; = \; \cdots \; = \; \mathbf{A}^n \vec{v}_1.$$

Thus, if we can compute high powers of a matrix quickly, we can jump ahead to whatever Fibonacci number we want. See Exercise 1.7.41 for another example.

Remark 1.6.1. *Whenever you see a problem in mathematics, you should look for generalizations or special cases where you have intuition. Here we are taking large powers of a matrix; this should recall our earlier investigations into large powers of a number. We saw an enormous advantage in writing the binary expansion of the exponent; we can do that here as well! Thus to compute A^{100} we would do $\boldsymbol{A}^{64}\boldsymbol{A}^{32}\boldsymbol{A}^{4}$.*

We conclude this chapter with one final remark: often one does not know how a math result will be useful, but only that there is a chance it may be so. Just because we can multiply matrices relatively quickly now does not mean we should proceed this way; perhaps there is another approach using more advanced ideas.

Given a matrix $\mathbf{A}$, $\mathbf{A}\vec{v}$ is typically a new vector with a different length and direction than $\vec{v}$; if, however, $\mathbf{A}\vec{v}$ is in the same direction as $\vec{v}$ (and is not the zero vector), we say $\vec{v}$ is an **eigenvector** of $\mathbf{A}$. We can thus write $\mathbf{A}\vec{v} = \lambda\vec{v}$ for some $\lambda \in \mathbb{C}$, and call λ the **eigenvalue** associated to $\vec{v}$; note λ could be zero, but $\vec{v} \neq \vec{0}$. The advantage of eigenvectors is that matrix multiplication just rescales their length, and we can thus compute high powers quickly.

For example, the Spectral Theorem asserts that if $\mathbf{A}$ is a real-symmetric matrix (the result holds more generally), then there is an orthonormal basis of eigenvectors. This means that we have eigenvectors $\vec{v}_i$ with eigenvalues λ_i such that given any $\vec{v}$ we have

$$\vec{v} \; = \; c_1\vec{v}_1 + \cdots + c_n\vec{v}_n,$$

and thus

$$\mathbf{A}^m\vec{v} \; = \; c_1\lambda_1^m\vec{v}_1 + \cdots + c_n\lambda_n^m\vec{v}_n.$$

The difficulty, of course, is in finding the eigenvalues and eigenvectors of the matrix. We know from Linear Algebra that we can find the eigenvalues of an $n \times n$

matrix $\mathbf{A}$ by finding the solutions to the equation

$$\det(\mathbf{A} - \lambda \mathbf{I}_n) = 0,$$

where $\mathbf{I}_n$ is the $n \times n$ identity matrix. As an aside, if you've never been warned about trying to use this approach for large matrices, consider yourself warned! There are stability issues, and once $n \geq 5$ we no longer have closed form expressions for the roots.

Fortunately, there are many situations in which eigenvalues and eigenvectors are easy to find. For instance, let's revisit the Fibonacci sequence. By finding the eigenvalues, eigenvectors, and matching the initial conditions, in Exercise 1.7.38 you will prove **Binet's formula** for the nth Fibonacci number:

$$F_n = \frac{1}{\sqrt{5}} \left(\frac{1 + \sqrt{5}}{2} \right)^n - \frac{1}{\sqrt{5}} \left(\frac{1 - \sqrt{5}}{2} \right)^n.$$

This expression is quite surprising: even though it's filled with fractions and irrational numbers, it always yields an integer result given a natural number n. Further, notice how instead of finding a large power of a 2×2 matrix we need only compute high powers of two real numbers, another great savings.

1.7. Exercises

Introduction:

Exercise 1.7.1. *Find an algorithm to factor an integer $N \geq 2$ that requires on the order of N divisions. Can you do $N/2$? What about $N/3$? What about N^δ for some $\delta < 1$?*

Exercise 1.7.2. *In the previous problem you hopefully came up with several approaches to factor a number. It's important to be able to compare them and see how much of an improvement the observations make. First assume that any division costs the same to perform as any other; how many divisions do each of your algorithms require? Now assume that the cost of a division depends on the number of digit multiplications (to keep things simple we'll assume there is zero cost for addition and carrying, though you are welcome to estimate the effects of these). Thus as 41446301 divided by 1701 is 24601, the cost of the division would be 20 digit multiplications. Now how expensive are the methods?*

Exercise 1.7.3. *In fourth grade my son had a homework assignment requiring him to count how often each letter of the alphabet appears in three sentences. While we did this with a computer program, imagine you have to do it by hand. Which is a better method:*

- *read letter by letter, and as you read increase the running tally for that letter, or*
- *read the entire sentence from start to finish and count the number of a's, then read from start to finish to count the number of b's, and so on*

(or will the two methods take the same amount of time)? Try this for several sentences, and if the two do not take the same amount of time see which factors affect which wins. Are there other approaches you could do by hand?

Babylonian Multiplication:

Exercise 1.7.4. *Another way to do multiplication is to note*

$$a \cdot b = \frac{(a+b)^2 - a^2 - b^2}{2};$$

how does this compare to the Babylonian method?

Exercise 1.7.5. *If you were a Babylonian, how would you efficiently compute* $a \cdot b \cdot c$ *and* $a \cdot b \cdot c \cdot d$?

There are many ways to create and use a look-up table. If we only have a small number of possibilities, as we did with multiplying base 60, we can just do an exhaustive enumeration. Most of the time, however, it's not possible to pre-compute every possibility, and we must somehow interpolate between entries. The next few problems explore some methods to do just that. An important part of the problems is to determine how to measure errors; for example, an error of size .01 is much worse if the values are around .2 than if they are around 20.

Exercise 1.7.6. *If we know the values of either* $\sin(x)$ *or* $\cos(x)$ *for* $0 \le x \le \pi/4$ *(or from 0 to 45 degrees if you prefer not to work in radians), then we can find the values of all trig functions using basic relations (such as* $\sin(x + \pi/2) = \cos(x)$*). Create a look-up table of* $\sin(x)$ *by finding its values for* $x = \frac{k}{45}\frac{\pi}{4}$ *with* $k \in \{0, 1, \dots, 45\}$. *Come up with at least two different ways to interpolate values of* $\sin(x)$ *for* x *not on your list, and compare their accuracies. For which values of* x *are your interpolations most accurate?*

A **continuous probability distribution** (or **density**) f is a non-negative function that integrates to 1; the area under f from a to b is the probability that we observe a value between a and b.

Exercise 1.7.7. *One of the most important densities in probability theory is the standard normal:*

$$f(x) = \frac{1}{\sqrt{2\pi}} e^{-x^2/2}.$$

The corresponding cumulative distribution function, $F(x)$, *gives the probability of obtaining a value at most* x:

$$F(x) = \int_{-\infty}^{x} f(t) dt;$$

note $F' = f$. *Unfortunately there is no simple closed form expression for the anti-derivative. We could try to find a series expansion for it by expanding the exponential; unfortunately, as one of the bounds is negative infinity we cannot integrate term by term. We are saved by noting that* $F(0) = 1/2$ *and thus*

$$F(x) = \frac{1}{2} + \int_{0}^{x} \frac{1}{\sqrt{2\pi}} e^{-t^2/2} dt;$$

the convention is to define the **error function** *(or* **erf***) by*

$$\text{erf}(x) = \frac{2}{\sqrt{\pi}} \int_{0}^{x} e^{-u^2} du = 2 \int_{0}^{x\sqrt{2}} \frac{1}{\sqrt{2\pi}} e^{-t^2/2} dt.$$

Create a look-up table for $F(x)$ for $x \in [0,5]$ (say in steps of size .1). Come up with at least two different ways to interpolate values of $\mathrm{erf}(x)$ for x not on your list, and compare their accuracies. For which values of x are your interpolations most accurate?

Exercise 1.7.8. *Imagine you want to create a look-up table to find square-roots of numbers from 0 to 100. If you could only have 10 entries in your table, what would you choose and why? What if you could have 20 values?*

You should have a great familiarity with **logarithms**, but as their uses extend to (and far beyond) this course, as well as provide an excellent example of look-up tables in action, we'll quickly go over their basics. We say $\log_b x = y$ if and only if $x = b^y$; note log is always the logarithm base e. We have the following, familiar **laws of logarithms**:

(1) $\log_b x_1 + \log_b x_2 = \log_b(x_1 x_2)$,

(2) $\log_b(x^r) = r \log_b x$,

(3) $b^{\log_b x} = x$,

(4) $\log_b x_1 - \log_b x_2 = \log_b(\frac{x_1}{x_2})$,

(5) $\log_b x = \frac{\log_c x}{\log_c b}$, the oft-forgotten change of basis formula!

Exercise 1.7.9. *Prove the five log laws. For example, for the first we have $\log_b x_i = y_i$, so $x_i = b^{y_i}$. Thus $x_1 x_2 = b^{y_1} b^{y_2} = b^{y_1+y_2}$. By definition, we now get $\log_b(x_1 x_2) = y_1 + y_2$, which finally yields $\log_b(x_1 x_2) = \log_b x_1 + \log_b x_2$.*

The change of basis formula is extremely useful because it allows us to compute logarithms in any base so long as we know how to compute them in a single base. Before we had calculators, the ability to do this was crucial since all values of functions had to be tabulated manually: **look-up tables**, which store pre-computed values, were vital in increasing efficiency. The change of basis formula meant that we only needed a single look-up table, for a single basis value, in order to compute logarithms of any kind, which allowed for immense savings in terms of storage, time, and efficiency.

Exercise 1.7.10. *Prove $\log(e^x) = x$ and $e^{\log x} = x$; thus the exponential and logarithm functions are inverses. This allows us to pass from the derivative of one to the derivative of the other via the Chain Rule: if $f(g(x)) = x$, then $f'(g(x))g'(x) = 1$. Using the derivative of e^x is e^x find the derivative of $\log x$, as well as the derivatives of b^x and $\log_b x$ for a fixed $b > 1$.*

Horner's Algorithm:

Exercise 1.7.11. *Consider a polynomial $f(x) = a_n x^n + \cdots + a_1 x + a_0$, and assume $a_i = 0$ whenever i is even. How can you modify Horner's Algorithm to efficiently compute $f(x)$?*

Exercise 1.7.12. *How would you generalize Horner's Algorithm to a polynomial in two variables? In d variables? For example, how would you efficiently compute*

$$f(x, y) = \sum_{m=0}^{M} \sum_{n=0}^{n} a_{mn} x^m y^n$$

at $(x, y) = (a, b)$?

Exercise 1.7.13. *Show that*

$$0 + 1 + 2 + \cdots + (n-1) + n = \frac{n(n+1)}{2}.$$

Hint: one way is to proceed by induction; another way is to write the numbers in reverse order, add, and then divide by 2.

Exercise 1.7.14. *Generalize the previous problem and find a formula for the sum of the squares of integers up to n. Find a formula for the sum of cubes.*

Exercise 1.7.15. *In the last two problems you should have noticed that your answer is always a polynomial, and perhaps you also noticed something interesting about the exponent of the highest degree term as well as its coefficient. Make some conjectures, test with other power sums, and try to prove. For more on this topic look up Bernoulli polynomials.*

Exercise 1.7.16. *Here is another way to get the result from Exercise 1.7.13. Let $S_k(n) = 1^k + 2^k + \cdots + n^k$. Then*

$$(n+1)^2 = S_2(n+1) - S_2(n) = \sum_{m=0}^{n} \left((m+1)^2 - m^2\right) = \sum_{m=0}^{n} (2m + 1).$$

Complete the argument. Can you generalize this approach and get a formula for $S_3(n)$ or $S_4(n)$?

Exercise 1.7.17. *In Exercise 1.7.13 you proved that the sum of the first n integers is $n(n+1)/2$, which is of approximately $n^2/2$. For many problems the actual value doesn't matter; all we need is a good approximation. The following provides a way to sketch that, in some sense, this sum grows at a similar rate as cn^2 for some constant c:*

- *Show that the sum is at most n^2.*
- *Show that the sum is at least $(n/2) \cdot (n/2)$ if n is even (if n is odd show it's at least $\frac{n-1}{2} \frac{n-1}{2}$).*

The previous problem shows that $1 + 2 + \cdots + n$ is at least $n^2/4$ and at most n^2; the actual answer is about $n^2/2$. Interestingly one obtains the correct approximation by taking the **geometric mean** of the two numbers, where the geometric mean of k numbers is $(a_1 a_2 \cdots a_k)^{1/k}$; this is different than the **arithmetic mean**, which would be $\frac{1}{k}(a_1 + a_2 + \cdots + a_k)$.

Exercise 1.7.18. *Let a_1 and a_2 be two positive numbers. Prove that their arithmetic mean is never smaller than their geometric mean. When are the two equal?*

Exercise 1.7.19. *More generally, given $a_1, \ldots, a_k > 0$ prove that their arithmetic mean is never smaller than their geometric mean. When are the two equal?*

Exercise 1.7.20. *Given $a_1, \ldots, a_n > 0$, define the kth symmetric mean $M_k(a_1, \ldots, a_n)$ by*

$$M_k(a_1, \ldots, a_n) = \left(\sum_{1 \leq i_1 < i_2 < \cdots < i_k \leq n} \frac{a_{i_1} a_{i_2} \cdots a_{i_k}}{\binom{n}{k}} \right)^{1/k};$$

*note $M_1(a_1, \ldots, a_n)$ is the arithmetic mean and $M_n(a_1, \ldots, a_n)$ is the geometric mean. Prove **Maclaurin's inequality**:*

$$M_1(a_1, \ldots, a_n) \geq M_2(a_1, \ldots, a_n) \geq \cdots \geq M_n(a_1, \ldots, a_n)$$

*(see [**B-AC**] if you are stuck).*

The next problem explores how we can take our initial crude bound and refine it. Of course, while for this problem we can easily determine the exact answer, our goal is to introduce a new method that is useful for a variety of other problems.

Exercise 1.7.21. *Let's revisit our estimation for $1 + 2 + \cdots + n$; in Exercise 1.7.17 we saw it's between $n^2/4$ and n^2. We highlight a powerful technique, called **dyadic decomposition**, which allows us to improve these bounds and close their gap. For simplicity you may assume $n = 2^k$ so we can divide n evenly by any power of 2.*

We first split $\{1, 2, \ldots, n/2, n/2 + 1, \ldots, n\}$ into two sets: $S_{11} = \{1, 2, \ldots, n/2\}$ and $S_{12} = \{n/2 + 1, n/2 + 2, \ldots, n\}$ and apply our bounds to the sum from each, remembering that $n/2$ is playing the role of n from before. Thus the sum of the terms in S_{11} is at least $(n/2)^2/4$ and at most $(n/2)^2$. For S_{12}, notice that each term is $n/2$ more than the corresponding term in S_{11}, and therefore its sum is $(n/2) \cdot (n/2) = n^2/4$ more than the sum from S_{11}. Combining we find

$$\frac{(n/2)^2}{4} + \frac{n^2}{4} + \frac{(n/2)^2}{4} \leq 1 + 2 + \cdots + n \leq \left(\frac{n}{2}\right)^2 + \frac{n^2}{4} + \left(\frac{n}{2}\right)^2$$

or

$$\frac{3}{8}n^2 \leq 1 + 2 + \cdots + n \leq \frac{6}{8}n^2.$$

Notice that the bounds are a lot closer and that the lower bound is less than $1/2$ and the upper bound exceeds $1/2$ (so our answers are reasonable).

Iterate this procedure one or two more times. Do the upper and lower bounds appear to converge to $1/2$? Can you prove that rigorously?

Exercise 1.7.22. *The **factorial** of a non-negative integer n, written $n!$, is the product of all integers at most n (with $0!$ set to be 1); note we can interpret this as the number of ways to order n objects when order matters. Find easy upper and lower bounds for $n!$, and then use the method of dyadic decompositions from the previous problem to improve your bounds and narrow the gap between them. For n large, **Stirling's formula** states that $n! \approx (n/e)^n \sqrt{2\pi n}$.*

Fast Multiplication:

Many encryption algorithms, such as RSA, use **modular (or clock) arithmetic**. We say $x \equiv y \bmod n$ (read x is equivalent or congruent to y **modulo** n) if $x - y$ is a multiple of n; thus $5 \equiv 17 \bmod 12$ (our choice of 12 for this example is to highlight the connection with a clock). For more on RSA see the problems beginning with Exercise 2.6.81.

Exercise 1.7.23. *Estimate how large 25^{100} is. Assume we only need to know $25^{100} \bmod 47$; use fast multiplication and congruences to compute this.*

Exercise 1.7.24. *Instead of using binary we could use ternary (base 3). How many operations would it take to compute x^{100} using the ternary expansion of 100? Is it ever more efficient to use ternary over binary? If yes, try to quantify how often ternary is better and how often binary is better. Note there are many ways you can compare the two, from average number of operations to worst-case scenarios.*

Exercise 1.7.25. *Show that any non-negative integer n has an unique binary expansion $n = \sum_{i=0}^{k} d_i 2^i$ with each $d_i \in \{0,1\}$.*

Exercise 1.7.26. *Prove that the number of multiplications required by fast multiplication to compute x^n is at most $2 \log_2(n)$.*

Exercise 1.7.27. *On average how many multiplications are needed to compute x^n for $n \in \{1, \ldots, N\}$? When investigating questions such as this it's often easier if we assume N is of a special form; in this case, perhaps it would be helpful to assume $N = 2^k$.*

Exercise 1.7.28. *Let $N = 2^k$. Which $n \in \{1, \ldots, N\}$ require the most multiplications to compute?*

Exercise 1.7.29. *Consider four algorithms: the first runs in e^N steps, the second in N^2 steps, the third in $N^{1/2}$, and the fourth in $\log N$ steps. Percentagewise how much of an increase is there in run-time going from an input of N to $2N$ (i.e., doubling the input)? Express your answer as a function of N, but for definiteness also do it for $N \in \{100, 10^{10}, 10^{100}\}$.*

Strassen's Algorithm:

Exercise 1.7.30. *Prove that Strassen's algorithm works as claimed (i.e., that we do recover the matrix product).*

Exercise 1.7.31. *Let $n \in \{10^4, 10^9, 10^{100}\}$. Compare how many multiplications are needed to compute the product of two $n \times n$ matrices using the brute force approach and using Strassen's algorithm.*

Exercise 1.7.32. *Implement Strassen's algorithm for 4×4 matrices.*

Exercise 1.7.33. *Discuss how to implement Strassen's algorithm if n is not a power of 2. What can you do? Which $n \in \{2^k, \ldots, 2^{k+1} - 1\}$ will lead to the least savings?*

Exercise 1.7.34. *Is there a fast way to multiply three matrices? Four? If yes how much better can you do than brute force?*

Exercise 1.7.35. *Let $n = 2$ and $\boldsymbol{A}$ be a 2×2 matrix. What is the most efficient way to compute $\boldsymbol{A}^m$ for $m = 2^\ell$, and how many steps does it take?*

Exercise 1.7.36. *Consider two 2×2 matrices $\boldsymbol{A}$ and $\boldsymbol{B}$ such that the lower left entry of $\boldsymbol{A}$ and the lower right entry of $\boldsymbol{B}$ are zero. Can you improve on Strassen's algorithm? What if now these are $2^k \times 2^k$ matrices and the lower left $2^{k-1} \times 2^{k-1}$ block of $\boldsymbol{A}$ and the lower right $2^{k-1} \times 2^{k-1}$ block of $\boldsymbol{B}$ are all zeros?*

Exercise 1.7.37. *If you know the sum of the first powers up to n, one can find the the sum of the first n cubes by noting*

$$\left(\sum_{k=1}^{n} k \right)^2 = \sum_{i=1}^{n} i^2 + 2 \sum_{j=2}^{n} \sum_{\ell=1}^{j-1} j\ell.$$

Prove this identity and deduce

$$\sum_{m=1}^{n} m^3 = \frac{n^2(n+1)^2}{4}.$$

Eigenvalues, Eigenvectors and the Fibonacci Numbers:

Exercise 1.7.38. *Find the eigenvalues and eigenvectors for the Fibonacci matrix*

$$\boldsymbol{A} = \begin{pmatrix} 1 & 1 \\ 1 & 0 \end{pmatrix},$$

and prove Binet's formula.

Exercise 1.7.39. *Prove that n is a Fibonacci number if and only if either $5n^2 + 4$ or $5n^2 - 4$ is a square. This problem was posed by Ira Gessel, Problem H-187 in the Fibonacci Quarterly **10** (1972), no. 4, page 417. Notice this provides a very fast test of whether or not an integer is a Fibonacci number, though it doesn't tell us which Fibonacci number it is; for more on fast tests see the Carmichael number exercises from §7.7.*

Exercise 1.7.40. *Generalize the previous problem to other recurrences; for example, is there a nice relation for the Tribonacci numbers, given by $T_{n+1} = T_n + T_{n-1} + T_{n-2}$ for appropriate choices of T_0, T_1 and T_2? For what a and b is there a nice recurrence relation describing numbers n such that n is in the sequence if and only if $an^2 \pm b$ is a square?*

Exercise 1.7.41. *Consider the following simplified model for the number of pairs of whales alive at a given moment in time. We make the following semi-reasonable assumptions:*

(1) Time moves in discrete steps of 1 year.

(2) The number of whale pairs that are 0, 1, 2, and 3 years old in year n are denoted by a_n, b_n, c_n, and d_n respectively; all whales die when they turn 4 (and at no other time).

(3) *If a whale pair is 1 year old it gives birth to two new pairs of whales, if a whale pair is 2 years old it gives birth to one new pair of whales, and no other pair of whales give birth.*

Let

$$v_n \; = \; \begin{pmatrix} a_n \\ b_n \\ c_n \\ d_n \end{pmatrix};$$

find a matrix $\boldsymbol{A}$ such that

$$v_{n+1} \; = \; \boldsymbol{A} v_n$$

*(these are called **Leslie matrices**, and are very important in mathematical biology). In Star Trek IV the crew of the Enterprise goes back in time and rescues a pair of humpback whales (George and Gracie), needed to repopulate the species and save the planet from an alien probe. How many whales will there be, according to this model, in 100 years? In 1000 years? In a million years?*

Exercise 1.7.42. *Compare how many multiplications are needed to find F_n using the Strassen algorithm versus Binet's formula.*

Exercise 1.7.43. *Show that for n large we do not need the second term in Binet's formula and can deduce F_n from an examination of just the first.*

   The next few problems concern real matrices; they can be generalized to complex matrices.

Exercise 1.7.44 (The Triangularization Lemma)**.** *Given any $n \times n$ matrix $\boldsymbol{A}$ there exists an invertible matrix $\boldsymbol{S}$ such that $\boldsymbol{S}^{-1} \boldsymbol{A} \boldsymbol{S}$ is an upper triangular matrix $\boldsymbol{T}$ (i.e., the entries of $\boldsymbol{T}$ below the main diagonal are all zero).*

Exercise 1.7.45. *Find the eigenvalues of an upper triangular matrix $\boldsymbol{T}$.*

Exercise 1.7.46. *Prove that if λ is an eigenvalue of $\boldsymbol{A}$, then there is a corresponding eigenvector. Is it possible that an eigenvvalue could have multiplicity r but not have r linearly independent eigenvectors? If yes give an example.*

Exercise 1.7.47. *Prove if $\boldsymbol{A}$ is a real symmetric matrix (so $\boldsymbol{A}^T = \boldsymbol{A}$), then A has an orthonormal basis of eigenvectors.*

Exercise 1.7.48. *A (real) $n \times n$ matrix $\boldsymbol{Q}$ is orthogonal if $\boldsymbol{Q}^T \boldsymbol{Q} = \boldsymbol{I}_n$. Prove that the eigenvalues of $\boldsymbol{Q}$ have absolute value 1 and that if $\boldsymbol{A}$ is a real-symmetric matrix, then we can find an orthogonal $\boldsymbol{Q}$ such that $\boldsymbol{Q}^T A \boldsymbol{Q} = \Lambda$, where $\Lambda = \mathrm{diag}(\lambda_1, \dots, \lambda_n)$ is a diagonal matrix of eigenvalues of $\boldsymbol{A}$. How many choices of Λ are there?*

Exercise 1.7.49. *We can use the Spectral Theorem to find eigenvalues, rather than the determinant formula. Assume for simplicity that A is an $n \times n$ real symmetric matrix, so it has an orthonormal basis of eigenvectors $\overrightarrow{v}_i$ with eigenvalues λ_i, ordered so that $|\lambda_1| \geq |\lambda_2| \geq \cdots \geq |\lambda_n|$. Consider a random vector $\overrightarrow{v}$; as the eigenvectors are a basis we have c_i such that*

$$\overrightarrow{v} \; = \; c_1 \overrightarrow{v}_1 + \cdots + c_n \overrightarrow{v}_n.$$

If $|\lambda_1| > |\lambda_2|$ (and thus non-zero), then if we are fortunate enough to have $c_1 \neq 0$, then

$$A^m \vec{v} \ = \ c_1 \lambda_1^m \vec{v}_1 + \cdots + c_n \lambda_n^m \vec{v}_n \ \approx \ c_1 \lambda_1^m \vec{v}_1.$$

Thus if we take the logarithms of the lengths of both sides,

$$\log \|A^m \vec{v}\| \ \approx \ m \log \lambda_1 + \log c_1 \|\vec{v}_1\|,$$

and sending m to infinity gives

$$\lambda_1 \ = \ \lim_{m \to \infty} \frac{\log \|A^m \vec{v}\|}{m}.$$

Justify the approximations above and discuss how we should send m to infinity and how we should compute the corresponding A^m. Also, the above assume $c_1 \neq 0$, but we don't know when we choose a vector whether or not the corresponding linear combination will or will not have $c_1 \neq 0$; what should we do to deal with this issue? What if $|\lambda_1| = |\lambda_2|$?

Chapter 2

Efficient Multiplication, II

In Chapter 1 we began our exploration by studying products. We were introduced to many of the techniques and ideas that will appear again and again in this book, from brute force approaches to pre-computing and look-up tables to searches for efficiency. In this chapter we build on these and study binomial coefficients and the Euclidean algorithm. While they are two very different topics, both provide another example of consequences of multiplication and are excellent springboards to showcase efficient techniques and the interplay between theory and applications.

2.1. Binomial Coefficients

We first quickly review some of the basics about factorials and binomial coefficients. The **factorial function** represents the number of ways to arrange n objects when order matters. We denote it by $n!$, and set $0!$ equal to 1 (there is one way to arrange nothing!). Note that $n! = n \cdot (n-1)!$; thus our function is recursively defined, and if we have a value the next is easily found.

If instead we want to choose k objects from n and arrange them, with order mattering, the number of ways is $n(n-1)\cdots(n-(k-1))$. Notice that this looks a lot like a factorial, except it does not go all the way down to 1. We can remedy this by using one of the most powerful techniques in mathematics: **multiplying by 1**. Thus,

$$n(n-1)\cdots(n-(k-1)) \ = \ \frac{n(n-1)\cdots(n-(k-1))\cdot(n-k)!}{(n-k)!} \ = \ \frac{n!}{(n-k)!}$$

(we define it to be zero if $k > n$).

Finally, if we want to choose k objects from n when order does not matter, the number of ways is the **binomial coefficient** $\binom{n}{k}$, which equals $\frac{n!}{k!(n-k)!}$. This follows immediately from there being $n!/(n-k)!$ ways to choose k objects from n when order matters, and there being $k!$ ways to arrange k objects.

While it would take us too far afield to review all the great properties of binomial coefficients, we state a few and give proofs, since they illustrate some great techniques.

The first is immediate: $\binom{n}{k} = \binom{n}{n-k}$. While this can be proved by brute force (expand both terms and see that they are equal), it's much better to **prove by story**. We can interpret $\binom{n}{k}$ as the number of ways to choose k objects from n when order doesn't matter; however, if we're choosing k that's the same as excluding $n-k$, and the claim follows.

The next property is far more important and will play a central role in much of the rest of this chapter:

$$\binom{n+1}{k+1} = \binom{n}{k} + \binom{n}{k+1}.$$

We could prove this by expanding each quantity, cross multiplying, and seeing that they are the equal, but again it's must easier to prove this by story. We'll show

$$\binom{n+1}{k+1} = \binom{1}{1}\binom{n}{k} + \binom{1}{0}\binom{n}{k+1};$$

as $\binom{1}{1} = \binom{1}{0} = 1$, this is equivalent.

Imagine we have $n+1$ people, say n physicists and 1 mathematician. There are $\binom{n+1}{k+1}$ ways to choose $k+1$ of these $n+1$ people; this is the left side. We can split the choice into two ways: either we choose the mathematician or we don't. There is $\binom{1}{1}$ way to choose the mathematician, and then we must take k of the remaining n physicists to get up to $k+1$ people. Similarly, there is $\binom{1}{0}$ way to not choose the mathematician, and then we must take $k+1$ of the n physicists. As these cases are mutually exclusive and exhaustive, the claim follows.

2.2. Pascal's Triangle

You may recall learning about Pascal's triangle in high school, which begins

$$
\begin{array}{ccccccccccccc}
 & & & & & & 1 & & & & & & \\
 & & & & & 1 & & 1 & & & & & \\
 & & & & 1 & & 2 & & 1 & & & & \\
 & & & 1 & & 3 & & 3 & & 1 & & & \\
 & & 1 & & 4 & & 6 & & 4 & & 1 & & \\
 & 1 & & 5 & & 10 & & 10 & & 5 & & 1 & \\
1 & & 6 & & 15 & & 20 & & 15 & & 6 & & 1
\end{array}
$$

Pascal's triangle has lots of nice properties, particularly from a computational point of view. The outer edges are two lines of 1's, and then each interior number in the triangle equals the sum of the number directly above and to the left and the number above and to the right; for instance, we see that the second number in the fourth row is 3, which is the sum of 1 and 2, which are the numbers in the third row directly above it. This property should remind you of our just-proved result on binomial coefficients: $\binom{n+1}{k+1} = \binom{n}{k} + \binom{n}{k+1}$. The proof of that identity follows by noting that the kth entry in the nth row (counting from 0) is $\binom{n}{k}$; this is the famous **Binomial Theorem** (which you are asked to prove in Exercise 2.6.16). Thus, we may view the $n+1$ terms in the nth row as the coefficients from expanding $(x+y)^n$; explicitly, the kth term in the row is the coefficient of $x^{n-k}y^k$, which turns out to be $\binom{n}{k}$.

Is it difficult, theoretically speaking, to understand and find $\binom{n}{k}$? Clearly not, as we can write it as $\frac{n!}{k!(n-k)!}$, and thus we have a simple, closed-form expression. Practically, however, it's a very different story: if you plug 100! into most calculators you'll get an overflow error, as 100! is too large to store its full value in a calculator's memory, much less display it on a small screen. Now imagine that we want to find $\binom{100}{39}$, which is 90139240300346304926343340800 or about 10^{28}. How can we manage the calculation to avoid the overflow? Before reading further, pause and think of how you would handle such a large computation.

We could write out Pascal's triangle until we reach row 100. This avoids having to do the enormous multiplication directly (as we are just adding numbers to move from row to row), but it's extremely time consuming. Notice the similarity between this brute force row approach and the recurrence relation approach to finding Fibonacci numbers; in both problems we end up computing many more terms than needed. We eventually found Binet's formula for the Fibonaccis, which allowed us to jump to the term we wanted without finding the earlier ones. In the current situation, we already have a formula for the term we want. The difference is that within the formula we need to deal with exceptionally large numbers and find a way to manage the calculations to avoid the overflow.

We start by writing it out to see what we might be able to simplify:
$$\binom{100}{39} = \frac{100!}{39!61!} = \frac{100 \cdot 99 \cdots 62 \cdot 61!}{39!61!} = \frac{100 \cdot 99 \cdots 62}{39 \cdot 38 \cdots 1} = \frac{100}{39} \cdot \frac{99}{38} \cdots \frac{62}{1}.$$

As binomial coefficients are integers, the above value must be an integer. Thus, the denominator divides the numerator, and we should be able to cancel terms; the difficulty of course is finding how the cancellation happens. Because we don't compute the entire triangle (we only work in the row we need), we can manage these cancellations very well by understanding new values in terms of previous values in a row.

Let's say we've already computed $\binom{n}{k}$ and want to calculate $\binom{n}{k+1}$. How might we go about this? Simple algebra gives
$$\binom{n}{k} = \frac{n!}{k!(n-k)!}, \quad \binom{n}{k+1} = \frac{n!}{(k+1)!(n-k-1)!} = \frac{n!}{k!(n-k)!} \cdot \frac{n-k}{k+1},$$

and thus we find
$$\binom{n}{k+1} = \frac{n-k}{k+1}\binom{n}{k}.$$

This is a terrific way to handle the fractions. Why is this helpful? This number $\frac{n-k}{k+1}\binom{n}{k}$ is always an integer. Not only that, but we also already know that $\binom{n}{k}$

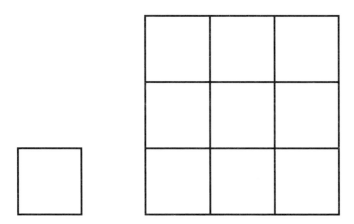

Figure 1. Increasing the sides of a square by a factor of 3 increases the area by a factor of 3^2, indicating the dimension of a square is 2.

is an integer, and we're multiplying it by $\frac{n-k}{k+1}$ to get an integer. Thus, $k+1$ has to divide $(n-k)\binom{n}{k}$; if we keep sliding along like this, we only have to deal with integers and the way to divide is clear, which is not the case for the earlier method of splitting into fractions.

This may be the longest amount of time in your college career that you spend on a computation that you'd normally just plug into your calculator, but studying this in depth can be very enlightening. We see that by simply rearranging the calculations that need to be done, we can completely change the computational dynamics involved. When computing, you want to be very careful to avoid rounding errors, which would happen if we shifted to fractions and decimals. We'll see an interesting application of fast approaches such as this to Pascal's triangle and fractals, but to do so we first need a quick review of dimension.

2.3. Dimension

There are many problems in which the dimension of the system controls the complexity or the difficulty of the calculation. For example, for polynomial evaluation the naive approach had complexity n^2, but we did much better by grouping terms with Horner's Algorithm, reducing the complexity to order n. We can view this as lowering the exponent from 2 to 1, leading to massive savings as n grows. For matrix multiplication the standard brute force approach took n^3 multiplications, while Strassen's algorithm reduced it to order $n^{\log_2 7}$.

This of course begs the question of what is the dimension of a system. Since our last example shows that it doesn't have to be an integer, we see that this question is harder than it might at first appear. Let's build intuition by looking at some familiar examples. Most people have no problem saying a line has dimension 1, a square has dimension 2, and a cube has dimension 3. For these shapes, if we scale each side by a factor of r, then the 'volume' (which is length in the first case, area in the second, and volume in the third) of a d-dimensional object is rescaled by a factor of r^d; see Figure 1.

Instead of taking this as an observation, we can use this to define the dimension of our object. Central in our definition will be the notion of a **dilation**; if $S \subset \mathbb{R}^n$, then its dilation by a factor of r replaces each point $(s_1, \ldots, s_n) \in S$ with $(rs_1, \ldots, rs_n)$.

Definition 2.3.1 ((Hausdorff) dimension). *Let $S \subset \mathbb{R}^n$ be a set. If dilating S by a factor of r yields c copies of S, then the **dimension** d of S satisfies $r^d = c$. (There are several different notions of dimension; this is essentially the **Hausdorff dimension**.)*

To understand how this definition works—and why we are defining it this way—it's helpful to examine how it operates in the cases we know well.

Example 2.3.2. *Let S be a line of length n. Dilating S by a factor of r yields a line of length rn. An example with $r = 3$ is shown in Figure 2.*

Figure 2. Tripling a line.

We can divide this new line into r segments of length n, meaning we have $c = r$ copies of S. We have $r = r^1 = c$, yielding $d = 1$ for the dimension of our line S.

Example 2.3.3. *If we now take a square of length 1 and dilate by a factor $r = 3$, we get the image from Figure 1. As we have 9 copies of the original square, we find $9 = 3^2$, and thus the dimension is $d = 2$.*

The reason we like this definition of dimension is that it allows us to handle cases we couldn't handle before. The first is the **Cantor set**, which is defined as follows:

- Let $C_0 = [0, 1]$, the unit interval.
- Given C_n, let C_{n+1} be the set formed by removing the middle third of each interval in C_n.

Thus, $C_1 = \{0, 1/3\} \cup \{2/3, 1\}$ and

$$C_2 = \{0, 1/9\} \cup \{2/9, 3/9\} \cup \{2/3, 7/9\} \cup \{8/9, 1\}.$$

If we dilate by a factor of 3, we end up with 2 copies of the Cantor set (see Figure 3). Thus, its dimension d satisfies $3^d = 2$ or $d = \log_3 2 \approx 0.63093$. While the actual value is surprising the first time you see it, it is at least reasonable. It's a subset of the unit interval, so it's dimension should be less than 1.

Figure 3. The first six iterations of the construction of the Cantor set. Image from Sarang (Wikimedia Commons).

There are many other fractal shapes we can consider. We end with one more, the **Sierpinski triangle**, as there is a wonderful connection between it and Pascal's triangle. It is defined as follows:

- Start with an equilateral triangle with sides of length 1, which we denote T_0.
- Given T_n, subdivide each of the equilateral triangles into four smaller equilateral triangles. Throw away the one in the center and keep the other three.

We show the first four stages of the construction in Figure 4. If we continue to subdivide and throw away, in the end it converges to the Sierpinski triangle.

Figure 4.   The construction process leading to the Sierpinski triangle; first four stages. Image from Wereon (Wikimedia Commons).

To find its dimension, notice that doubling results in three copies; thus, its dimension d satisfies $2^d = 3$ or $d = \log_2 3 \approx 1.58$. For the Cantor set, we obtained a dimension less than 1; here we have $1 < d < 2$. This is reasonable: due to its deleted sections, the Sierpinski triangle doesn't seem to have as great of a dimension as, say, a square. It does contain many lines, which, being 1-dimensional themselves, indicate its dimension should be greater than 1.

In the next section we'll see connections between the Sierpinski and Pascal triangles.

2.4. From the Pascal to the Sierpinski Triangle

We recall the first few rows of Pascal's triangle, where the kth entry in the nth row is $\binom{n}{k}$.

$$
\begin{array}{ccccccccccccccc}
&&&&&&& 1 &&&&&&& \\
&&&&&& 1 && 1 &&&&&& \\
&&&&& 1 && 2 && 1 &&&&& \\
&&&& 1 && 3 && 3 && 1 &&&& \\
&&& 1 && 4 && 6 && 4 && 1 &&& \\
&& 1 && 5 && 10 && 10 && 5 && 1 && \\
& 1 && 6 && 15 && 20 && 15 && 6 && 1 & \\
1 && 7 && 21 && 35 && 35 && 21 && 7 && 1
\end{array}
$$

Let's modify Pascal's triangle. Instead of writing $\binom{n}{k}$ in the kth entry of the nth row, we place a $\bullet$ if $\binom{n}{k}$ is odd and leave it blank if $\binom{n}{k}$ is even. Thus if we have just one row we would see •, if we have four rows we would see

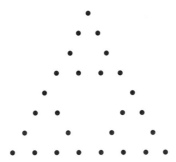

while for eight rows we find

As this section follows our discussion of dimensions and fractals, it's natural to conjecture that in the appropriate limit Pascal's triangle converges to the Sierpinski triangle. However, maybe we haven't gone far enough and need to collect more points. Collecting more points, we soon run into a variety of computational issues. Specifically, how do we compute the rows of Pascal's triangle modulo 2 quickly, to determine if $\binom{n}{k}$ is even or odd? Determining this value for a given n and k is very different from determining the values of a given row, which is also very different from determining the values of all rows up to a fixed level. Any one of these situations involves performing many calculations. Rather than taking the time to perform them all, we might be able to do only a few and reuse the result. This should remind you of the incredible gains we obtained with Strassen's algorithm. The gains there are not a one-time trick, but rather one of the many examples of a powerful technique. We further explore these issues in the exercises.

Another issue, which is one of the reasons why I wanted to discuss this topic, is that it highlights the importance and difficulty of presenting data well enough to highlight the limiting behavior. If we just keep adding more and more rows our image quickly grows too large to fit on a page. If we want to investigate what happens in the limit, we need to standardize. A good way to do this is to fix a size and rescale our image so that it's always displayed in a box of the same size; this is done in Figure 5. Notice that while we do see more detail at row 2^{10} than we do at row 2^8, the thickness of the dots is starting to cause a problem. These images were created in Mathematica; due to the pre-defined graphics functions, it was easier to rotate the triangle so that it starts at the origin.

It's a nice computational challenge to create an animated movie; you can hear a short lecture on the subject and see the resulting movie at https://www.youtube.com/watch?v=tt4_4YajqRM (I created this short because I was unable to find one online.)

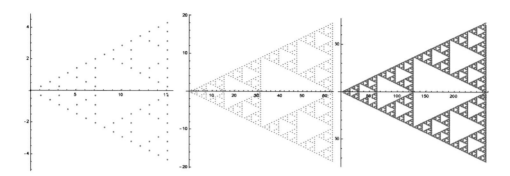

Figure 5. Plot of Pascal's triangle modulo 2 for 2^4, 2^8 and 2^{10} rows.

2.5. The Euclidean Algorithm

The final entry in our multiplication discussion is the Euclidean algorithm, though perhaps it's more appropriate to view this as division. Given two positive integers x and y, their **greatest common divisor** (**gcd**), often denoted $\gcd(x, y)$, is, as the name suggests, the largest integer dividing both. Thus $\gcd(12, 66) = 6$ and $\gcd(8, 27) = 1$. When the greatest common divisor is 1, we say the numbers are **co-prime** or **relatively prime**. Another notation for the gcd is (x, y), and thus $(x, y) = 1$ means x and y numbers are relatively prime. The gcd is extremely useful both in theoretical mathematics (it plays a key role in group theory) and applied mathematics (especially in RSA encryption and decryption; we briefly describe that method in a series of problems beginning with Exercise 2.6.81).

We first describe the brute force approach. While it trivially finds the gcd of two numbers, it is horribly inefficient, as should be expected by now. In the worst-case scenario it will take x trials.

Brute Force GCD Algorithm: Start with two positive integers $x < y$.

(1) Let $d = x$.

(2) If $d|x$ and $d|y$, then stop, and d is the greatest common divisor of x and y.

(3) Let d equal $d - 1$ and return to Step (2).

The above procedure works but is quite inefficient. In the worst-case scenario it will take x trials to find $\gcd(x, y)$. The Euclidean algorithm takes less. Far less. We won't say how much less now to keep some suspense.

While the Euclidean algorithm is a gem and worth studying in its own right, we'll explore it in depth here primarily because it's different from the other algorithms we've seen so far. Understanding it introduces many of the concepts we'll meet throughout the semester. Horner's algorithm, fast multiplication, and Strassen's algorithm all ran faster than the naive, brute force algorithm, but we had to do something clever. The Euclidean algorithm is in some sense the exact opposite of that: it's fairly straightforward, has a trivial way to estimate its run-time, and yet manages to run a lot faster than we might expect.

2.5.1. The Euclidean Algorithm and Clock Arithmetic. The Euclidean algorithm is more than just a fast way to find the greatest common divisor of two numbers; it provides a very efficient way to find the numbers a and b in the theorem below.

Theorem 2.5.1. *If x, y are relatively prime positive integers (so $(x, y) = 1$), then there exist integers a and b such that $ax + by = 1$. More generally, there are a and b such that $ax + by = \gcd(x, y)$.*

This theorem might at first seem out of place in an applied mathematics course; you might remember it from an Abstract Algebra class (which, don't worry, is not a prerequisite for this book). As mentioned earlier, one reason we're discussing it is its application in RSA encryption and decryption. RSA is used so much not only because it's secure, but also because it's fast; it needs to operate with huge numbers and find multiplicative inverses relatively quickly.

While Abstract Algebra is not required, its language and terminology are helpful in explaining the true import of this result. In the Fast Multiplication exercises from §1.7 we quickly introduced some notation on **clock arithmetic**. We say $x \equiv y \bmod z$ if $x - y$ is a multiple of z. As the name suggests, this is math you've done all your life. If it's 10 o'clock in the morning, what time will it be in 5 hours? The answer, of course, is 3pm. This doesn't mean that $10 + 5 = 3$; what we're saying is that $10 + 5$ leaves 3 after we take away integer multiples of 12. This is what it means to say that $10 + 5 \equiv 3 \bmod 12$: it's the result that you get if you're using a clock, where you reset to 0 every time you hit 12. We say 3 and 15 are in the same **equivalence class** (modulo 12); another notation is to say 3 and 12 are in the same **residue class** (modulo 12).

Using this language, we can revisit our theorem above: if $ax + by = 1$, then $ax \equiv 1 \bmod y$. This means that the difference between ax and 1 is a multiple of y (in this particular case, we can see that the multiple is $-by$).

Whenever we are presented with something in mathematics, there is a checklist, of sorts, of questions we need to ask about it. The first question is: *Does it exist?* As far as a and b are concerned, Theorem 2.5.1 gives us the answer: yes they do.

The second question we should ask follows the first quite naturally: *Is it unique?* If it isn't unique, we can ask: *How many are there?* Finally, we should ask: *Is there a canonical representation?* By **canonical representation** we mean some representation that seems more natural than all other representations.

So what do you think? If a and b exist, are they unique? Or do you think that, if a and b exist, we can find other a and b that work? We'll first explore properties of the a and b in Theorem 2.5.1, and then describe how to efficiently find them.

2.5.2. Existence and Uniqueness. Let's now investigate the existence and uniqueness of the a and b from Theorem 2.5.1. We tackle uniqueness first. Before trying to find whether or not there is *a* solution (and if so what it is), let's first see how many solutions there could be.

Uniqueness and Canonical Solutions:

Imagine $a_0 x + b_0 y = 1$. Now consider

$$(a_0 + \alpha)x + (b_0 + \beta)y = 1.$$

Are there any non-zero α, β that satisfy this? From that equation, we get

$$(a_0 x + b_0 y) + (\alpha x + \beta y) = 1.$$

As we already know that $(a_0 x + b_0 y) = 1$, we're left with $(\alpha x + \beta y) = 0$. Are there any solutions to this? Yes there are! Just take $\alpha = y$ and $\beta = -x$ to get a solution. In fact, $\alpha = ny$ and $\beta = -nx$ is a solution for all integers n, meaning there are infinitely many solutions. (This last bit is a very general argument strategy – if you're looking for a solution, see if you can find a trivial one that works.)

This leads us to the canonical representation question we mentioned earlier. We've shown that if there is a solution there are infinitely many; we want a way to single out one 'special' choice (i.e., a *canonical* choice). One great choice is to see what happens if we insist a and b lie in a certain region; specifically, we don't want them to be too large. A similar restriction arises often in linear programming: we will have problems with multiple solutions, and it will be important to move between solutions and to find one that has a certain desirable property.

Based on our discussion above, a natural restriction might be $|a| \le y$ and $|b| \le x$; we can't assert both a and b are positive as our linear combination would always exceed $x + y$, but our arguments show it's reasonable to expect that we can move from a solution to one with this additional property. You're asked to prove this in Exercise 2.6.64.

To recap, we've shown that *if* a solution to $ax + by$ exists, then there are infinitely many, and in particular at least one has $|a| \le y$ and $|b| \le x$. Now we have to show that there is at least one solution, and then we have to somehow find it.

Existence:

We now give a procedure that will find a canonical solution. We don't claim that this is the best procedure, but it's guaranteed to work; later we'll find a better method.

Consider all $ax + by$, with $0 \le a < y$ and $0 \le b < x$. We first solve the easier problem $ax + by \equiv d \bmod xy$, because once we have a solution to this related problem we can convert it to obtain a strict equality. For example, imagine $ax + by = d + mxy$. Then $(a - my)x + by = d$, and we have a new a and b that satisfy the strict equality. Thus we need only solve $ax + by \equiv d \bmod xy$. We'll do the case $d = 1$, and leave the general case to you as Exercise 2.6.65.

From our ranges for a and b we see that we have xy possibilities for $ax + by$. We also have xy residues modulo xy (specifically, $0, 1, 2, \ldots, xy - 1$). We now have two cases:

(1) One of the xy pairs of a, b give $ax + by \equiv 1 \bmod xy$.

(2) None of the xy pairs of a, b give $ax + by \equiv 1 \bmod xy$.

If the first case happens, our argument above shows how we can pass from such a solution to a solution of our original problem. Thus, we're reduced to proving the second case cannot happen.

We see that in this case, we have xy possible values of $ax + by$ mod xy, but at most $xy - 1$ residues modulo xy are realized, as the residue 1 is forbidden. This is the standard set-up to apply the **Pigeonhole Principle** (if we have N items and $N - 1$ boxes, then at least one box has at least two elements). For us, our boxes are the equivalence classes modulo xy that are realized.

As we're trying to fit xy numbers into $xy - 1$ residue classes, there must be at least one class which has at least two of our pairs in it. So we have $a, b, \tilde{a}, \tilde{b}$ such that

$$ax + by \equiv \tilde{a}x + \tilde{b}y \text{ mod } xy,$$

which implies that

$$ax + by = \tilde{a}x + \tilde{b}y + mxy$$

for some integer m. Without loss of generality, we may assume that $b > \tilde{b}$. We now get $(a - \tilde{a})x + (b - \tilde{b})y = mxy$, and we know that $0 < b - \tilde{b} < x$. Thus

$$(b - \tilde{b})y = [my - (a - \tilde{a})] x,$$

and x must divide $(b - \tilde{b})y$. But we know that x doesn't divide $(b - \tilde{b})$, since $0 < (b - \tilde{b}) < x$; thus x must share a factor with y. This contradicts our assumption that x and y are relatively prime. Therefore the second case never occurs, and a solution to $ax + by = \gcd(x, y)$ exists.

2.5.3. Efficiency of the Euclidean Algorithm. Our discussion of Theorem 2.5.1 gives more than just a constructive proof to find the greatest common divisor of two positive integers x and y; it gives us two integers a and b so that $ax + by = \gcd(x, y)$ and $|a| \leq y$ and $|b| \leq x$. Thus there are finitely many cases to check, and worst case we check xy values. Unfortunately, for applications in cryptography we can easily have $x, y \approx 10^{200}$. In that case the number of steps could be as large as 10^{400}.

Let's put this in perspective. There are $31,536,000$ seconds in a year, and a single super computer (being very generous) can perform 10^{20} operations in a second. There are at most 10^{100} objects in the entire universe, and if we assume that each of these is a supercomputer devoted exclusively to our needs, then in 1 year we can do about 10^{127} operations. This means that with every object in the universe running the algorithm you'd need 10^3 years to complete in the worst-case scenario. If each person on our planet had one supercomputer devoted exclusively to this computation, the universe would reach its heat death long before the algorithm finishes.

Our brute force algorithm is unfortunately useless for *practical* purposes, though it is useful for *theoretical* ones (see for example Exercise 2.6.60). We clearly need a more efficient procedure.

The Euclidean Algorithm: Let $x < y$ be positive integers. Set $r_0 = y$, $r_1 = x$ and given $r_n > 0$ let a_{n+1} and r_{n+1} be the unique integers satisfying

$$r_{n-1} = a_{n+1}r_n + r_{n+1}, \quad 0 \le r_{n+1} < r_n.$$

Then if m is the smallest index such that $r_m = 0$, we have $\gcd(x,y) = r_{m-1}$. Further, there are (easily) computable a and b (in terms of the a_i's and r_i's) such that $ax + by = \gcd(x,y)$.

At first glance, the Euclidean algorithm appears to have a comparable run-time to our brute force approach. The reason is that the worst-case scenario each time is that $r_{n+1} = r_n - 1$. If that happens, it'll take $x - 1$ steps to terminate. As our brute force approach found the gcd by trying all integers at most x, there are no savings here, though it's potentially faster at finding the a and b in the linear combination than trying all xy pairs.

In the course of proving why the algorithm works we'll also prove it runs significantly faster than we might fear; it takes on the order of $\log x$ steps, not on the order of x. This gain in efficiency is similar to the difference between brute force multiplication and repeated squaring.

The idea is to keep using the division algorithm, writing the division as a quotient and a remainder, and then considering a related problem involving the remainder. We start by expressing $r_0 = y$ as a multiple of $r_1 = x$ and a remainder $r_2 \in \{0, 1, \ldots, r_1 - 1\}$:

$$r_0 = a_2 r_1 + r_2.$$

If $r_2 > 0$ we write r_1 as a multiple of r_2 plus a remainder:

$$r_1 = a_3 r_2 + r_3, \quad r_3 \in \{0, 1, 2, \ldots, r_2 - 1\}.$$

We continue, and thus at the nth stage we have

$$r_{n-1} = a_{n+1}r_n + r_{n+1}, \quad r_{n+1} \in \{0, 1, \ldots, r_n - 1\}.$$

This method is often called **lather, rinse, repeat** after shampoo instructions: we keep doing the same steps again and again. Each step reduces the size of the problem, and after finitely many steps it successfully terminates. We first show it finds the gcd, then that it yields a and b, and finally analyze the run-time.

Finding the gcd.

Let m be the smallest index such that $r_m = 0$. We claim that $r_{m-1} = \gcd(x,y)$. This is a consequence of the following observation:

$$\gcd(r_{i+1}, r_i) = \gcd(r_i, r_{i-1});$$

in other words, the gcd is unchanged as we march down the line.

To prove the claim, let's look at the relevant expansions; we write more than we need here as these relations will be useful when we analyze the run-time:

$$
\begin{aligned}
r_{i-2} &= a_i r_{i-1} + r_i, \\
r_{i-1} &= a_{i+1} r_i + r_{i+1}, \\
r_i &= a_{i+2} r_{i+1} + r_{i+2}.
\end{aligned}
$$

(2.1)

Let $d_i = \gcd(r_i, r_{i-1})$ and $d_{i+1} = \gcd(r_{i+1}, r_i)$. We need to show $d_{i+1} = d_i$. We use the observation that any number which divides two numbers also divides their greatest common divisor (see Exercise 2.6.59). A nice way to do this is to show that $d_{i+1}|d_i$ and $d_i|d_{i+1}$, which implies they must be equal; a similar idea is often used to show two sets are equal (sets A and B are equal if $A \subset B$ and $B \subset A$):

- We first note $d_{i+1}|d_i$. The reason is that by definition d_{i+1} divides r_{i+1} and r_i; thus if we can show d_{i+1} also divides r_{i-1} we will have $d_{i+1}|\gcd(r_i, r_{i-1})$. From the middle equation in (2.1) we see that anything dividing both r_i and r_{i+1} divides r_{i-1}, completing the proof of this direction.

- We now must show $d_i|d_{i+1}$. Again from the definition we have d_i divides r_i and r_{i-1}, and thus we just need to show d_i divides r_{i+1}. We rewrite the middle equation in (2.1) as

$$r_{i+1} = r_{i-1} - a_{i+1}r_i;$$

thus as d_i divides both terms on the right hand side, d_i divides r_{i+1}, completing the proof.

The result of our analysis is that the gcd does not change as we march down the line. We continue until we reach $r_m = 0$, and since $d_m = \gcd(r_m, r_{m-1}) = r_{m-1}$, we find the gcd is r_{m-1} as claimed.

Determining a and b.

We leave the details of finding a and b to Exercise 2.6.68, and sketch the approach. From the above we know

$$r_{m-3} = a_{m-1}r_{m-2} + r_{m-1}, \quad r_{m-1} = d = \gcd(x, y).$$

Rewriting we get

$$d = r_{m-3} - a_{m-1}r_{m-2}.$$

Now we use

$$r_{m-4} = a_{m-2}r_{m-3} + r_{m-2},$$

which allows us to write r_{m-2} as a linear combination of r_{m-3} and r_{m-4}: $r_{m-2} = r_{m-4} - a_{m-2}r_{m-3}$. Substituting yields

$$d = r_{m-3} - a_{m-1}(r_{m-4} - a_{m-2}r_{m-3}) = -a_{m-1}r_{m-4} + (1 + a_{m-1}a_{m-2})r_{m-3}.$$

We just continue marching down the line, and eventually we'll find integers a and b such that

$$d = ar_1 + br_0 = ax + by.$$

Efficiency of the Euclidean algorithm.

All that remains is to show that the Euclidean algorithm runs in logarithmic time. While it's possible that $r_{i+1} = r_i - 1$, it is *not* possible that $r_{i+2} = r_i - 2$; in fact, as the following lemma shows it must be significantly less.

Lemma 2.5.2. *At every two iterations, the remainder in the Euclidean algorithm is decreased by at least a factor of 2:*

$$r_{i+2} \leq \frac{1}{2}r_i.$$

This implies that the number of steps is at most $2\log_2 x$.

Proof. We break the analysis into two cases; this is a common technique, as doing so gives us more information to use in each setting. The first case has r_{i+1} at most half of r_i, which is trivial to analyze; the second explores what happens if r_{i+1} is large relative to r_i.

Case 1: Assume $r_{i+1} \leq \frac{1}{2}r_i$. Not surprisingly, our claim immediately follows as $r_{i+2} < r_{i+1}$.

Case 2: Assume $r_{i+1} > \frac{1}{2}r_i$. From (2.1) we have

$$r_i = a_{i+2}r_{i+1} + r_{i+2}.$$

By assumption $r_{i+1} > \frac{1}{2}r_i$, and by construction we have $r_{i+2} < r_{i+1}$. Thus the only possibility is $a_{i+2} = 1$, which then gives

$$r_{i+2} = r_i - r_{i+1} \leq \frac{1}{2}r_i,$$

completing the proof.   □

We end with a few brief comments to put the results of our two introductory chapters in the proper perspective.

The examples of these first two chapters introduce many of the ideas and techniques that we'll use in our future investigations. It's fine if you don't understand line by line everything above. We won't be using the Euclidean algorithm in the rest of our studies, but it's very useful for getting a sense of the calculations we're going to do, and it's a great transition between the theoretical mathematics you've seen before and the more applied mathematics we'll start to examine. It also provides one final example of the benefits of revisiting preconceived notions on how to do something to find better ways.

Additionally, the analysis of the run-time of the Euclidean algorithm shows that while it may be hard to show that a fixed step yields a significant reduction, sometimes it's easy to show that a finite sequence of steps has the desired reduction. In other words, bad things cannot happen all the time. Instead of trying to show each step is 'good', we show that at least one step out of several must be 'good'. The advantage of such an approach is that we are breaking into cases, and thus have more facts at our disposal. For example, either the first step is 'good' or it is not; if not, we know certain facts about where we now are, and we can use those in analyzing what happens next.

2.6. Exercises

Binomial Coefficients

Whenever you see a product, you should have a Pavlovian response and think logarithms. The reason is that logarithms convert products to sums, and we have great familiarity with sums. We've seen Riemann sums in calculus class and have techniques to estimate them; very few students have taken a class on handling products! Of course, there's really no difference between analyzing a product $\mathcal{P} = p_1 \cdots p_n$ or a sum $\mathcal{S} = s_1 + \cdots + s_n$ (where $s_i = \log p_i$), as $\mathcal{P} = \exp(\mathcal{S})$ (or $\mathcal{S} = \log(\mathcal{P})$; however, it's often more convenient to take logarithms and convert as the resulting expression is more familiar. The first few problems below provide excellent illustrations of this.

Exercise 2.6.1. *Use the integral test from calculus to estimate $n!$ for n large.*

Exercise 2.6.2. *Use the integral test from calculus to estimate $\binom{n}{k}$ for n and k large. For what range of k (relative to n) do you expect this to be a good approximation?*

Exercise 2.6.3 (The Birthday Problem). *Assume in any group of N people that everyone is equally likely to be born on any day of the year, independent of whomever is in the room. The **Birthday Problem** asks how large N must be so that there is at least a 50% chance that two people share a birthday. If we assume no one is born on February 29th, then the probability anyone is born on a given day is 1/365. Show that the probability that the N people have distinct birthdays is*

$$P_N = \prod_{n=0}^{N-1} \frac{365 - n}{365} = \prod_{n=0}^{N-1} \left(1 - \frac{n}{365}\right).$$

Take logarithms and estimate what N is needed so that $P_N \approx 1/2$.

Exercise 2.6.4. *Generalize the Birthday Problem to the case where there are D days in a year. As a function of D, approximately how many people are needed before there is at least a 50% chance that two share a birthday?*

Exercise 2.6.5. *Another generalization of the Birthday Problem is to require three people share a birthday. Give some upper and lower bounds on the number of people needed to ensure there is at least a 50% chance this happens. What do you think the true answer is? Explore this by writing a computer program.*

Exercise 2.6.6. *In the previous problem we required three people to share a birthday. Another generalization asks for there to be at least two days with at least two people having those days as their birthday; note if three people were born on April 19th and two on October 17th, that would count. Give some upper and lower bounds on the number of people needed to ensure there is at least a 50% chance this happens. What do you think the true answer is? Explore this by writing a computer program.*

Exercise 2.6.7. *If $x, y \geq 0$ and $x + y = S$, prove xy is maximized when $x = y = S/2$.*

Exercise 2.6.8 (From Peter Schumer). *Find all n and k such that $\binom{n}{k}$, $\binom{n}{k+1}$, and $\binom{n}{k+2}$ are in a 1 to 2 to 3 ratio.*

Exercise 2.6.9. *Generalize the previous problem so that the three consecutive binomial terms are in a 1 to a to b ratio, where a and b are given integers. For what a and b is this possible? Or instead require the ratios to be 1:2:3, but now the terms are in an arithmetic progression with common distance d (so $\binom{n}{k}, \binom{n}{k+d}, \binom{n}{k+2d}$).*

Exercise 2.6.10. *We introduced the method of **dyadic decompositions** in Exercise 1.7.21, and you were asked to use it in Exercise 1.7.22 to estimate $n!$. Here is a nice way to improve that bound by using the previous exercise. Let $n = 2^k$. While we could bound the product $a(a+1)\cdots(a+k-1)$ below by a^k and above by $(a+k-1)^k$ (or $(a+k)^k$), we can do better by matching terms in pairs. Thus $a(a+k-1) \leq \left(a + \frac{k-1}{2}\right)^2$, $(a+1)(a+k-2) \leq \left(a + \frac{k-1}{2}\right)^2$, and so on. Use this idea to obtain a better upper bound for $n!$. Can you get a good lower bound? Compare this bound to the bound we get by trivially bounding such products by $(a+k-1)^k$.*

Exercise 2.6.11. *Define the **Gamma function** $\Gamma(s)$ by*

$$\Gamma(s) = \int_0^\infty e^{-x} x^{s-1} dx$$

for the real part of s greater than zero. Prove this integral converges, and if n is a non-negative integer, then $\Gamma(n+1) = n!$; more generally, prove $\Gamma(s+1) = s\Gamma(s)$ (this is called the functional equation of the Gamma function).

Exercise 2.6.12. *Show that $\Gamma(1/2)\sqrt{2} = \int_{-\infty}^\infty e^{-x^2} dx$, and thus the normalization constant of the standard normal is related to the Gamma function; recall that the density of the standard normal is $e^{-x^2/2}/\sqrt{2\pi}$. More generally, the kth moment of a continuous probability distribution $p(x)$, M_k, is defined by*

$$M_k = \int_{-\infty}^\infty p(x) dx.$$

Express the even moments of the standard normal in terms of the Gamma function.

Stirling's formula says that for n large, $n! \approx n^n e^{-n} \sqrt{2\pi n}$; this means that as $n \to \infty$, the ratio of those two quantities tends to 1. More precisely, we have the following series expansion:

$$n! = n^n e^{-n} \sqrt{2\pi n} \left(1 + \frac{1}{12n} + \frac{1}{288n^2} - \frac{139}{51840n^3} - \cdots\right).$$

Our work above provides bounds towards this important result; see also the bounds arising from the integral test (Exercise 2.6.1).

Exercise 2.6.13. *Using Stirling's formula, estimate $\binom{2n}{n}$, $\binom{3n}{n}$, and $\binom{bn}{an}$ (where $1 \leq a < b$ are integers) as $n \to \infty$.*

Exercise 2.6.14. *Let $m < n$ be non-negative integers. Find upper and lower bounds for $\binom{n}{m}$ in terms of n and m.*

Exercise 2.6.15. *Prove* $\sum_{k=0}^{n}(-1)^n \binom{n}{k} = 0$.

Exercise 2.6.16. *Prove the **Binomial Theorem**:*

$$(x+y)^n = \sum_{k=0}^{n} \binom{n}{k} x^k y^{n-k}.$$

Exercise 2.6.17. *Generalize the previous problem and find a formula for $(x+y+z)^n$, or more generally $(x_1 + \cdots + x_\ell)^n$. The multinomial coefficients might be useful: $\binom{n}{n_1,\ldots,n_\ell} = \frac{n!}{n_1! \cdots n_\ell!}$ (with $n_1 + \cdots + n_\ell = n$).*

Exercise 2.6.18. *Prove that the number of ways to divide c identical cookies among p distinct people is $\binom{c+p-1}{p-1}$; this can also be interpreted as counting the number of non-negative integer tuples that solve $x_1 + \cdots + x_p = c$.*

Exercise 2.6.19. *Prove $\sum_{c=0}^{C} \binom{c+p-1}{p-1} = \binom{C+p}{p}$.*

Pascal's Triangle

Exercise 2.6.20. *Without computing factorials, find $\binom{24}{11}$.*

Exercise 2.6.21. *Find a formula for $\binom{n}{k}$ in terms of entries from row $n-2$.*

Exercise 2.6.22. *Find a formula for $\binom{n}{k}$ in terms of entries from row $n+1$.*

Exercise 2.6.23. *Prove $\sum_{k=0}^{n} \binom{n}{k}^2 = \binom{2n}{n}$. Hint: $\binom{n}{k}^2 = \binom{n}{k}\binom{n}{n-k}$. Now tell a story.*

Exercise 2.6.24. *Prove that $\binom{2n}{n}$ is divisible by the product of all primes between n and $2n$. Use this and dyadic decompositions to get a bound on the product of all primes up to $2n$.*

Exercise 2.6.25. *Let T_m be the number of dots in an equilateral triangle with m rows and k dots in row k. Prove $T_{n-1} = \binom{n}{2} = n(n-1)/2$.*

Exercise 2.6.26. *Reprove the formula for T_m from the previous problem by deforming the equilateral triangle to an isosceles right triangle and using two such triangles to make a square. Hint: be careful about double counting.*

Exercise 2.6.27. *The outside diagonals of Pascal's triangle are all 1's, the next diagonal are the integers as $\binom{n}{1} = 1$, and the diagonal after that are the triangular numbers as $T_{n-1} = \binom{n}{2} = n(n-1)/2$. Prove that every positive integer is a sum of at most 3 triangular numbers.*

Exercise 2.6.28. *In the previous problem you showed every positive integer N is a sum of at most 3 triangular numbers. What if you want to write N as a sum of a fixed number of distinct triangular numbers? Is that possible? If so how many triangular numbers are needed?*

A **number pyramid** of level n and base $x_1, \ldots, x_n$ starts with those numbers, in that order, as the base, and then adds up to reach the top. For example, the number pyramid of level 4 and numbers 14, 11, 9, 17 would be

$$91$$

$$45 \quad 46$$

$$25 \quad 20 \quad 26$$

$$14 \quad 11 \quad 9 \quad 17$$

In my son's challenge math problems one day they were given problems such as this, with the four numbers for the bottom and the pyramid empty save for the number at the top. The goal was to figure out the order of the numbers to place on the bottom.

Exercise 2.6.29. *How many ways are there to place the n numbers in the bottom row of the pyramid?*

Exercise 2.6.30. *Up to symmetry, how many ways are there to place the n numbers in the bottom row of the pyramid?*

Exercise 2.6.31. *How many additions does it take to fill in an n-level pyramid once the bottom is given?*

Exercise 2.6.32. *If $n = 4$, how many additions will it take to check each of the possibilities? If you use symmetry how many additions will it take?*

Exercise 2.6.33. *Imagine now our goal is not to fill in the pyramid, but just to determine if the order on the bottom leads to the number given at the top. One solution of course is to fill in the entire pyramid, but that is costly. If $n = 4$ find a simple condition which is easily checked which tells us if the order on the bottom leads to the number on the top. How much of a savings is this if we use it for all possibilities for the bottom (with and without symmetry)?*

Exercise 2.6.34. *Generalize the previous problem to $n = 5$.*

Exercise 2.6.35. *Generalize Exercise 2.6.33 to arbitrary n.*

Exercise 2.6.36. *Generalize Exercise 2.6.33 to a three-dimensional pyramid, where now each cell is the sum of four numbers below it.*

Exercise 2.6.37. *Another type of pyramid is **Pascal's pyramid** (sometimes called **Pascal's tetrahedron**). The numbers come from the expansion of $(x + y + z)^n$. Prove that at level n we have*

$$(x + y + z)^n = \sum_{\substack{0 \le i,j,k \le n \\ i+j+k=n}} \frac{n!}{i!j!k!} x^i y^j z^k;$$

*the $n!/i!j!k!$ are called the **trinomial coefficients**. How should these be placed to form the pyramid? What would the generalization be to four dimensions?*

Exercise 2.6.38. *We have seen the power of parentheses and grouping; we can use this to obtain a formula for the trinomial coefficients by writing $(x + y + z)^n$ as $((x + y) + z)^n$ and using the binomial theorem twice. Flesh out this idea and express the trinomial coefficients as a product of binomial coefficients.*

Dimension

Exercise 2.6.39. *Prove that C_n is the union of 2^n intervals of length $1/3^n$, and thus the length of C_n tends to zero. In particular, its dimension must be less than 1.*

Exercise 2.6.40. *Consider a modified Cantor set (a **fat Cantor set**) where, instead of removing the middle third each time, we instead remove the middle $1/a$; thus $C_1(a) = \{0, \frac{1}{2a}\} \cup \{1 - \frac{1}{2a}, 1\}$. What is the length of $C_n(a)$ as $n \to \infty$? Is there a choice of where that length is positive? What is the dimension?*

Exercise 2.6.41. *Generalize the previous problem so that at the nth stage we remove the middle $1/a_n$. For what sequences a_n are we left with a set of positive length?*

Exercise 2.6.42. *Generalize the Cantor set to a two-dimensional space; thus we start with a unit square and throw away the middle of nine squares. What is the dimension of the limiting set?*

Exercise 2.6.43. *Generalize the previous problem to a hypercube in $\mathbb{R}^n$. What is the dimension of the limiting set?*

Exercise 2.6.44. *Show $1/4$ and $3/4$ are in C, but neither is an endpoint.*

Exercise 2.6.45. *Show the Cantor set is also the set of all numbers $x \in [0,1]$ which have no 1's in their ternary (base three) expansion. If we have a rational such as $1/3$, we write it by using repeating 2's: $1/3 = .02222\ldots$ in base three. By considering base two expansions, show there is a one-to-one and onto map from $[0,1]$ to the Cantor set.*

Exercise 2.6.46. *Use the previous exercise to show that every $x \in [0,2]$ can be written as a sum $y + z$ with $y, z \in C$.*

Exercise 2.6.47. *Another interesting fractal set is the **Koch snowflake**. Start with an equilateral triangle. At each stage split every edge into thirds, and draw an equilateral triangle at the middle segment. At stage n how many segments are there on the boundary? What is the perimeter at stage n? What is the area? What do you think is the limiting dimension? Note that our previous notion of Hausdorff dimension is not appropriate here.*

For more on the Cantor set, including dynamical interpretations, see [**Dev, Edg, Fal**].

From the Pascal to the Sierpinski Triangle

Exercise 2.6.48. *Find as fast of a way as you can to determine $\binom{n}{k}$ modulo 2 (i.e., is it even or odd?).*

Exercise 2.6.49. *Given $\binom{n}{k}$ mod 2 find as fast a way as you can to determine $\binom{n}{k+1}$. Notice this can be used to quickly construct a row of Pascal's triangle modulo 2, as $\binom{n}{0} = 1$.*

Exercise 2.6.50. *Given row $n-1$, find as fast a way as you can to determine row n.*

Exercise 2.6.51. *Use recursion to find an efficient way to determine large numbers of entries of Pascal's triangle modulo 2.*

Exercise 2.6.52. *Investigate what happens if instead of looking at the entries of Pascal's triangle modulo 2 we instead look at them modulo m. Is there different behavior for m composite versus prime?*

Exercise 2.6.53. *For a prime p, the **Legendre symbol** $\left(\frac{n}{p}\right)$ is defined to be 1 if n is a non-zero square modulo p, 0 if $n \equiv 0 \bmod p$, and -1 otherwise. Apply the Legendre symbol to each $\binom{n}{k}$ in Pascal's triangle and investigate the result.*

Exercise 2.6.54. *Look at Pascal's pyramid (see Exercise 2.6.37) modulo 2; what can you say about the shape? If we constantly rescale so that it lives in the unit cube, what is its dimension?*

The Euclidean Algorithm

Exercise 2.6.55. *Prove that the brute force approach to finding the greatest common divisor of x and y takes at most x steps. For a fixed y, how long on average does it take to find the gcd using this method? What if we started at 1 instead of at x? Would that be a better choice?.*

Exercise 2.6.56. *Here is another way to find the gcd of $x < y$: factor x and y and use the factorizations to determine the gcd. If $x = p_1^{r_1} \cdots p_k^{r_k}$ and $y = p_1^{s_1} \cdots p_k^{s_k}$, where $r_i, s_i \in \{0, 1, 2, \ldots\}$ and $p_1 = 2, p_2 = 3, p_3 = 5, \ldots$ are the primes, find $\gcd(x, y)$ in terms of the p_i, r_i, and s_i. How long does this method take? Why is this not used in practice?*

Exercise 2.6.57. *Find $\gcd(1776, 24601)$, $\gcd(1701, 1793)$, $\gcd(377, 610)$, and $\gcd(123456789, 987654321)$.*

Exercise 2.6.58. *Find $\gcd(F_n, F_m)$ where F_k is the kth Fibonacci number (normalized so that $F_0 = 0$, $F_1 = 1$, and $F_{n+1} = F_n + F_{n-1}$). Your answer should depend on n and m and the Fibonacci numbers.*

Exercise 2.6.59. *Prove that if $d = \gcd(x, y)$ and $\widetilde{d}$ divides x and y, then $\widetilde{d}$ divides d.* Hint: the Fundamental Theorem of Arithmetic, which states every integer can be written uniquely as a product of prime powers, might be useful.

Exercise 2.6.60. *Let $(\mathbb{Z}/N\mathbb{Z})^*$ be the subset of numbers in $\{0, 1, 2, \ldots, N\}$ which are relatively prime to N. Use Theorem 2.5.1 to prove that this is a multiplicative group; the difficulty is showing that given an x in the set there is an element, which we denote by x^{-1}, such that $xx^{-1} \equiv 1 \bmod N$.*

Exercise 2.6.61. *What is the relation between $\gcd(x^2, y^2)$ and $\gcd(x, y)$? More generally, if k, ℓ are positive integers is there a nice relation between $\gcd(x^k, y^\ell)$ and $\gcd(x, y)$?*

Exercise 2.6.62. *Prove that $x^2 \bmod N$ is the same as $(x \bmod N) \cdot (x \bmod N) \bmod N$; in other words, we can multiply and then reduce modulo N, or we can first reduce modulo N and then multiply and reduce. Generalize to x^r. If N has d digits, what is the largest number we need to be able to compute and store to find $x^r \bmod N$?*

Exercise 2.6.63. *Find all* (α, β) *such that* $\alpha x + \beta y = 0$; *of course,* α *and* β *will depend on* x *and* y.

Exercise 2.6.64. *Let* x, y *be positive numbers and assume we have integers* a, b *such that* $ax + by = \gcd(x, y)$. *Prove that* a *and* b *are either of opposite sign or one of them is zero. Use this to show that if a solution exists, then we can always find a solution with* $|a| \le y$ *and* $|b| \le x$.

Exercise 2.6.65. *Prove that there is always a solution to* $ax + by = \gcd(x, y)$. *How many have* $|a| \le y$ *and* $|b| \le x$?

Exercise 2.6.66. *Prove that the Euclidean algorithm terminates with some* $r_m = 0$.

Exercise 2.6.67. *Complete the proof sketched in our analysis of the Euclidean algorithm that we can pass from the* a_i's *and* r_i's *to finding* a *and* b *such that* $ax + by = \gcd(x, y)$.

Exercise 2.6.68. *Using the Euclidean algorithm, find* a *and* b *in terms of the* a_i's *and* r_i's.

Exercise 2.6.69. *How many multiplications are needed to compute* a *and* b *using the formulas from Exercise 2.6.68?*

Exercise 2.6.70. *Find* a *and* b *such that* $ax + by = \gcd(1776, 24601)$.

Exercise 2.6.71. *Investigate the Euclidean algorithm for various choices of* x *and* y. *What values cause it to take a long time? A short time? For problems like this you need to figure out what is the right metric to measure success. For example, if* $x < y$ *and it takes* s *steps, a good measure might be* $s / \log_2(x)$.

Exercise 2.6.72. *We proved that* $r_{i+2} \le r_i/2$ *in the Euclidean algorithm; thus the worst case for us is if* $r_{i+2} = r_i$. *Can this hold for say all even indices* i? *If so, for what pairs of numbers does this happen. If it cannot always hold, how many times can we have it?*

Exercise 2.6.73. *In the spirit of the previous exercise, we analyzed three consecutive steps in the Euclidean algorithm to show* $r_{i+2} \le r_i/2$. *Would we get a better bound if instead we looked at four steps? Five? Six?*

Exercise 2.6.74. *Generalize the Euclidean algorithm to find* $\gcd(x, y, z)$. *How quickly can we do this? What about finding* $\gcd(x_1, \ldots, x_n)$; *what is the fewest steps you need to compute that?*

Exercise 2.6.75. *The* **least common multiple (lcm)** *of two integers* x *and* y *is the smallest integer both divide. Prove that* $\mathrm{lcm}(x, y)$ *exists. What would be the brute force algorithm to find it?*

Exercise 2.6.76. *Building on the previous problem, find the most efficient algorithm you can to compute* $\mathrm{lcm}(x, y)$.

Exercise 2.6.77. *Generalize Exercise 2.6.56 to use the factorizations of* x *and* y *to quickly compute their lcm.*

Exercise 2.6.78. *Find as efficient an algorithm as you can to compute* $\mathrm{lcm}(x_1, \ldots, x_n)$.

Exercise 2.6.79. *Let* $\mathbb{Z}[i] = \{a + bi : a, b \in \mathbb{Z}\}$ *be the set of* **Gaussian integers**, *where* $i = \sqrt{-1}$. *Define the norm of a number* $z = a + ib$ *by* $||z|| = \sqrt{a^2 + b^2}$. *Generalize the Euclidean algorithm to this setting.*

Exercise 2.6.80. *Let* $\mathbb{Z}[\sqrt{d}] = \{a + b\sqrt{d} : a, b \in \mathbb{Z}\}$ *for a square-free integer* d. *Can you find choices of* d *where a Euclidean algorithm exists?*

The following exercises give a sense of how **RSA encryption** works and the role the Euclidean algorithm and fast multiplication play in it; for more details see [**CM**]. Briefly, Bob wants to send a secure message to Alice without having to previously meet with her and agree upon a password. It suffices to send numbers, as we can have 00 represent a space, 01 an A, 02 a B and so on; thus transmitting the number 130120 sends the message CAT.

Alice chooses two large, distinct primes p and q and finds their product $N = pq$; you should think of these numbers as having at least 200 digits to provide a reasonable sense of security. She then computes $\phi(N)$ (called the **Euler totient function**), which is the number of integers in $\{1, 2, \ldots, N\}$ relatively prime to N (see Exercise 2.6.60), and chooses an e (for encrypt) and d (for decrypt) such that $ed \equiv 1 \bmod \phi(N)$. She then publishes the pair (N, e) but keeps p, q and d secret. Bob chooses a message $M < N$ and sends $X = M^e \bmod N$ to Alice, who then computes $Y = X^d \bmod N$; amazingly, Y equals the original message M!

Exercise 2.6.81. *Discuss how fast multiplication (repeated squaring) is used in encrypting and decrypting the message.*

Exercise 2.6.82. *Prove* $\phi(N) = \phi(pq) = (p-1)(q-1)$. *More generally, show* $\phi(p^k) = p^k - p^{k-1}$, *and if* m *and* n *are relatively prime, then* $\phi(mn) = \phi(m)\phi(n)$; *use this property to find* $\phi(n)$ *for arbitrary input.*

Exercise 2.6.83. *Alice needs to find a pair* e *and* d *such that* $ed \equiv 1 \bmod \phi(N)$. *She decides to randomly choose* $e \in \{2, \ldots, \phi(N) - 1\}$. *If her choice is relatively prime to* $\phi(N)$ *show the Euclidean algorithm yields the desired* d *(and quickly too!).*

Exercise 2.6.84. *Let's revisit the previous problem and assume Alice is unlucky and her first choice of* e *shares a factor with* $\phi(N)$. *She decides to keep incrementing* e *by 1 until she reaches a number relatively prime to* $\phi(N)$. *Bound how many numbers she must test before she is successful.*

Exercise 2.6.85. *The reason RSA works is that* $Y = M$ *(the reason it is useful is that repeated squaring and the Euclidean algorithm allow us to perform the needed constructions, encrypting and decrypting quickly). Prove that* $Y = M$. *Hint:* **Fermat's little Theorem**, *which states that if* a *is relatively prime to* m, *then* $a^{\phi(m)} \equiv 1 \bmod m$ *might be useful.*

Exercise 2.6.86. *Prove Fermat's little Theorem (see the previous exercise).*

Exercise 2.6.87. *We discussed above how to find* e *and* d, *but implicit in our arguments above is that Alice can find two large primes. Discuss how she might do this efficiently.*

Exercise 2.6.88. *Imagine Bob chooses a message* M *that is not relatively prime to* N. *Show that Bob can deduce all of Alice's secret information and thus read all her messages.*

The problems above are meant to give a brief sense of the importance of what we have studied so far. While we are concentrating on these algorithms in part because they highlight important techniques, they also play central roles in some of the most important applied mathematics problems of our age: the need to securely send and read secure information.

Exercise 2.6.89. *A fundamental problem in cryptography is the threat that some-one might be masking as a trusted friend; for example, just because Alice receives a message from someone claiming to be Bob does not mean Bob sent the message! Or perhaps the person who says she is Alice is not! It's thus essential that Alice and Bob have a way to verify each other's identities. Generalize the RSA algorithm to allow Bob to 'sign' his message so that Alice knows it must come from him.*

Part 2

Introduction to Linear Programming

Chapter 3

Introduction to Linear Programming

In this chapter we study one of the earliest successes of *linear programming*: the Diet Problem. Before doing so, we quickly review some of the problems linear algebra can handle, and then discuss some natural generalizations which lead to linear programming. The general framework is that we have a set of constraints and some objective function; as the name suggests, the variables and constraints are linear relationships. From this we get a set of feasible solutions, which are solutions that satisfy our constraints. We want to choose one of these that maximizes (or minimizes if we're so inclined) our objective function. There are a variety of issues that can arise, from determining the existence of solutions to searching large spaces efficiently to find an optimal choice (or at least get close to one).

The subject is also called *linear optimization*; while this name fits better with the title and theme of our book, we use linear programming below as that highlights our technique of writing programs to find the desired solutions. The following quote, from *Linear Programming* by Dantzig and Thapa [**DaTh**], gives Dantzig's recollection of the origin of some of the terminology:

> Here are some stories about how various linear programming terms arose. The military refer to their various plans or proposed schedules of training, logistical supply and deployment of combat units as a program. When I first analyzed the Air Force planning problem and saw that it could be formulated as a system of linear inequalities, I called my paper Programming in a Linear Structure. Note that the term 'program' was used for linear programs long before it was used as the set of instructions used by a computer. In the early days, these instructions were called codes.
>
> In the summer of 1948, Koopmans and I visited the Rand Corporation. One day we took a stroll along the Santa Monica beach. Koopmans said: "Why not shorten 'Programming in a Linear Structure' to 'Linear Programming'?" I replied: "That's it! From now on that will be its

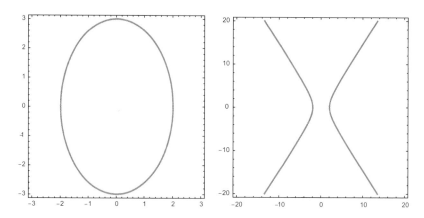

Figure 1. Left: ellipse with $a = 2, b = 3$. Right: hyperbola with $a = 2, b = 3$.

name." Later that day I gave a talk at Rand, entitled "Linear Programming"; years later Tucker shortened it to Linear Program.

3.1. Linear Algebra

One of the central problems in this subject is to find solutions to a matrix equation. Specifically, given an $n \times m$ matrix $\mathbf{A}$ (n rows and m columns) and a vector $\vec{b}$ with n components, can we find a vector $\vec{x}$ with m entries such that

$$\mathbf{A}\vec{x} = \vec{b}.$$

Here are just a few instances where this problem emerges:

- We saw this in the matrix formulation of the Fibonacci problems in §1.6:

$$\begin{pmatrix} F_{n+1} \\ F_n \end{pmatrix} = \begin{pmatrix} 1 & 1 \\ 1 & 0 \end{pmatrix} \begin{pmatrix} F_n \\ F_{n-1} \end{pmatrix}, \quad \begin{pmatrix} F_1 \\ F_0 \end{pmatrix} = \begin{pmatrix} 1 \\ 0 \end{pmatrix}.$$

This, of course, can be generalized to other difference equations.

- Another example is solving a system of linear equations. Given n equations in m unknowns, say

$$a_{r1}x_1 + a_{r2}x_2 + \cdots + a_{rm}x_m = b_r, \quad 1 \leq r \leq n,$$

is there a simultaneous solution $(x_1, \ldots, x_m)$?

- You may remember conic sections from high school algebra, which almost surely concentrated on the case when the figures are aligned with the coordinate axes. Such an **ellipse** is the set of solutions to $(x/a)^2 + (y/b)^2 = 1$, while a hyperbola would be either $(x/a)^2 - (y/b)^2 = 1$ or $(y/b)^2 - (x/a)^2 = 1$; see Figure 1.

  We can unify these as the solutions to the following equation:

$$\vec{x}^T \mathbf{A} \vec{x} = 1,$$

where

$$\mathbf{A}_{\text{ellipse}} = \begin{pmatrix} 1/a^2 & 0 \\ 0 & 1/b^2 \end{pmatrix},$$

$$\mathbf{A}_{\text{hyperbola}} = \begin{pmatrix} 1/a^2 & 0 \\ 0 & -1/b^2 \end{pmatrix} \text{ or } \begin{pmatrix} -1/a^2 & 0 \\ 0 & 1/b^2 \end{pmatrix}.$$

The matrix perspective is powerful and suggests a solution to writing down equations when our objects are not parallel to the coordinate axes (see Exercise 3.7.1).

- Many quantum mechanics problems are equivalent to solving $\mathbf{H}\psi_n = E_n \psi_n$, where $\mathbf{H}$ is the Hamiltonian and E_n is the energy level of the eigenstate ψ_n. See [**BFMT-B**] for an introduction to the role random matrices play in nuclear physics (and number theory!).

Now that we're motivated to solve $\mathbf{A}\vec{x} = \vec{b}$, how might we do this? You should have learned techniques to do this in linear algebra; one powerful one is **Gaussian elimination**, where we reduce to an equivalent problem $\mathbf{T}\vec{x} = \vec{b}'$ with $\mathbf{T}$ an upper triangular matrix. In Exercise 3.7.2 you're asked to investigate the efficiency of this process.

The questions below serve as useful guides here and throughout the book. Interestingly, sometimes the easiest way to prove that a solution exists is to show that there are many and thus there must be at least one. Examples of this in pure math include finding primes in arithmetic progressions (for a general arithmetic progression $an + b$ with $\gcd(a, b) = 1$ the only way we know to prove there is at least one prime of this form is to show that there are infinitely many) and the ternary Goldbach problem (to show that an odd number can be written as a sum of three primes we show it can be written as such a sum in many ways).

General Questions about Solutions: Given an equation, the following are natural questions to ask.

(1) Is there a solution?
(2) How many solutions are there?
(3) How can we (efficiently) find the solutions (or at least a solution)?

Returning to $\mathbf{A}\vec{x} = \vec{b}$, the only possibilities are that it has 0, 1, or infinitely many solutions; see Exercise 3.7.14. Which case do you think is going to be the most interesting for us as we move to optimization problems?

Right now we're trying to solve an equation; when we turn to linear programming we're going to additionally require our solution to minimize or maximize a desired quantity. Clearly, in the case of just 0 or 1 solution our analysis stops as soon as we find the solution (or prove none exist). If there's at most one answer, we can't look at different ones and see which is 'best'; however, if there are infinitely many solutions then we can search the solution space and try to find one with additional properties.

3.2. Finding Solutions

In the last section we isolated three general questions: (1) is there a solution, (2) how many solutions are there, and (3) how do we find them? As all three of these will play a key role in our linear programming explorations, we take a few moments to discuss some simpler problems and methods.

One of the best ways to find a solution is the **method of Divine Inspiration**. Briefly, the way it works is you look at the problem, think for a few moments, and then write down the solution. Of course, in general it's hard to know just when that flash of divine insight will happen; however, for many problems it's easy to quickly find at least one solution.

For example, consider the polynomial equation

$$x^3 y^3 z^3 - 12 x^4 y^4 z^4 + (x^2 + y^2 + z^2)(x + y + z) + 1701(xy + yz + zx) - 24601 xyz \;=\; 0.$$

It's quite difficult to write down all the solutions; however, it's very easy to find one: take $x = y = z = 0$. We find this by noticing that every term is a multiple of at least one of the x, y, or z, and thus if they are all zero then the expression vanishes. Furthermore, while it's hard to write down all solutions we can easily see some properties the set of solutions has. For example, if (α, β, γ) is a solution, then so too are

$$(\alpha, \gamma, \beta), \; (\beta, \alpha, \gamma), \; (\beta, \gamma, \alpha), \; (\gamma, \alpha, \beta), \; (\gamma, \beta, \alpha);$$

to see this, note that any permutation of (x, y, z) leaves the polynomial unchanged. Thus, without solving the problem we can deduce some properties of the set of solutions.

Let's take a more interesting example: **Pell's equation**. Given a square-free integer d, consider the equation

$$x^2 - dy^2 \;=\; 0;$$

we want to find all the integer solutions. For definiteness, let's take $d = 2$ and study $x^2 - 2y^2 = 1$. There's a beautiful theory describing these equations (and wonderful connections with continued fractions). The method of divine inspiration quickly gives $(1, 0)$ as a solution, but let's see if we can find something more interesting. Fortunately a little trial and error yields $(3, 2)$, and as the equation only involves the squares of the numbers, we also have $(3, -2), (-3, 2)$, and $(-3, -2)$. Are there any others? Staring at it a bit longer doesn't seem to help, but it's very easy to write a simple program to check pairs; for many problems **brute force searching** is a great way to proceed. Without loss of generality we can assume x and y are positive integers. The following simple Mathematica code finds all the solutions with $1 \le x, y \le 10000$.

```
For[x = 1, x <= 10000, x++,
  For[y = 1, y <= 10000, y++,
   If[x^2 - 2 y^2 == 1, Print["(", x, ", ", y, ")"]
    ]]].
```

While the code above is easy to write, with just a little work we can make it more efficient. Clearly $x > y$, so rather than having y run up to 10000 we might as well stop at x.

```
For[x = 2, x <= 10000, x++,
  For[y = 1, y < x, y++,
    If[x^2 - 2 y^2 == 1, Print["(", x, ", ", y, ")"]
    ]]].
```

Of course, this is still quite inefficient. There's no need to loop on two variables; once we choose x or y the other is forced. The shorter code below assumes we have a simple way to check if a number is an integer; if your environment doesn't have such a function you'll have to code slightly differently.

```
For[x = 2, x <= 10000, x++
  {
    y = Sqrt[(x^2 - 1)/2];
    If[IntegerQ[y] == True, Print["(", x, ", ", y, ")"]];
  }].
```

While in this range all three (fortunately!) return the same set of solutions,

$$(3, 2), \quad (17, 12), \quad (99, 70), \quad (577, 408), \quad (3363, 2378),$$

there is a marked difference in run-times. On my laptop the first took 234.9 seconds, the second 118.8 seconds, and the third 0.187 seconds!

3.3. Calculus Review: Local versus Global

As many of the optimization ideas from calculus transfer over to linear programming, it's worth quickly reviewing constrained **extrema** (i.e., maxima and minima). Typically we're given a function f on a compact set S with boundary $C = g(x_1, \ldots, x_n)$, and we want to find the maxima and minima of f. How do we proceed?

If all we know is f is continuous, we're often stuck; however, if f and g are differentiable, then there are powerful techniques. We break the problem into two cases which are solved differently. We first do interior extrema, points inside S, and then turn to extrema on S's boundary.

For the interior extrema, we look for points at which the **gradient** ∇f vanishes:

$$(\nabla f)(a_1, \ldots, a_n) := \left(\frac{\partial f}{\partial x_1}(a_1, \ldots, a_n), \ldots, \frac{\partial f}{\partial x_n}(a_1, \ldots, a_n) \right) = \vec{0}.$$

This is the generalization of finding **critical points** (points where the derivative vanishes) in one dimension; if there was a direction where the gradient was non-zero, flowing one way in that direction would increase the value of f while flowing in the opposite direction would decrease it. Thus, the only candidates for interior extrema are the critical points; however, it is not the case that every critical point is an extremum.

Finding extrema on the boundary is more interesting. Here we use **Lagrange multipliers**. We look for points $(a_1, \ldots, a_n)$ that simultaneously satisfy

$$(\nabla f)(a_1, \ldots, a_n) \;=\; \lambda (\nabla g)(a_1, \ldots, a_n), \quad g(a_1, \ldots, a_n) = C$$

for some $\lambda \neq 0$. The reason this works is that the gradient of g is in the direction of the normal to the surface S, and if the gradient of f is not in the same direction as the gradient of g, then there is a direction tangent to S to move where the function f is increasing (and in the opposite direction f is decreasing). As the derivative has n components, we see we have $n+1$ equations in $n+1$ variables (the a_i's and λ), and thus there is a hope we can solve this. The solutions to this system give us *candidate* points; some of these may not actually be extrema, but there can't be extrema anywhere else on the boundary.

We won't be using Lagrange multipliers much in this course, but they highlight a very powerful idea which we *will* use, time and time again. Initially we have infinitely many points on the boundary to check, but typically after applying Lagrange multipliers we reduce to a small, finite set of candidates. It's a lot easier to check a handful of points along a boundary than it is to check an infinite amount of them.

Before we move on, however, there is another important lesson from Lagrange multipliers. It's fine to say that the method yields candidate points, but how do we actually *find* these points? For example, say we start at a point $\vec{a} = (a_1, \ldots, a_n)$ where $(\nabla f)(\vec{a}) \neq \lambda (\nabla g)(\vec{a})$ for any non-zero λ. This point can't be an extremum, so we need to keep looking. What point should we try next? Since the two gradients are not parallel, the gradient of f at $\vec{a}$ has some component that is not in the same direction as the normal vector to the boundary. Which way along the boundary should we go to find, say, the maximum? Our gut instinct would be to go in the direction of this additional component of $(\nabla f)(\vec{a})$ (or the opposite direction). By doing this, we can use Lagrange multipliers to obtain a fairly efficient algorithm for finding extrema along a boundary.

We need to be careful, though: just because the direction of the component of $(\nabla f)(\vec{a})$ is the best way to go to increase (or decrease) the value of f *locally* doesn't mean that it's necessarily the best way to maximize (or minimize) f *globally*. This is an example of a more general phenomenon: being greedy locally isn't necessarily the best way to operate globally, as we will now see.

The example below is from a beautiful series of papers [**MiPe, Pe**] about whether or not Pennings' dog, Elvis, knows calculus. Imagine you and Elvis are playing in the water, a short distance from the shore. Say you throw Elvis a ball in a straight line roughly parallel to the shore, and assume Elvis isn't allowed to start going after it until it's landed in the water (see Figure 2). What's the most efficient way for Elvis to get to the ball?

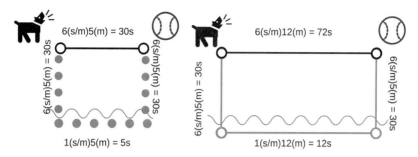

Figure 2. Left: Where it's more efficient for Elvis to just swim (30s) rather than use the land (65s), the local max is the global max. Right: Where both paths are equally efficient (72s).

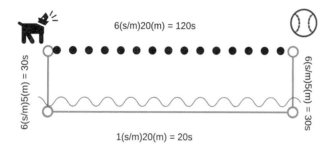

Figure 3. Where it is more efficient for Elvis to use the land (80s) rather than just swim (120s).

The first and most natural option is for Elvis to just swim (or doggie paddle) straight to the ball. For short throws, this will be the most efficient way to reach it (see Figure 2), but now say the ball has been thrown a long distance. Will this still be the fastest way for Elvis to reach his ball? No! Instead of swimming all the way there, it's more efficient for Elvis to make his way towards the shore, run most of the way to the ball, then swim in to get it, as Elvis can move much faster on land than in the water; see Figure 3. Note that the pictures are not accurately drawn; if Elvis does not have an infinite speed on land, then he will angle towards land, run on land, then angle out. As a nice exercise, determine what angle he should take; the answer is related to Snell's law and how light propagates through different media.

For long distances, the most efficient path for Elvis *globally* is to swim to the shore, then run, then swim to the ball; but *locally*, if we look only at his progress a short while after he's begun moving, it's clear that he's closed more distance by swimming straight towards the ball than by swimming diagonally towards the shore.

What's the lesson? If you take as your strategy "I will always do what is locally best", you can sometimes get stuck in a suboptimal outcome. You can reach a local extremum and not the global one. This is a danger we must always be wary of when designing algorithms.

3.4. An Introduction to the Diet Problem

It's time to get started with the meat of the course, with true operations research and solving real problems. We'll start with the **Diet Problem**: what is the cheapest way to keep people alive? Later we'll revisit this and take care of other concerns, such as variety and quality; doing so will lead us to multi-objective linear programming. This is one of the most important problems you might face if, say, you were to consider a career as a dictator. You'd want your followers to be fed enough so that they don't die or revolt, and maybe have energy left over to construct monuments in your honor. Or, in a more historical situation, maybe you are in charge of supplies for the U.S. Army and want a diet that keeps soldiers alive and in good fighting shape, yet allows you to use your budget for other purchases. There are many variants and applications, both to this problem and natural generalizations to other industries. See [**St**] for one of the first papers in the subject and [**Da2**] for an entertaining account of early attempts at solving it through linear programming's simplex method (a small input mistake had the optimal diet for one problem include 500 gallons of vinegar – a day!).

So, we're (finally!) ready for our first foray into linear programming. Let's say we're shopping for food, and we have two options. For simplicity we'll assume they're spoo and gagh (two favorites in the sci-fi community), and all we care about is iron and protein.

- Spoo costs \$20 per unit, with each unit of spoo containing 30 units of iron and 15 units of protein.

- Gagh costs \$2 per unit, with each unit of gagh containing 5 units of iron and 10 units of protein.

Our goal is to live cheaply, but we need to have 60 units of iron and 70 units of protein in our diet or we die. Right now nothing matters but the cost and the minimum daily requirements; taste is not important, though an interesting generalization adds that as a secondary constraint.

Say we buy x_1 units of spoo and x_2 units of gagh. We can start by writing our constraints mathematically:

$$30x_1 + 5x_2 \geq 60 \text{ (iron)},$$
$$15x_1 + 10x_2 \geq 70 \text{ (protein)},$$
$$x_1, x_2 \geq 0.$$

The first two constraints reflect how much iron and how much protein we consume, and ensure that we meet the minimum requirements. The third wasn't directly mentioned, but is quite important: we are not allowed to eat negative quantities! In many problems we'll have constraints like this that are implicitly in the problem; it's essential that we make them explicit.

Next we write out our goal. We want to minimize the cost C, which is given by the expression

$$C = 20x_1 + 2x_2.$$

The nutritional values of our two foods were deliberately chosen to make this particular problem easy to solve. A little inspection gives the best choice is an all-gagh diet. Although 1 unit of gagh is subsumed by 1 unit of spoo, 2 units of gagh already has more protein, and by the time we reach 10 units of gagh the price is the same as that of 1 unit of spoo and yet we have both more iron and more protein. Thus, no advanced mathematics is needed to solve this realization!

What would we need to do to make this a reasonable problem? One way is to just increase the cost of gagh; but if we increase the cost of gagh too much, we're just going to buy spoo. We can see that as we vary the cost of gagh, we'll go from an all-gagh diet to a mixed diet, and finally to an all-spoo diet. In the next section we'll see how to solve this problem no matter what the costs and nutritional values may be. Of course, we're making assumptions here; what if we don't know the costs beforehand? This leads to dynamic programming, which we'll discuss later; for now, however, we assume that we have perfect knowledge of all the prices.

Finally, it's worth collecting and generalizing what we've done.

Diet Problem: two foods. Imagine we have two foods and two nutrients, where one unit of the first has a_{11} units of the first nutrient and a_{21} of the second, while our second food has a_{12} units of the first nutrient and a_{22} of the second. Assume we need b_i units of the nutrient i each day to survive, and the cost of food k is c_k dollars per unit. Our goal is to minimize

$$c_1 x_1 + c_2 x_2$$

subject to the condition that

$$\begin{pmatrix} a_{11} & a_{12} \\ a_{21} & a_{22} \end{pmatrix} \begin{pmatrix} x_1 \\ x_2 \end{pmatrix} \geq \begin{pmatrix} b_1 \\ b_2 \end{pmatrix}, \quad x_1, x_2 \geq 0.$$

We can rewrite this as

$$\text{Goal}: \text{ minimize } \vec{c}^{\,T} \vec{x} \text{ subject to } \mathbf{A}\vec{x} \geq \vec{b} \text{ and } \vec{x} \geq \vec{0}.$$

Notice how similar this is to the linear algebra problems we've previously studied. There are a few important differences. The first is that instead of $\mathbf{A}\vec{x} = \vec{b}$ we now have $\mathbf{A}\vec{x} \geq \vec{b}$, although we'll later see that by adding additional variables we can replace the inequality laden problem with an equivalent one involving only equalities. Next, we have constraints on the variables; here the x_i's must be non-negative. Finally, and most importantly, we want more than just a pair (x_1, x_2) that satisfy our constraints; we want a pair that leads to a minimal cost.

There are a lot of similarities with this problem and the Lagrange multipliers. In both cases we have a constrained optimization. Lagrange multipliers helped us move from having to examine all points on the boundary to a small finite set of candidate points; fortunately something similar is possible here.

3.5. Solving the Diet Problem

A good maxim to go by is to always draw a picture. It will help to plot our constraints on a graph, with x_1 on the horizontal axis and x_2 on the vertical axis.

- We start with the constraint $30x_1 + 5x_2 \geq 60$. If we had an equality this would be a line, thus the inequality means it will be all pairs of points (x_1, x_2) above and to the right of this line. When $x_1 = 2$ we hit the horizontal axis,

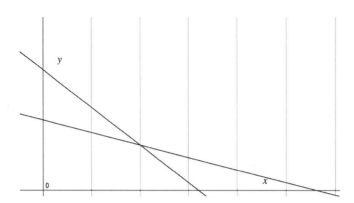

Figure 4.   Plot of constraints for Diet Problem.

and when $x_2 = 12$ we hit the vertical. Thus any points above this line satisfy our iron constraint.

- Now we look at the second constraint: $15x_1 + 10x_2 \geq 70$. We get a vertical intercept at $x_2 = 7$ and a horizontal intercept somewhere between $x_1 = 3$ and $x_1 = 4$; any points above this second line satisfy our protein constraint.

- We also have the constraints that $x_1, x_2 \geq 0$: this means we must stay in the first quadrant of the graph, above the horizontal axis and to the right of the vertical axis.

We now have a graphical representation of the feasible diets. We get a convex region with three special points, which we show in Figure 4 (see (3.1) for $\mathbf{A}$, $\vec{x}$, and $\vec{b}$).

Before finding an optimal solution, we first find all possible feasible solutions. Note that if (x_1, x_2) is a feasible solution, then it must be in the region in the first quadrant above both lines. Thus there are infinitely many candidates for the optimal solution (or solutions!).

Fortunately, we can greatly winnow down the candidate list for optimal solutions. The idea is somewhat similar in spirit to the optimization problem from calculus; here we first show that optimal solutions *cannot* occur in the interior and therefore *must* occur on the boundary of the polygon, and then we show that they *must* occur at a vertex.

We're trying to minimize the cost $20x_1 + 2x_2$. Consider the linear function $\text{Cost}(x_1, x_2) = 20x_1 + 2x_2$. We look at contours where the function is constant: $\text{Cost}(x_1, x_2) = c$; we sketch some constant cost contours in Figure 5.

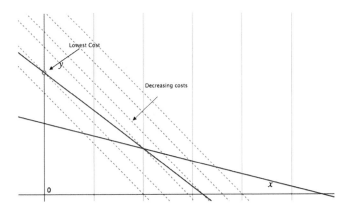

Figure 5. Solving the diet problem.

Note that the contours of constant cost are parallel; this is clear, as they are of the form $20x_1 + 2x_2 = c$. Further, the smaller the value of c, the smaller the cost. Thus, given the choice between two lines, to minimize cost we choose the *lower* line.

Therefore, if we start at any point *inside* the polygon (the set of feasible solutions), by flowing on a line of constant cost we may move to a point on the boundary with the same cost. Thus, to find *an* optimal solution, it suffices to check the feasible solutions on the boundary. This is a general feature of linear programming problems, and it is very helpful in limiting our search to the points along the boundary.

Notice the above looks a lot like the Lagrange multipliers we saw earlier: we're trying to find some way to maneuver along the boundary. Similar to our work there, we can reduce from having to check infinitely many points to just finitely many. It's enough to check just the *vertices* of the polygon. This is because if we're on one of the edges of the polygon, if we move to the left the cost decreases. Thus, it suffices to check the three vertices to find the cheapest diet which contains the minimum daily requirements.

As we keep sliding along the boundary, we eventually hit one of the three vertices. Say we hit the one on the horizontal axis first. Can we keep going left? This now depends on the slope of our cost lines; depending on this, we might be able to keep going until the intersection vertex, or even all the way until the vertex on the vertical axis. Those are the three candidate points.

Because of the cost of a unit of spoo (\$20) and gagh (\$2), the slope of the line is such that the cheapest diet has *no* spoo, and only gagh. This is why we chose such unreasonable numbers for the cost of spoo and gagh, so that we could check our first solution with our intuition.

Let's now consider a more reasonable set of prices, where the cost function is $\mathrm{Cost}(x_1, x_2) = x_1 + x_2$ (so the two products both cost \$1 per unit). We plot some cost lines in Figure 6.

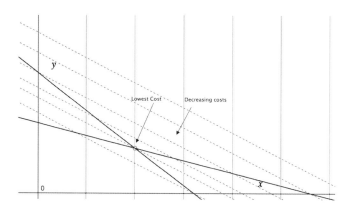

Figure 6. Using new costs, solving the diet problem.

Note now that the optimal solution, while still one of the vertex points, is *not* $x_1 = 0$.

There is one technical issue which we didn't cover. What if the equi-cost lines are parallel to one of the constraint lines? In this case, however, we simply pick one of the vertices on this constraint line—if we're indifferent among the points, we can simply choose one. In real life, we would probably choose the mixed-diet vertex over either of the other two in a situation like this; this is what is helpful about getting a situation where there are many remaining points we are indifferent amongst, as it allows us to bring in additional condition we might want to optimize.

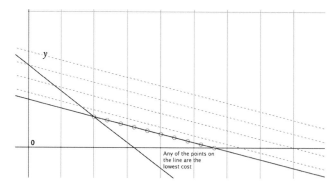

Figure 7. Using costs that parallel a constraint, solving the diet problem.

Our discussion above is a good example of how linear programming works. For a lot of problems, we'll end up having a huge, many-dimensional system that is convex. If we're ever at an interior point, we're not at an extremum, so we flow to the boundary, and continue flowing until we reach a vertex. All that needs to be done is to verify the finite number of vertices at the intersections of the hyperplanes in our many-dimensional space. This is a huge improvement: we've made the problem finite!

Even though we've reduced it to a finite calculation, this may still not be a calculation we can or want to do. We need a way to search the candidate points. This isn't much of a problem in our simplified diet problem example, but this example was just that—simplified and artificial. In the real world, there can be dozens if not hundreds of additional constraints and parameters we will have to manage. The diet problem is very useful and helps to illustrate some key features of linear programming, but we are still going to need a way to quickly go from vertex to vertex in an efficient way. The idea here will be similar to the Euclidean algorithm: we will develop something called the simplex method which will tell us how to navigate the vertices surprisingly efficiently.

3.6. Applications of the Diet Problem

We have gone over the diet problem's solutions graphically, but we have yet to see a way to solve it using other means. Microsoft Excel offers an add-on to quickly and easily solve the Diet Problem, although there are many more powerful programs available to attack problems like this (many of which are freely available). Much of the business world uses Microsoft Excel, so it's worth seeing how to use it to solve linear programming problems.

Diet Problem				
	Spoo	Gagh		
	0	0		
Cost	$ 20.00	$ 2.00	0	Constraints
Iron	30	5	0	60
Protein (g)	15	10	0	70

Figure 8. Setting up the Diet Problem in Excel.

Using Excel, we will set up a problem similar to the one in Figure 8. The green cells will be our quantities of each of the foods, spoo and gagh, which can be filled in by the user. The costs are independent of any other variables, just like the nutrients each food provides. Spoo, in this case, costs 20 dollars, gives 30 units iron and 15 g of protein.

The red cell depends on the quantity of spoo and gagh multiplied by their respective prices. In other words, the cell depends on $Q_{spoo} * P_{spoo} + Q_{gagh} * P_{gagh}$. The blue cells are the sum of the total quantities of the iron and protein for the given combination of spoo and gagh. The orange constraint cells are the how much of each nutrient is needed to survive and are filled in independently of other variables. The end result should be an Excel sheet that changes costs and nutrient quantities if the user alters the green cells.

Using the "Solver Tool" (under Data) we can solve the problem. Figure 9 shows the interface. From top to bottom, our objective function is minimizing the cost of spoo and gagh, represented in cell E6, the red cell. Notice maximums or particular values can also be calculated.

The variable cells are the amounts of spoo and gagh, found in cells C5 and D5. The constraints are more difficult to code. In this case, C5 and D5 (the amounts

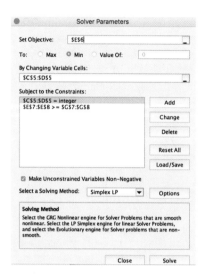

Figure 9. Using Excel to solve the Diet Problem.

of spoo and gagh) both must be integers; because we cannot have 'parts' of spoo or gagh, we must purchase discrete amounts. The second constraint signifies that E7 (the sum of the iron, the top blue cell) $\geq$ G7 (60, the top constraint) and E8 (the sum of protein, the bottom blue cell) $\geq$ G8 (70, the bottom constraint). We want to keep unconstrained variables non-negative (we cannot purchase negative amounts of spoo or gagh) and we want to use the Simplex method to solve the system. We will learn more about the Simplex Method later, but this is a good introduction to it.

If working correctly, the program should solve the system, giving 12 spoo, shown in Figure 10. We also see that the minimum cost is 24, with 60 being the limiting factor.

Diet Problem						
	Spoo	Gagh				
	0	12				
Cost	$ 20.00	$ 2.00	24		Constraints	
Iron	30	5	60			60
Protein (g)	15	10	120			70

Figure 10. Finished using Excel to solve the Diet Problem.

3.7. Exercises

Linear Algebra

Exercise 3.7.1. *Consider an ellipse where the major axis has length $2a$, the minor axis length $2b$, and the major axis makes an angle of $\theta \in [0, \pi/2)$ with the x-axis. Write a matrix equation for the ellipse.*

Exercise 3.7.2. *Let A be an $n \times n$ matrix. How many multiplications are needed to solve the matrix equation $A\vec{x} = \vec{b}$ using Gaussian elimination?*

Exercise 3.7.3. *A matrix is **invertible** if its determinant is zero; equivalently, no linear combination of its row (or its columns) is the zero vector. Let M_n be the $n \times n$ matrix where we fill in the numbers 1 through n^2, with the first row $1, 2, \ldots, n$, the next row $n + 1, n + 2, \ldots, 2n$, and so on. For what values of n is this matrix invertible?*

Exercise 3.7.4. *Solve*

$$\begin{pmatrix} 1 & 3 & 3 \\ 2 & 3 & 5 \\ 2 & 1 & 3 \end{pmatrix} \vec{x} = \begin{pmatrix} 4 \\ 0 \\ 1 \end{pmatrix}.$$

Exercise 3.7.5. *Recall λ is an **eigenvalue** of an $n \times n$ square matrix A with **eigenvector** $\vec{v} \neq \vec{0}$ if $A\vec{v} = \lambda\vec{v}$. Further, if B satisfies $AB = BA$, then we say A and B **commute**. Investigate the relationship between eigenvectors of two commuting matrices.*

Exercise 3.7.6. *Prove that if $\det(A - \lambda I) = 0$, then there is an eigenvector with eigenvalue λ.*

Exercise 3.7.7. *Prove any $n \times n$ matrix has n eigenvalues, not necessarily distinct. If A is real must the eigenvalues be real?*

Exercise 3.7.8. *Obtain bounds for the absolute value of the eigenvalues of an $n \times n$ matrix A in terms of n and its entries a_{ij}. If the eigenvalues of A are real, prove there exists some constant C such that $A + CI$ has positive eigenvalues (here I is the $n \times n$ identity matrix).*

While it may not be criminal to leave you with the impression that the determinant equation from Exercise 3.7.6 is a good way to find eigenvalues, at the very least it's a bad idea and goes against the spirit of this class. There are numerous issues with this method, especially numerical stability. Fill in the details of the following alternative approach.

Exercise 3.7.9. *Let A be a real symmetric matrix, so it has an orthonormal basis of eigenvectors $\vec{v}_1, \ldots, \vec{v}_n$ with real eigenvalues $\lambda_1, \ldots, \lambda_n$; we may order so that $|\lambda_1| \geq |\lambda_2| \geq \cdots \geq |\lambda_n|$. Thus, given any $\vec{v}$ there exist c_i such that*

$$\vec{v} = c_1 \vec{v}_1 + \cdots + c_n \vec{v}_n.$$

Assume for simplicity that the eigenvalues are distinct and $\lambda_1 > 0$. Let $\|\vec{w}\|$ denote the length of the vector $\vec{w}$. Show that as $N \to \infty$, if $c_1 \neq 0$, then

$$\|A^N \vec{v}\| \approx |c_1| \lambda_1^n \|\vec{v}_1\|;$$

thus taking logarithms yields

$$\lambda_1 \approx \frac{1}{n} \log \|A^N \vec{v}\|.$$

Exercise 3.7.10. *Building on the previous problem, once you know λ_1 and $\vec{v}_1$ (or at least approximations for them), how would you approximate λ_2 and $\vec{v}_2$?*

Exercise 3.7.11. *In Exercise 3.7.9 we need to compute $\boldsymbol{A}^N \vec{v}$ for a sequence N tending to infinity; what would be a good choice for N?* Hint: think back to earlier chapters and methods in the book!

Exercise 3.7.12. *One way to solve matrix equations is to calculate the inverse: if $\boldsymbol{A}\vec{x} = \vec{b}$, then $\vec{x} = \boldsymbol{A}^{-1}\vec{b}$ (if the inverse exists). How many multiplications does it take to compute $\boldsymbol{A}^{-1}$?*

Exercise 3.7.13. *The previous problem explored using inverses to solve matrix equations. The formula for a 2×2 matrix is well known:*

$$\text{if } \boldsymbol{A} \;=\; \begin{pmatrix} a & b \\ c & d \end{pmatrix}, \quad \text{then } \boldsymbol{A}^{-1} \;=\; \frac{1}{\det(\boldsymbol{A})} \begin{pmatrix} d & -b \\ -c & a \end{pmatrix},$$

where $\det(\boldsymbol{A}) = ad - bc$. Find a formula for the entries of $\boldsymbol{A}^{-1}$ if $\boldsymbol{A}$ is an invertible 3×3 matrix. Repeat for $\boldsymbol{A}$ an $n \times n$ matrix.

Exercise 3.7.14. *Prove that $\boldsymbol{A}\vec{x} = \vec{b}$ has 0, 1, or infinitely many solutions.* Hint: it suffices to show that if there are two solutions, then there are infinitely many.

Exercise 3.7.15. *We showed that there are either 0, 1, or infinitely many solutions to a matrix equation. Consider now the simultaneous system, where we want to find all $\vec{x}$ such that $\boldsymbol{A}\vec{x} = \vec{b}_1$ and $\boldsymbol{B}\vec{x} = \vec{b}_2$. Now how many solutions can there be? Give examples of the different possibilities.*

Exercise 3.7.16. *Find all solutions to*

$$\begin{pmatrix} 1 & 2 \\ 2 & 4 \end{pmatrix} \begin{pmatrix} x \\ y \end{pmatrix} \;=\; \begin{pmatrix} 1 \\ 2 \end{pmatrix}.$$

Among all the solutions, find the one with smallest $x^2 + y^2$ (if possible) and the one with largest $x^2 + y^2$ (if possible).

Finding Solutions

Exercise 3.7.17. *Let $\boldsymbol{A} = \begin{pmatrix} 1 & 2 \\ 2 & 1 \end{pmatrix}$. Find a solution to $\boldsymbol{A}\vec{x} = \begin{pmatrix} 3 \\ 3 \end{pmatrix}$ and to $\boldsymbol{A}\vec{x} = \begin{pmatrix} 0 \\ 0 \end{pmatrix}$. Use those solutions to find infinitely many solutions to $\boldsymbol{A}\vec{x} = \vec{33}$.*

Exercise 3.7.18. *Calculate how many multiplications are needed in the three different programs from the book to find solutions to $x^2 - 2y^2 = 1$ in the range studied.*

Exercise 3.7.19. *Given a solution to Pell's equation one can construct infinitely many more. Notice we can write the solution $(3, 2)$ of $x^2 - 2y^2 = 1$ as*

$$(3 + 2\sqrt{2})(3 - 2\sqrt{2}) \;=\; 1.$$

If we raise both quantities to the nth power we get

$$(3 + 2\sqrt{2})^n (3 - 2\sqrt{2})^n \;=\; 1 \quad \text{or} \quad (a_n + b_n\sqrt{d})(c_n + d_n\sqrt{2}) \;=\; 1.$$

Find a_n, b_n, c_n, and d_n in terms of $3, 2, \sqrt{2}$, and n. Investigate the solutions found in the chapter; are all of the solutions with both coordinates positive obtainable through an appropriate choice of n?

Exercise 3.7.20. *Study $x^2 - 3y^2 = 1$. Find a non-trivial solution (i.e., a solution where neither x nor y is zero). Can you generate new solutions from your old one? What is the 'best' choice to take for your first solution?*

Exercise 3.7.21. *Study $x^2 - dy^2 = 1$. Does this always have a non-trivial solution (i.e., a solution where neither x nor y is zero) for d square-free? If yes how can you find it? Can you generate new solutions from such a solution?*

Calculus Review: Local versus Global

Exercise 3.7.22. *Find a differentiable function $f : \mathbb{R} \to \mathbb{R}$ such that f does not have an extremum at its critical point.*

Exercise 3.7.23. *Prove that if a differentiable function f has an interior maximum or minimum at $(a_1, \ldots, a_n)$, then its gradient ∇f vanishes at that point.*

Exercise 3.7.24. *Prove that if f and g are differentiable, then the extrema of f restricted to $g(x_1, \ldots, x_n) = C$ occur when the gradient of f and g are parallel; explicitly, show there must be a λ such that*

$$(\nabla f)(a_1, \ldots, a_n) = \lambda (\nabla g)(a_1, \ldots, a_n)$$

if $(a_1, \ldots, a_n)$ is an extremum of f subject to the boundary of the surface $g(x_1, \ldots, x_n) = C$.

Exercise 3.7.25. *Generalize the previous problem to find the extrema of a differentiable f subject to $g_i(x_1, \ldots, x_n) = C_i$ for differentiable functions $g_1, g_2, \ldots, g_r$ with $r < n$.*

Exercise 3.7.26. *Use Lagrange multipliers to find the largest and smallest values of $3x + 4y$ on the ellipse $(x/5)^2 + (y/2)^2 = 1$. Are you surprised by your answer? In hindsight could you have predicted it?*

Exercise 3.7.27. *Redo the previous problem by parametrizing the boundary and thus reducing to a one-dimensional problem. Specifically, the boundary of the ellipse is given by $\gamma(t) = (5\cos t, 2\sin t)$ for $0 \le t \le 2\pi$, and we can convert our two-dimensional problem to a one-dimensional optimization.*

Exercise 3.7.28. *Maximize $\sqrt[n]{x_1 x_2 \cdots x_n}$ subject to $x_i \ge 0$ and $x_1 + \cdots + x_n = n$. Interpret your result in terms of the arithmetic mean - geometric mean inequality.*

Exercise 3.7.29. *Consider a triangle with sides a, b and c.* **Heron's formula** *(also called* **Hero's formula***) states that the area equals*

$$\sqrt{s(s-a)(s-b)(s-c)},$$

where $s = (a + b + c)/2$ is the semi-perimeter. Prove Heron's formula and find the triangle that has the largest area for a given perimeter. Is there a triangle with smallest area for a given perimeter?

Exercise 3.7.30. *Continuing the previous problem, find the triangle with the smallest perimeter for a given area. Is there a triangle with largest perimeter for a given area?*

Exercise 3.7.31. *Given a positive integer S, which decompositions $a_1 + \cdots + a_n = S$ with the a_i positive integers have the largest product $a_1 \cdots a_n$? What if the a_i are only required to be real? Note you have freedom to have n depend on S.*

Exercise 3.7.32. *Assume Elvis swims at a rate of v_w in the water and runs at a rate of v_ℓ on land; without loss of generality we can normalize so that one of the two rates is 1 (let's take $v_w = 1$). What is the optimal path if $v_\ell = 1$? What is the optimal path if $v_\ell = \infty$?*

Exercise 3.7.33. *Fix $x_1, x_2 > 0$. Building on the previous problem, show that for any speed v_ℓ if Elvis starts off x_1 units from the infinitely long horizontal shore and the stick is thrown d units, landing x_2 units from shore, then there is some value of d such that the straight line path to the stick takes as long as the optimal path swimming to shore. Show this also holds if $v_\ell = \infty$, and use this to deduce the arithmetic mean - geometric mean inequality: $(x_1 + x_2)/2 \geq \sqrt{x_1 x_2}$.*

An Introduction to the Diet Problem

Exercise 3.7.34. *Imagine we could 'consume' negative amounts of foods. What interpretation would you give for such behavior in the Diet Problem?*

Exercise 3.7.35. *Assume all quantities in the statement of the diet problem are positive. Prove there must be at least one cheapest diet.* Hint: some results from real analysis might be helpful here.

Exercise 3.7.36. *Modify the Diet Problem so that now one cannot consume too much in a day; note part of this problem is figuring out exactly what constraint you want for this.*

Exercise 3.7.37. *Can any matrix linear algebra problem $A\vec{x} = \vec{b}$ be recast as a Diet Problem? If yes, you need to find a related matrix A and related vectors $\vec{x}, \vec{b}, \vec{c}$ such that the solution to $A\vec{x} = \vec{b}$ is the optimal solution to $\min \vec{c}^T \vec{x}$ subject to $A\vec{x} \geq \vec{b}$ with $\vec{x} \geq \vec{0}$.*

Solving the Diet Problem

Exercise 3.7.38. *Find the optimal solution to the Diet Problem when the cost function is $\mathrm{Cost}(x_1, x_2) = x_1 + x_2$.*

Exercise 3.7.39. *There are three vertices on the boundary of the polygon (of feasible solutions). We have seen two choices of cost functions that lead to two of the three points being optimal solutions; find a linear cost function which has the third vertex as an optimal solution.*

Exercise 3.7.40. *Generalize the Diet Problem to the case when there are three or four types of food, and each food contains one of three items a person needs daily to live (for example, calcium, iron, and protein). The region of feasible solutions will now be a subset of $\mathbb{R}^3$. Show that an optimal solution is again a point on the boundary.*

Exercise 3.7.41. *Generalize the Diet Problem to f foods and n nutrients. Assume all quantities are positive. Prove there are only finitely many vertices.*

Exercise 3.7.42. *Using notation as in the problem above, how many vertices do you expect? Try small values of f and n to build your intuition.*

Chapter 4

The Canonical Linear Programming Problem

In the previous chapter we introduced the notion of linear programming. We also saw, through the example of the Diet Problem, how one can go about setting up and solving a linear programming problem. It's now time to shift our focus to **canonical forms**. Developing the theory of canonical forms will provide us with a convenient and simple way of dealing with arbitrary linear programming problems. In mathematics, a canonical form (also sometimes referred to as a standard or normal form) of some mathematical object is simply a standard way of representing said object as a mathematical expression.

As a first example, we can look at the canonical form of any k digit positive integer n in base-10 as just a finite sequence of digits $a_0 a_1 \cdots a_k$, where $a_0, a_1, \ldots, a_k \in \{0, 1, 2, 3, 4, 5, 6, 7, 8, 9\}$ and $a_0 \neq 0$. Often there are many ways of representing a quantity (in our previous example, if we drop the requirement that $a_0 \neq 0$, then a given integer could be written in many ways); by a canonical form we mean a preferred choice. We've already seen another example of a canonical representation in the **Fundamental Theorem of Arithmetic**: every integer $n \geq 2$ can be written uniquely as a product of primes in the form $n = p_1^{r_1} \cdots p_k^{r_k}$ with $1 < p_1 < \cdots < p_k$. For yet another example, you may remember having found parametrizations of curves in calculus; it's common to parametrize by unit speed, as this allows us to find the length of a curve based on its parametrization in a simple way. Finally, a book about linear programming would be incomplete without an example from linear algebra. In an introductory linear algebra class, Gaussian elimination is sometimes used to solve a system of equations. The idea here is to reduce a system of equations to an 'equivalent' upper triangular system and then solve the new system using back-substitution. In this case the upper triangular matrices serve as canonical representations of systems of equations.

There are advantages and disadvantages to standardizing notation. One of the greatest advantages is that it facilitates different people conversing, since both now know how the other is viewing it. It's of course important to be aware of other people's conventions. A certain degree of care must be taken when using canonical representations of a mathematical object. For example, some people define the exponential density with parameter λ to be equal to $\lambda^{-1} \exp(-x/\lambda)$ for $x \geq 0$ and 0 otherwise, while others use $\lambda \exp(-x\lambda)$ (and thus the parameters have a reciprocal relationship). Either definition is fine as long as you are cognizant of how the mathematical analysis changes when you switch from one to the other, but if you are unaware of the normalization convention used you could make a mistake. A powerful real world warning comes from NASA, which in 1999 lost a \$325 million dollar probe because some systems were working with metric units and others with English units (see `http://www.cnn.com/TECH/space/9909/30/mars.metric/` for example). Their mistake serves as a wonderful illustration of the need for standard notation. When several people work on the same problem, a common language facilitates communication and helps everyone understand each other's work.

Since real analysis provides some very useful tools in the study of linear programming, we start by briefly reviewing some of its useful terminology. We also provide a brief overview of certain results from real analysis that are important in the study of canonical linear programming. Following this, to introduce the concept of canonical forms, we first cover their use in deriving the formula for the roots of the general quadratic equation, and then move on to studying canonical forms of linear programming problems.

4.1. Real Analysis Review

When we formulate the canonical linear programming problem, we'll exclude strict inequalities. The reason is that our objective function is continuous, and there are powerful results from analysis (see for example Exercise 4.9.6) which say that a continuous function on a closed, bounded set attains its maximum and minimum; if we have strict inequalities we'll have open sets and these results will be unavailable. We thus take a few moments to quickly recall some basic notation from real analysis and give some exercises on many of the key results.

Our problems come from the real world and usually deal with real-valued quantities, and as a result, all of them involve subsets of $\mathbb{R}^n$ as the domain and range of the functions that describe these problems. Therefore, in our definitions, we assume our sets live in $\mathbb{R}^n$. While the results are not given in the greatest possible generality, the formulations suffice for our purposes and highlight what is going on. The following discussion is supposed to serve as a primer of real analysis and only introduces the most basic ideas. We assume that the reader is somewhat familiar with elementary set theory and real analysis. For a comprehensive treatment of the subject, see [**Rud1**].

We begin by introducing one of the most fundamental objects in real analysis: the **ball** of radius r.

Definition 4.1.1. *The **ball** (or **disk**) about $\vec{a} = (a_1, \ldots, a_n) \in \mathbb{R}^n$ of radius $r > 0$, denoted by $B(\vec{a}; r)$, is the set of all points less than r units from a:*

$$B(\vec{a}; r) = \{(x_1, \ldots, x_n) : (x_1 - a_1)^2 + \cdots + (x_n - a_n)^2 < r^2\}.$$

*Sometimes we call this the **open ball**. The set of points exactly r units from the center $\vec{a}$ is the boundary of the disk, and the union of the open ball and its boundary is the **closed ball**.*

Note that since the definition of $B(\vec{a}; r)$ requires $r > 0$, every ball (whether open or closed) contains $\vec{a}$ itself. A unit ball (open or closed) about a point is simply a ball of radius 1. In $\mathbb{R}$, an open ball is called an interval. Using this notion of open and closed balls, we can now introduce the concept of open and closed sets.

Definition 4.1.2. *A set $S \subset \mathbb{R}^n$ is **open** if and only if given any $\vec{x} \in S$ there is some $r > 0$ (which can depend on $\vec{x}$) such that the ball of radius r centered at $\vec{x}$ is entirely contained in S, i.e. $B(\vec{a}; r) \subset S$.*

Definition 4.1.3. *A set $S \subset \mathbb{R}^n$ is **closed** if and only if its **complement** is open (the complement of S is the set of all points not in S).*

Note that a set $S \subset \mathbb{R}^n$ can be both open and closed. For example, $\mathbb{R}^n$ and the empty set $\emptyset$ are both open and closed in $\mathbb{R}^n$. In $\mathbb{R}$, open intervals are open and closed intervals are closed, just like their names suggest. Furthermore, using the above definitions one can easily check the following properties of open and closed sets (the reader is advised to prove these results as an exercise):

- An arbitrary union of open sets is open, and a finite intersection of open sets is open.

- An arbitrary intersection of closed sets is closed, and a finite union of closed sets is closed.

Definition 4.1.4. *A set $S \subset \mathbb{R}^n$ is **bounded** if there is some R such that every $\vec{x} \in S$ is at most R units from the origin. Equivalently we say that a subset S of $\mathbb{R}^n$ is bounded if it's contained in a ball of finite radius.*

It is easy to see that any interval (open or closed) in $\mathbb{R}$ with finite end-points is bounded. The set $S = \{1/n : n \in \mathbb{N}\}$ is another example of a bounded set. See Exercise 4.9.3 for examples of bounded and unbounded sets.

Definition 4.1.5. *A set $S \subset \mathbb{R}^n$ is **compact** if and only if it's closed and bounded.*

An equivalent definition of *compactness* can be given in terms of *open covers* (see Exercise 4.9.1). It's easy to verify the two definitions are equivalent. An example of a compact set in $\mathbb{R}^2$ is $S = \{(x, y) \in \mathbb{R}^2 : x^2 + y^2 \leq 1\}$.

We end this review section with one of the most important definitions in analysis.

Definition 4.1.6. *A function $f : \mathbb{R}^n \to \mathbb{R}$ is **continuous** at $\vec{x} \in S$ if given any $\epsilon > 0$ there is some $r > 0$ (which may depend on $f, \vec{x}$, and r) such that for all $\vec{y} \in S \cap B(\vec{x}; r)$ we have $|f(\vec{x}) - f(\vec{y})| < \epsilon$. Note that it's possible for $f : [0, 1] \to \mathbb{R}$*

to be continuous at the endpoints, as we only require closeness with points in the set. If f is continuous at every point in S we say f is **continuous on** S.

4.2. Canonical Forms and Quadratic Equations

We start our discussion on canonical forms with an example you've seen many times, solving quadratic equations. We build up to these by first studying linear equations, such as

$$ax + b = 0.$$

It's easy to find the solution; it's just $x = -b/a$. Therefore, we only need to understand how to solve this canonical form to find a solution for any other linear equation.

Linear equations, however, are very simple to solve regardless of whether we put them in canonical form or not. Let's try something more complicated. We can move on to a quadratic equation; let's start with the simplest one:

$$ax^2 + c = 0.$$

Solving this equation is not much harder; in fact it's almost the same as the first one. We get that $x^2 = -c/a$, which then yields $x = \pm\sqrt{-c/a}$.

Building on our success, let's generalize a bit more and consider an arbitrary quadratic:

$$ax^2 + bx + c = 0.$$

How should we solve this? Assume, of course, that we don't know the quadratic formula (the point of this discussion is, in fact, to find that formula!). One idea is to somehow use the simpler case discussed above to solve the general case. What do we do? We want to convert our equation to an equivalent one that has the same form as the simpler quadratic, as we know how to solve that. Thus we want to reduce it so that it looks like $\alpha z^2 + \gamma = 0$, where α and γ are constants depending on a, b and c, and z depends on x, a, b, and c. While we could write $\alpha(x + \beta)^2 + \gamma$ and then expand and set this equal to $ax^2 + bx + c$, we instead jump straight to the answer by using an important technique called **complete the square**:

$$a\left(x^2 + \frac{b}{a}x\right) + c = 0,$$

$$a\left(x^2 + \frac{b}{a}x + \frac{b^2}{4a^2} - \frac{b^2}{4a^2}\right) + c = 0,$$

$$a\left(x + \frac{b}{2a}\right)^2 - \frac{b^2}{4a} + c = 0.$$

Now we have something that looks like our simpler quadratic. We can thus use our solution from before and find the roots for the general quadratic. Simple algebraic manipulations can be employed to find that the roots are

$$x = -\frac{b}{2a} \pm \sqrt{\frac{b^2 - 4ac}{4a^2}},$$

which is just our familiar **quadratic formula**.

By writing any quadratic in the form $ax^2 + bx + c = 0$ we can write down the solution immediately in terms of a, b, and c using the formula we just derived; this is the advantage of choosing a canonical representation for these equations. We can do the same thing for the even more difficult cases of cubic and quartic equations; however, there's a deep theorem in a fascinating area called Galois Theory which states that it's impossible to write a general solution to quintic or higher polynomials in terms of the coefficients using only the arithmetic operations and taking roots.

The cubic and quartic cases are quite difficult, but they are feasible. Doing them requires a similar approach to this one: examining "easy" cases first, and trying to find ways of reducing the more complex ones to the more simple cases we know how to work with. As interesting as these higher-order equations are, however, they lie outside the scope of this course, so we will leave the topic at the quadratic case.

4.3. Canonical Forms in Linear Programming: Statement

In this section we discuss the notion of canonical form in the context of Linear Programming problems. Our goal now is to write down a good, general form for Linear Programming problems. In particular, we'll standardize the way we set up such problems. A benefit of this is that we only need to write one type of program, as we can always assume our input is formatted in a particular way. Now, from a practical point of view, do we always want to convert everything into the same format? The answer is fairly clearly "no"; it can often be very convenient, depending on local situations, to write things in different ways. In fact we will see in §4.7 that canonical forms are not always the best way of looking at a given problem. From a theoretical point of view, however, it's very good to have a standard form that we can always use to study a given problem.

We consider the class of problems of the following form. We have some matrix $\mathbf{A}$ such that $\mathbf{A}\vec{x} \bowtie \vec{b}$, where $\bowtie$ indicates that the relationship for each equation is drawn from any of $\{=, \geq, \leq\}$ (and we do not require it to be the same relation for each component of $\vec{b}$), though we *forbid the strict inequalities $<$ and $>$*. We also assume there are constraints on the x_i's; say $x_i \geq m_i$ where the m_i could be negative or even negative infinity. Finally we want to maximize or minimize a function $\vec{c}^{\,T}\vec{x} = c_1 x_1 + \cdots + c_n x_n$.

A large class of problems fall under this framework. While they are structurally different, we have three places where we can have variation: in the relations, in the constraints on the x_i's, and in minimizing or maximizing. It would be nice to convert any of these variations to a canonical one. Our goal is to convert any problem in this form to one specific, canonical form. We first state the answer and then discuss the structure of solutions and how to do the conversion.

Definition 4.3.1 (Canonical Linear Programming problem). *The canonical Linear Programming problem has the following form:*

(1) We have variables $x_j \geq 0$ for $j \in \{1, \ldots, N\}$.

(2) *The variables satisfy linear constraints, which we can write as* $\mathbf{A}\vec{x} = \vec{b}$ *(where* $\mathbf{A}$ *is a matrix with* M *rows and* N *columns, and* $\vec{b}$ *is a column vector with* M *components).*

(3) *The goal is to minimize a linear function of the variables:* $\vec{c}^{\,T}\vec{x} = c_1 x_1 + \cdots + c_N x_N$.

If $\vec{x}$ satisfies the constraints ($\mathbf{A}\vec{x} = \vec{b}$, $\vec{x} \geq \vec{0}$), then we call $\vec{x}$ a **feasible solution** to the canonical Linear Programming problem; if further $\vec{x}$ minimizes the linear function $\vec{c}^{\,T}\vec{x}$, then $\vec{x}$ is called an **optimal solution** to the canonical Linear Programming problem. We can see how not all linear programming problems have this form. We can have a problem where not all variables are non-negative. Certainly, we cannot expect for all of our relationships to be equalities either. Finally, we assume that our canonical form has a linear objective function. In the following section we will discuss how we can deal with some of these standardization issues.

We discuss some pathological cases. Consider the following canonical Linear Programming problems:

(1) We want to minimize $10x_1$ where the constraints are $x_1 = -2018$, with $x_1 \geq 0$. There are no feasible solutions; thus there are no optimal solutions.

(2) We want to minimize $-17x_1$ where the constraints are $2x_1 - 5x_2 = 0$, with $x_1, x_2 \geq 0$. There are infinitely many feasible solutions: any (x_1, x_2) works with $x_1 = 2.5x_2$; however, there is no optimal solution (send $x_1 \to \infty$).

(3) We want to minimize $x_1 + x_2$ where the constraints are $x_1 + x_2 = 1$, with $x_1, x_2 \geq 0$. Here there are infinitely many feasible solutions, and each feasible solution is also an optimal solution.

The above examples show some care is required. A general Linear Programming problem need not have a feasible solution. If it does have a feasible solution, it need not have an optimal solution. Further, even if it does have an optimal solution, it need not have a unique optimal solution.

We can also discuss some issues in the language of linear algebra. If our system $\mathbf{A}\vec{x} = \vec{b}$ is overdetermined or if it has exactly one solution, then we cannot have more than one feasible solution and sometimes we won't have any feasible solutions. Moreover, in the case that we have one solution, we do not have any options to maximize our objective function. Therefore, it's desirable to have an under-determined system that has multiple feasible solutions so that we have many options for finding optimal solutions.

We end by revisiting our canonical formulation. If we have strict inequalities we can have open regions, and as shown in the exercises this can lead to optimization issues; we avoid this problem by using $\geq$ and $\leq$ (though as discussed above it's still possible to have situations where there are no feasible or no optimal solutions).

If, however, we really want a strict inequality, what we can do is add a very small quantity to the corresponding component of $\vec{b}$, say $10^{-1000!}$ (ok, we probably don't need something quite so small!), and for many problems this will lead to a negligible change. Later we'll study problems where the variables are restricted to be integers, and thus a restriction to be strictly greater can easily be adjusted (i.e., being strictly greater than 4 is the same as being greater than or equal to 5).

4.4. Canonical Forms in Linear Programming: Conversion

4.4.1. Dealing with the relationships. First off, let's deal with the issue of the relationship. We want to show that we can convert our problem to an equivalent one where all the relationships are equalities. Remember that we're dealing with matrices and column vectors. This means that a different relation may hold for each row of $\mathbf{A}\vec{x}$ and $\vec{b}$. To narrow down the possible relations, we can try targeting the rows for which a certain relation holds. Specifically, note that in the case of a row for which $\geq$ or $\leq$ holds, we can obtain the opposite relation by simply negating the corresponding row of our matrix A.

So by negating specific rows of $\mathbf{A}$ we can assume that we have no $\leq$ relations, and we are left with just $=$ and $\geq$ relations. Of course, we could have done the algebra to eliminate the $\geq$ relations; the choice is ours. Which of the two we choose won't matter much, however, as we still need to reduce the possibilities further (down to just equalities).

Our goal now is to find a way of removing all $\geq$ relations. This is significantly more difficult, and it's important to point out that this is a case in which obtaining a canonical form requires us to make some sacrifices that may not always be advisable, as we will soon see.

Let's take a row for which the $\geq$ relation holds. We have

$$a_{i,1}x_1 + \cdots + a_{i,N}x_N \geq b.$$

To turn this into an equality we introduce a new variable, call it z_i, with the constraint that $z_i \geq 0$. Introducing this new variable gives us the power to remove the inequality. If we now subtract z_i from the left side of the relation, we can now balance the scales and obtain an equality:

$$a_{i,1}x_1 + \cdots + a_{i,N}x_N - z_i = b.$$

By introducing a new variable to pick up the slack in these relations, we can reduce all cases to equalities. This is what we wanted for a canonical form, but is this always a good idea?

Think back to the Diet Problem that we studied earlier. When we worked through it, we obtained certain *feasibility regions* that satisfied our nutritional restrictions. By eliminating all inequalities, these will be lost: we will be limited to a search for some exact amount of nutrition. When dealing with something like the Diet Problem, we may not *want* to put the problem in canonical form, but it's *possible* to do so, and that's what is important here.

Before moving on to the second element we need to narrow down for our canonical form; let's survey the damage that's been done so far. We've added this new variable z_i; this can potentially add up to one new variable for each row of our matrix $\mathbf{A}$. If we have a lot of constraints, this amounts to a great many additional variables. If we were to write a program with all these additional variables, we are sure to get slower run times and increased memory use. Have we changed the objective function? No, as these new variables do not appear in the objective function.

We've now settled our first issue: we've reduced the possible relations in our original set-up to equalities only, and we can safely assume that we have $\mathbf{A}\vec{x} = \vec{b}$. There are two other elements of the problem left for us to standardize: we have the constraints on the x_i's as $x_i \geq m_i$, and then the issue of whether we are maximizing or minimizing $\vec{c}^T\vec{x}$. Let's continue in order and deal with the constraints of the x_i's next.

4.4.2. Dealing with bounds on the x_i's. In many real-world problems, there are two types of constraints: either x_i has to be greater or less than a certain amount, or x_i has to be integer-valued—you can't fly half a plane to Salt Lake City or buy half a set of hot dog rolls (unless you're Steve Martin in *Father of the Bride*, though he was incarcerated for such misbehavior). That being said, can you buy, say, half a pound of a given commodity? In many cases the answer is yes. The more serious issue seems to be the lower bounds and upper bounds. Can you think of situations in which there might be hard bounds on the quantity of certain products, for instance when making a sale or a purchase?

Upper- or lower-bound constraints are fairly common in real-world problems: we can't buy more of a product than there is supply, for example. In problems of distribution, concerns about fairness might require that everybody receive at least some baseline amount. How do we deal with these kinds of constraints?

Take a given constraint $x_i \geq m_i$. For our canonical form the constraint should be that each x_i is non-negative. We can do this, but there are several cases depending on the value of m_i.

- *Case I: $m_i \geq 0$.* This case is straightforward. We simply move the condition $x_i \geq m_i$ into a new row in our matrix $\mathbf{A}$ (it has a $\geq$ relation, but we can deal with that as we just saw).

- *Case II: $m_i < 0$ and finite.* In this case, we use a change of variables: let $y_i = x_i - m_i$. This leaves us with $y_i \geq 0$, as desired, and all that is needed is to make the substitution $x_i = y_i + m_i$ wherever it appears. Unfortunately if m_i is not finite this method does not work.

- *Case III: $m_i = -\infty$.* This is the completely unrestricted case: the only constraint is that x_i be real. We want to replace it with a variable that is greater than or equal to zero. How can we do this? The solution is quite beautiful. We write $x_i = u_i - v_i$, where $u_i, v_i \geq 0$. This allows us to write x_i in terms of non-negative variables, regardless of what real value x_i is taking. However, this change of variables is also quite wasteful: we introduce two new variables just to deal with this one "constraint" (or lack thereof, as it were).

4.4.3. Dealing with maximizing/minimizing. As before, not all real-world problems are created equal. We cannot always expect ourselves to go around maximizing all quantities. For example, an investment banker maximizing losses will get fired almost instantly. Similarly another investment banker who goes around minimizing profit won't fare much better. So we are left with a single issue left to resolve: maximization or minimization? As it turns out, this is the easiest of the three issues to resolve. It suffices to just consider minimization, since maximizing $\vec{c}^T \vec{x}$ is equivalent to minimizing $-\vec{c}^T \vec{x}$. Done!

Recap.

We can put any Linear Programming problem into canonical form, using the three methods from above. From now on we'll assume problems are given in canonical form, for instance when we prove the Simplex Method works.

Of course, we should reiterate that putting problems into canonical form is not always the most efficient approach; in fact it can at times be quite costly. Though the benefits of having a canonical form are great, and though we will make use of them quite often, there are cases in which it's simply the wrong approach to take, particularly in cases where we already have a given structure. This kind of situation is what we'll investigate next.

4.5. The Diet Problem: Round 2

Let's revisit the Diet Problem that was formulated in the previous chapter and see how we can convert it to the canonical form! We will add some new information to make the conversion fun. As before, let's say we are shopping for food, and we have two options: spoo and gagh. The nutrients that we care about are iron and protein. Furthermore, we have been informed that the Romulans have put cyanide in all the spoo and gagh, so we have to look out for that too.

- Spoo costs \$20 per unit. Each unit of spoo contains 5 units of iron, 3 units of protein, and 4 units of grade-A Romulan cyanide.
- Gagh costs \$10 per unit. Each unit contains 3 units of iron, 2 units of protein, and 1 unit of grade-A Romulan cyanide.

We want to live cheaply but we need at least 15 units of iron and 10 units of protein. Additionally, we cannot consume more than 10 units of Romulan cyanide or we would die. Let's say we buy x_1 units of spoo and x_2 units of gagh. Also, the store that we are shopping at doesn't let people buy less than 2 units of gagh. Therefore, we get the following constraints:

$$
\begin{aligned}
5x_1 + 3x_2 &\geq 15, \\
3x_1 + 2x_2 &\geq 10, \\
4x_1 + x_2 &\leq 10, \\
x_1 &\geq 0, \\
x_2 &\geq 2.
\end{aligned}
$$

We want to maximize the number of units of food that we consume, so our objective is to maximize $x_1 + x_2$.

Let's follow the process described in the previous section. First, we want to turn all inequalities into $\geq$. Therefore we multiply the third equation by -1 to get

$$-4x_1 - x_2 \geq -10.$$

Now let's turn all inequalities into equalities. For this purpose, we introduce three slack variables $z_1, z_2, z_3 \geq 0$. Our constraints now become

$$
\begin{aligned}
5x_1 + 3x_2 - z_1 &= 15, \\
3x_1 + 2x_2 - z_2 &= 10, \\
-4x_1 - x_2 - z_3 &= -10.
\end{aligned}
$$

Now we need to deal with x_2. We make a new constraint $x_2 \geq 0$, and introduce a new slack variable $z_4 \geq 0$ to get rid of the inequality that we added. Thus, our system in canonical form is

$$
\begin{aligned}
5x_1 + 3x_2 - z_1 &= 15, \\
3x_1 + 2x_2 - z_2 &= 10, \\
-4x_1 - x_2 - z_3 &= -10, \\
x_2 - z_4 &= 2,
\end{aligned}
$$

where we minimize the function $-x_1 - x_2$ and we have $x_1, x_2, z_1, z_2, z_3, z_4 \geq 0$.

4.6. A Short Theoretical Aside: Strict Inequalities

We began our discussion of canonical forms in linear programming by saying that we will exclude strict inequalities from our formulations. From a practical standpoint, it's not unreasonable to make this assumption. In everyday real-life problems, we can often replace strict inequalities with non-strict inequalities without changing the problem too much. For example if someone says that they want to cover strictly less than 10000 miles in their cross-country national park road trip, we can safely assume that they wouldn't be averse to travelling exactly 10000 miles. Problems, however, can arise sometimes when we do away with strict inequalities in our formulation of linear programming problems, but they are almost always easy to fix. Why is it then that we don't want to include strict inequalities? The answer is pretty simple: we want our feasible regions to be closed and bounded! To explain why, let's state the following important result from real analysis, sometimes called the extreme value theorem.

Theorem 4.6.1. *If f is continuous on a compact set $S \subset \mathbb{R}^n$, then f is bounded, and f attains its maximum and minimum.*

Now, the idea is that the continuous function in our Linear Programming problems is our objective function. Since we are usually interested in minimizing (or maximizing) our objective function, it's of paramount importance that the minimum (or maximum) of our objective function exists in our feasible region. Let's consider a very simple example to illustrate what happens when we have strict inequalities in our constraints.

Suppose we want to maximize the objective function $f(x) = x$ subject to $x^2 + y^2 < 1$, $x, y \geq 0$. Then the feasible region is given in Figure 1. Now we

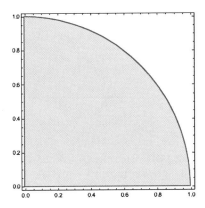

Figure 1. Feasible region for the constraints $x^2 + y^2 < 1$, $x, y \geq 0$.

can see that our objective function will never hit its maximum on this feasible region. Note that we have

$$\lim_{x \to 1} x = 1.$$

Now suppose we picked some x_0 in the feasible region as our candidate for an optimal solution. Then, since we know that for some $\epsilon > 0$, $B(x_0; \epsilon)$ is completely contained in the feasible region, we have that $x_0' = x_0 + \frac{\epsilon}{2}$ is a feasible solution and clearly

$$f(x_0') > f(x_0).$$

Every time we have a candidate for an optimal solution, we can construct a solution that is better than it. Therefore the fact that we have an open set as our feasible region (which is a result of having strict inequalities in our constraints) makes it impossible for us to have any optimal solutions! This is why we require our constraints to not have strict inequalities in them, so that we get a nice feasible region (that is closed and bounded). Once we have a closed and bounded feasible region, we can appeal to the extreme value theorem to say that maximum and minimum values of our objective function exist (given that it's continuous on the feasible region).

4.7. Canonical is Not Always Best

Consider the following sum:

$$
\begin{array}{r r}
 & 14 - 1 \\
+ & 112 - 14 \\
+ & 1170 - 112 \\
+ & 1701 - 1170 \\
+ & 24601 - 1701 \\
\hline
 & ?
\end{array}
$$

Can you find the result without computing term by term? You can, if you notice the pattern in the sums: each intermediate line adds a new term while subtracting the one that was added at the previous line. By cancelling all these redundant terms, we can easily find the answer:

$$
\begin{array}{rl}
& \cancel{14} - 1 \\
+ & \cancel{112} - \cancel{14} \\
+ & \cancel{1170} - \cancel{112} \\
+ & \cancel{1701} - \cancel{1170} \\
+ & 24601 - \cancel{1701} \\
\hline
& 24600
\end{array}
$$

This is called a **telescoping sum** (or series), because, like a pirate's telescope, all the intermediate lines collapse down to leave only the first and last terms.

Now, if we were to write an algorithm to evaluate general sums, we might calculate results line by line and then sum them up. In this case, however, it's best to add the terms vertically and exploit the telescoping effect. In general, does it save time to add the terms vertically? No, but sometimes it will. We have a case of incredible cancellation; most of the time we won't see cancellation of this sort and should add as we would normally, but in cases where we have it and can recognize it we would be foolish not to exploit the additional structure. This should remind you of the differences we saw between Horner's Algorithm, which is a great way to evaluate a general polynomial, and fast multiplication, which is magnitudes faster for pure powers (see §1.3 and §1.4).

The above is a great warning to carefully study the structure and symmetry of your system before embarking on a calculation. Often there is an opportunity to use certain algorithms. For example, there's a whole class of methods for dealing with *sparse* matrices—those in which almost all entries are zero. These algorithms all run much more efficiently than the best general algorithms but are confined to the restricted case of these sparse matrices.

4.8. The Oil Problem

Now that we've covered canonical forms and their benefits and drawbacks, let's end this chapter by looking at a famous and important linear programming problem. Imagine you run an oil corporation and are trying to ship oil from your refineries to the markets available to you. Your goal, which is the goal in almost any problem where you're asked to imagine yourself running a business, is to maximize profits.

Parameters.

- R refineries, which we denote with $r \in \{1, 2, \ldots, R\}$.
- M markets, which we denote with $r \in \{1, 2, \ldots, M\}$.
- p_m is price oil sells for at market m.
- d_m is the demand for oil at market m.
- s_r is the oil supply available from refinery r.
- $c_{r,m}$ is the cost of shipping 1 unit of oil from refinery r to market m.

There are a lot of issues with our set-up and a lot of questions that can be asked. What about competition for the markets? What about different methods of shipping? Perhaps most importantly, can we really have an accurate, static estimate of the demand at a given market? All these parameters have many underlying factors, especially demand, which can be affected by countless items, can change rapidly, and is very difficult to predict.

While these issues can and should be dealt with when solving real-life problems, for a first look at a Linear Programming problem in an academic course it's best to keep things simple: we assume that all factors are independent and that we have perfect information regarding all these parameters.

Variables.

- $x_{r,m}$ is the amount of oil refinery r ships to market m.

Constraints.

- $x_{r,m} \geq 0$: we don't have markets shipping oil back to the refineries!

- $\sum_{m=1}^{M} x_{r,m} \leq s_r$: we can't ship more oil from a refinery than the refineries have.

- $\sum_{r=1}^{R} x_{r,m} \geq d_m$: we must at least meet the demand at all markets.

We've succeeded in converting our approximation to the real-world problem to a Linear Programming problem; now it's time to develop techniques to solve it!

4.9. Exercises

Real Analysis Review

The following problems provide a review of some key results from real analysis.

Exercise 4.9.1. *Let* $\{\mathcal{O}_i\}_{i=1}^{\infty}$ *be a collection of open sets such that* $S \subset \bigcup_{i=1}^{\infty} \mathcal{O}_i \subset \mathbb{R}^n$; *we call these sets an **open cover** of* S. *Prove that if* S *is compact, then it must be contained in the union of finitely many of the* $\mathcal{O}_i$ *(this is called a **finite subcover**).*

Exercise 4.9.2. *A function is **uniformly continuous** on* S *if it's continuous and there is an* r *that works for all* $\vec{x} \in S$. *Prove that a continuous function on a compact set is uniformly continuous.*

Exercise 4.9.3. *Consider the following subsets of* $\mathbb{R}^2$:

$$A = \left\{ (x, y) \in \mathbb{R}^2 : x^4 + 8x^2y^6 + y^8 = 1 \right\} \quad and$$

$$B = \left\{ (x, y) \in \mathbb{R}^2 : x^2 - y^2 = 1 \right\}.$$

Prove that A *is bounded while* B *is not.*

Exercise 4.9.4. *A set* S *is sequentially compact if given any sequence of points in* S, *there is a subsequence converging to a point in* S. *Prove that the open interval* $(0, 1)$ *is not sequentially compact but the closed interval* $[0, 1]$ *is.*

Exercise 4.9.5. *Prove that $S \subset \mathbb{R}^n$ is compact if and only if it's sequentially compact.*

Exercise 4.9.6. *Prove that if f is continuous on a compact set $S \subset \mathbb{R}^n$, then f is bounded, and f attains its maximum and minimum (i.e., there are values $\vec{x}_{\max}$ and $\vec{x}_{\min}$ in S such that for all $\vec{x} \in S$ we have $f(\vec{x}_{\min}) \leq f(\vec{x}) \leq f(\vec{x}_{\max})$).*

Exercise 4.9.7. *A set $S \subset \mathbb{R}^n$ is **connected** if there do not exist disjoint open sets U and V such that $S = (S \cap U) \cup (S \cap V)$. Let $f : S \to \mathbb{R}$ be an integer valued function. Prove that if f is continuous, then S is connected.*

Canonical Forms and Quadratic Equations

Exercise 4.9.8. *Prove that any cubic $ax^3 + bx^2 + cx + d = 0$ can be written as $x^3 + px + q = 0$ (i.e., we can rewrite so that the coefficient of the x^2 term vanishes and the coefficient of the x^3 term is 1); this is called the **depressed cubic** associated to the original one.*

Exercise 4.9.9. *Building on the previous problem, introduce two new variables u and v so that $x = u + v$. Prove the following statements, which lead to a formula for the roots of the depressed cubic:*

- *$u^3 + v^3 + (3uv + p)(u + v) + q = 0$.*

- *If we add the additional constraint that $3uv + p = 0$ (which implies $u^3 v^3 = -p^3/27$), then the depressed cubic becomes $u^3 + v^3 = -q$.*

- *Substitute $v^3 = -(u^3 + q)$ into $u^3 v^3 = -p^3/27$; setting $z = u^3$ we obtain the equation $z^2 + qz - p^3/27 = 0$. This last is a quadratic equation which we can solve for z, and then u^3, and then v^3, and then u and v, and then x.*

Complete the analysis and determine the cubic formula.

Exercise 4.9.10. ***Polar coordinates*** *are usually written $x = r\cos\theta$ and $y = r\sin\theta$ with $r \geq 0$ and $\theta \in [0, 2\pi)$. There are many conventions for **spherical coordinates**; a common one is*

$$x = \rho \sin\theta \cos\phi, \quad y = \rho \sin\theta \sin\phi, \quad z = \rho \cos\theta,$$

with $\rho \geq 0$, $\phi \in [0, 2\pi)$, and $\theta \in [0, \pi]$. Prove these equations have the claimed properties. Explicitly, prove that if $\rho > 0$ is fixed, then this parametrizes the points on the sphere of radius ρ; thus, given an (x, y, z) on that sphere there is a unique choice of (θ, ϕ) giving that point.

Exercise 4.9.11. *Building on the previous problem, write down coordinates for a sphere in 4-dimensional space; remember to include the ranges of each variable. More generally, do for an arbitrary-dimensional space.*

Exercise 4.9.12. *Generalize the previous problem to find a formula for roots of a quartic. Warning: this is a hard problem, but there are plenty of write-ups online.*

Canonical Forms in Linear Programming: Statement

Exercise 4.9.13. *Can you construct a canonical Linear Programming problem that has exactly two feasible solutions? Exactly three? Exactly k where k is a fixed integer?*

Exercise 4.9.14. *Can you construct a canonical linear programming problem that has exactly two optimal solutions? Exactly three? Exactly k where k is a fixed integer?*

Canonical Forms in Linear Programming: Conversion

The following problems show why we cannot have strict inequalities in the canonical formulation of our Linear Programming problems. It is essential that we work in a compact space.

Exercise 4.9.15. *Find a continuous function defined in the region $(x/2)^2+(y/3)^2 < 1$ (i.e., the interior of an ellipse) that has neither a maximum nor a minimum but is bounded.*

Exercise 4.9.16. *Find a continuous function defined in the region $(x/2)^2+(y/3)^2 < 1$ (i.e., the interior of an ellipse) that has neither a maximum nor a minimum and gets arbitrarily large and positive and arbitrarily large and negative.*

Exercise 4.9.17. *Find a continuous function defined in the region $(x/2)^2+(y/3)^2 > 1$ (i.e., the exterior of an ellipse) that has neither a maximum nor a minimum and gets arbitrarily large and positive and arbitrarily large and negative.*

Exercise 4.9.18. *Find a continuous function defined in the region $(x/2)^2+(y/3)^2 \geq 1$ (i.e., the exterior and the boundary of an ellipse) that has neither a maximum nor a minimum and gets arbitrarily large and positive and arbitrarily large and negative.*

Canonical Forms in Linear Programming: Conversion

Exercise 4.9.19. *Consider the problem*

$$\max(4x_1) - 2x_2 \ \ \text{subject to} \ \begin{pmatrix} 2 & 3 \\ 3 & 5 \end{pmatrix} \begin{pmatrix} x_1 \\ x_2 \end{pmatrix} \geq 21, \ \ x_1 \geq -4, \ \ x_2 \leq 5.$$

Convert this to canonical form.

Exercise 4.9.20. *Given an $n \times m$ matrix, worst-case scenario how many equations and how many variables must be added to convert it to canonical form?*

Exercise 4.9.21. *When converting a Linear Programming problem to canonical form, is it always possible to do so in such a way that the resulting matrix has at least as many positive components as negative components?*

Canonical is Not Always Best

Exercise 4.9.22. *How should you arrange the summands in*

$$\sum_{n=-N}^{N} \sin\left(\frac{2n^3 + n}{4n^2 + 6}\pi\right)$$

in order to compute it as efficiently as possible?

Exercise 4.9.23. *How should you perform the integration in*

$$\int_0^{\pi} \left(\cos^5 x - 3\cos^3 x + 7\cos x\right) dx$$

in order to compute it as efficiently as possible?

The Oil Problem

Exercise 4.9.24. *Make up numbers for the oil problem with two cities and two refineries. Can you use the methods from our Diet Problem investigation to solve?*

Exercise 4.9.25. *Prove that the optimal solution for the oil problem never has more oil sent to a city than demanded.*

Exercise 4.9.26. *Give the weakest possible condition you can for the parameters in the oil problem which ensures there is a feasible solution, and show that if your condition is not met, then there is no solution.*

Exercise 4.9.27. *Modify the oil problem so that certain refineries cannot ship to certain cities (perhaps the corresponding countries are now at war). Does your condition from the previous problem still ensure a feasible solution exists? If not, what condition would you now need?*

Exercise 4.9.28. *Prove that if there is a feasible solution to the oil problem, then there is an optimal solution.*

Exercise 4.9.29. *Modify the oil problem so that instead of shipping oil directly from the refineries to each city it's now possible for oil to first be sent to a city and then either kept there or have some of it shipped further to other cities. What new parameters, variables, and constraints are needed? Write down the corresponding Linear Programming problem.*

Chapter 5

Symmetries and Dualities

We've covered how to formulate a Linear Programming problem, and seen some examples of important problems that can be expressed in this framework; now it's time to start learning how to solve them. Solving problems is often quite hard. In some cases it can even be impossible to solve a problem perfectly, as we saw with roots of quintic polynomials. Often we're reduced to approximating a solution. Since we're approximating, we want to find (or at least bound) the error we're making and, hopefully, keep it to an acceptable size.

This is quite a challenge; how are we to find the distance our approximation is from the actual result without ever knowing what the actual result is? The idea is that often we can determine a tight range where the solution must live and then use that to bound our error. For many linear programming problems, this is accomplished by looking at a Dual Problem.

In this chapter we discuss duality and how it can be used. Before doing that, we first explore how to use symmetries to reduce the size of the space of candidate solutions. In addition to being of use in linear programming, these ideas are valuable for a variety of other problems and worth mastering.

5.1. Tic-Tac-Toe and a Chess Problem

Imagine that due to budget cuts we can only afford a 5×5 chessboard, and for some reason it's essential that we place 5 queens in such a way that we can put 3 pawns down and have none of the queens attacking any of the pawns (remember queens can attack anything in the same row, column or diagonal). While this may seem like an arbitrary busywork problem, it turns out that the way we solve it illustrates one of the most powerful ideas in mathematics, that of duality.

Let's start off by attacking the problem through brute force. While in general this is a bad way to solve a problem efficiently, it will often at least get us to the right solution. Then, once we know whether or not it can be done, we can work on trying to do it well.

The brute force approach is quite grueling; we just go through every possible set-up of 5 queens and 3 pawns on a 5×5 board, and find the set-ups that work. This method involves testing a grand total of $\binom{25}{5}\binom{20}{3}$ possible configurations, where the first factor is how many ways there are to place 5 queens, and the second is how many ways there are to choose 3 of the remaining squares for our pawns. This involves examining over 60 million different set-ups!

How can we improve this? As a first step, we can remove the $\binom{20}{3}$ term entirely: we need only place the 5 queens to be able to find whether or not there are 3 safe spots for pawns to be placed. We don't need to place the pawns, as we can just see how many squares are not attacked by the queens. This leaves us with $\binom{25}{5} = 53,130$ configurations to examine.

Fortunately, there's a better approach. We can learn a great lesson from **Tic-Tac-Toe** and exploit some symmetries from the problem. Consider the standard Tic-Tac-Toe grid:

How many possible first moves are there? There are 9 squares available to the first person, so there are 9 possible first moves. This leaves 8 responses. The pattern continues: 7 next moves, then 6, then 5. We get $9 \cdot 8 \cdot 7 \cdot 6 \cdot 5$ possible paths for the first 5 moves. After this point, things get a bit more complicated, as the game may already have completed, like the game below (in order to make it easy to follow how this game evolved, the subscript of the letter indicates the move corresponding to its placement on the board):

X_1	O_2
X_3	O_4
X_5	

Analyzing the possibilities in full detail gets a little hairy after this, as once there are three of the same letter in a row, column, or diagonal, the game ends; however, we can bound the total number of games by $9! = 362,880$ possible games. If you're attempting to become a Tic-Tac-Toe grandmaster by remembering each possible game, it should be clear that this is not a desirable approach.

Fortunately we can reduce the load by exploiting symmetry. There are in effect only 3 opening moves: we can go in a corner (we denote this by 1, and note there are four ways to do this), in the middle of a side (we denote this by 2, and there are four of these as well), or in the center (we denote this by 3, and this time see there is only one way to choose this case):

1	2	1
2	3	2
1	2	1

Thus, up to symmetry, there are but three opening moves. We've saved a factor of three, a terrific start. We can continue to exploit symmetry to further reduce the computations. We'll look one move further (it's of course possible to keep going, but the enumeration becomes more involved).

If the first move was in the corner, there are 5 responses up to symmetry, not 8: adjacent middle, corner in same row or column, center, further middles, diagonally opposite corner:

X	1	2
1	3	4
2	4	5

If instead the first move was on a middle, there are again 5 responses:

1	2	4
X	3	5
1	2	4

Finally, if the opening move was in the center, there are only 2 responses:

1	2	1
2	X	2
1	2	1

Whereas before we had $9 \cdot 8 = 72$ possibilities for the first 2 moves, we now have $5+5+2 = 12$ possibilities. We've reduced the number of cases by a factor of 6.

We now apply this same technique to the chess problem. Originally we had $\binom{25}{5}$ possible set-ups for the queens. Here again, we reduce this number by exploiting symmetry.

How many possible opening moves are there for the first queen's placement? Up to symmetry there are 6:

1				
2	3			
4	5	6		

The rest follow from symmetry. Even if we cover all remaining cases naively, this reduces the number of cases to check to $6\binom{24}{4} = 10,626$. We could continue to argue as we did before, and for each of the five openings look at the possible responses. As soon as we resort to the brute force method we lose the savings from symmetry, which means we have a lot more cases to examine and thus a lot more work to do.

Is there ever a reason why we might want to do more work? There can be: if we were to continue to use symmetry to reduce the number of cases, the paths get increasingly complex and difficult to code. This means we lose time elsewhere, in the process of writing the code for the more efficient method. Sadly, it also makes it far more likely that there will be a mistake somewhere in the code, while at the same time making any such mistakes harder to identify.

In this situation, $10,626$ cases are too many to do by hand, but a computer can exhaust all of these in a heartbeat. In fact, the $\binom{25}{5}$ cases were already quite manageable from a computer's perspective. The savings begin to make a difference if the computation needs to be repeated multiple times, but as this is a one-and-done computation, it's faster to write slightly inefficient code quickly than to take a long time to write efficient code.

Even if a problem is a one-and-done, it's also important to consider the amount of time efficient code can save over the course of solving a single problem. Consider a generalization of the chess problem to an $n \times n$ board with n queens. On a home computer, brute force will get you up to around $n = 6$ or $n = 7$. A few optimizations for efficiency can get you to $n = 10$ or $n = 11$. The world record is around $n = 17$. If it takes a few hours on a home computer for a reasonably skilled programmer to get this close to the world record, then that speaks to the rate at which the problem grows in complexity. In cases like this, brute force methods quickly become ineffective and efficient code is mandatory, despite the fact that there is only one problem to solve.

To recap, we've tried a few approaches to solving our chess problem, covering the brute force method as well as a more refined approach exploiting symmetry. There's another technique we can use here, which is called **duality**. Duality can be summarized as follows: *I don't want to solve your problem. I want to solve a related problem, and if I can do that, I can do your problem, too.* Thus we want to find a different and simpler problem that is equivalent to that of placing 5 queens on a board such that 3 pawns can be placed safely.

What is an equivalent, easier problem? Instead of placing 5 queens so that 3 pawns are safe, we place 3 queens so that 5 pawns are safe. If we can do this, we simply perform an exchange operation, replacing all queens with pawns and all pawns with queens, obtaining a solution to the original problem. Alternatively, we can just place 3 pawns directly and see if there are 5 squares that do not attack them.

Why is this helpful? This reduces the number of cases to $\binom{25}{3} = 2300$, which is already better than our mix of symmetry and brute force above; if we use symmetry, we can reduce this even further to $6\binom{24}{2}$, taking us all the way down to $1,656$ configurations to check. Sadly, we cannot always use symmetry, though when it's available it's often a great aid in winnowing the options. (For example, I recently tried to rent a car and even though I had a reservation, they didn't have one available when I arrived. I had to wait till they negotiated one from another company. I wanted to inform them when I returned that sadly I didn't have a car for them, but perhaps the next person would be able to help them.) In the next section we see how duality arises in linear programming.

5.2. Duality and Linear Programming

Converting a problem into a dual one with an easier approach to finding a solution is an extremely helpful tool, and it is used quite frequently. For instance, when solving differential equations we often use Fourier or Laplace transforms to convert one system of differential equations into another one that's simpler to solve (though often we pay the price when we try to convert back). When computing multiple integrals, changing the order of integration can at times turn integrals that are impossible to compute into relatively simple computations.

What does duality have to do with linear programming? Remember that our goal is to find an optimal solution if possible, and if not to at least find a feasible solution which is close to the optimal value. It turns out that given any canonical problem we can associate a dual problem, and the solutions to this problem will help us with our two goals.

Recall that our Canonical Linear Programming problem has the following form.

- Constraints: $\mathbf{A}\vec{x} = \vec{b}$, $x_i \geq 0$.
- Objective function: minimize $\vec{c}^{\,T}\vec{x}$.

Dual Problem: Given a canonical Linear Programming problem, its Dual Problem is the following:
- Constraints: $\vec{y}^{\,T}\mathbf{A} \leq \vec{c}^{\,T}$, $y_j \in \mathbb{R}$.
- Objective function: maximize $\vec{y}^{\,T}\vec{b}$.

It's fairly easy to see that these two problems are related; in fact it's a good exercise (Exercise 5.4.17) to prove that the Dual Problem to the Dual Problem is the original problem. At this point, though, it should be a bit of a mystery as to how the Dual Problem can assist us. It turns out that if we have information about solutions to the Dual Problem, then we obtain information about the solutions to the original problem. The following lemma illustrates this. Here **feasible** means the solution satisfies the constraints, but it may or may not be optimal as far as maximizing or minimizing the desired function.

Lemma 5.2.1. *Let $\hat{x}$ be a feasible solution to the Canonical Linear Programming problem, and let $\hat{y}$ be a feasible solution to the dual problem. Then*

$$\hat{y}^{\,T}\vec{b} \ \leq \ \vec{c}^{\,T}\hat{x},$$

and if equality holds, then $\hat{x}$ is optimal.

Simply put, this lemma means that if we have feasible solutions to the canonical problem and the Dual Problem, then these solutions place upper and lower bounds, respectively, on the optimal solution to the canonical problem. Thus, the Dual Problem helps us approximate the optimal solutions to the canonical problem and estimates the error.

Proof. While the proof is essentially just simple algebra, the difficulty is that we must first identify what terms or expressions we need to study in order to be successful. Looking back at our formulations of the canonical and Dual Problems, we see that both are fairly similar, but they employ different variables, $\vec{x}$ and $\vec{y}$. Thus, it might be promising to study $\hat{y}^T \mathbf{A} \hat{x}$, which combines both problems' constraint expressions. We have

$$\hat{y}^T \left(\mathbf{A} \hat{x} \right) \; = \; \left(\hat{y}^T \mathbf{A} \right) \hat{x}.$$

Since $\hat{y}$ and $\hat{x}$ are feasible solutions to their problems, they satisfy the constraints, meaning we have

$$\hat{y}^T \mathbf{A} \; \leq \; \vec{c}^{\,T} \quad \text{and} \quad \mathbf{A} \hat{x} \; = \; \vec{b}.$$

Substituting these back into the previous expression yields

$$\hat{y}^T \left(\mathbf{A} \hat{x} \right) \; = \; \hat{y}^T \vec{b} \; = \; \left(\hat{y}^T \mathbf{A} \right) \hat{x} \; \leq \; \vec{c}^{\,T} \hat{x}.$$

The desired bound $\hat{y}^T \vec{b} \leq \vec{c}^{\,T} \hat{x}$ now follows immediately.

All that remains is to show that if we have equality above, then $\hat{x}$ is optimal. If we assume equality, we see that

$$\hat{y}^T \vec{b} \; = \; \left(\hat{y}^T \mathbf{A} \right) \hat{x} \; = \; \vec{c}^{\,T} \hat{x}.$$

If $\vec{c}^{\,T} \hat{x}$ is not optimal, then there exists a feasible x such that

$$\vec{c}^{\,T} x \; < \; \vec{c}^{\,T} \hat{x} \; = \; \hat{y}^T \vec{b}.$$

However, we now have

$$\vec{c}^{\,T} x \; < \; \hat{y}^T \vec{b},$$

whereas we just proved that

$$\hat{y}^T \vec{b} \; \leq \; \vec{c}^{\,T} x$$

for any feasible solution x, a contradiction. Thus we've shown that if there's equality, then $\hat{x}$ is optimal. $\qquad\qquad\qquad\qquad\qquad\qquad\qquad\square$

We've succeeded – we've found a test for optimality of solutions to the canonical problem! However, at this point it's not clear whether or not this test is *useful*.

5.3. Appendix: Fun Versions of Tic-Tac-Toe

Our interest in Tic-Tac-Toe was due to the fact that it's a simple, easily explainable game which illustrates some important concepts. Unfortunately, it's not very interesting, as every game between two skilled players always ends in a tie, and this level of proficiency is not very hard to attain. There are, however, other versions of Tic-Tac-Toe that *are* interesting.

One way to make the game more fun is to implement bidding: both players start with the same, arbitrary amount of funds, and bid for the right to make each move. The winning bidder gets to make the given move, but the losing bidder receives the funds from the winner's bid (alternatively, winning bids could be discarded instead of transferred). This version of Tic-Tac-Toe makes strategy much more complex: is it worth giving up the two first moves in hopes of getting three consecutive moves further down the line? How much is the starting move worth? The added depth from the bidding mechanism can make the game much more fun. See [**DP**] for more on this.

Another more entertaining version of the game is "Gobble", or Russian doll Tic-Tac-Toe. Each player has 2 large, 2 medium, and 2 small pieces, and a larger piece can be placed over any smaller piece on the board. This allows players to overwrite each other's pieces, giving the game a much greater tactical dimension.

If we abandon the restriction to the traditional 3×3 grid, another fun modification of Tic-Tac-Toe is Tic-Tac-Toe Cubed. The Tic-Tac-Toe Cubed gameboard is one large 3×3, each square of which contains a smaller 3×3 board, as shown:

$$\begin{array}{|c|c|c|} \hline \# & \# & \# \\ \hline \# & \# & \# \\ \hline \# & \# & \# \\ \hline \end{array}$$

We will refer to the position on the larger board as an upper level position and the position on a smaller board as a lower level position. On the first move of the game, Player 1 plays at any upper level position and at any lower level position. Subsequently, each player can play at any lower level position, but must play at the higher level position that matches the lower level position of the previous player's move. As an example, consider the following sequence of possible plays (where to make things easier to see we have expanded our board from before):

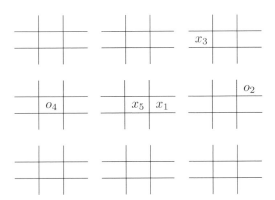

This covers the basics of Tic-Tac-Toe Cubed. Edge cases occur once the board starts to fill up and options for play become limited, but the game still works no matter how you choose to handle edge cases, so we won't discuss them here.

5.4. Exercises

Tic-Tac-Toe and a Chess Problem

Exercise 5.4.1. *In our Tic-Tac-Toe analysis we used the fact that the first move must be an X and the second move an O, but then ignored the fact that the players alternate in our crude upper bounds. Can you use this information to improve your estimate? Does the order in which the X's are placed matter?*

Exercise 5.4.2. *Up to symmetry, how many possible ways are there to place two queens on a 5×5 board? Note that this is a bit different than the Tic-Tac-Toe problem, as the two queens are undistinguishable (which is not the case with the first two moves in Tic-Tac-Toe).*

Exercise 5.4.3. *Imagine we want to place n queens on an $n \times n$ board in such a way as to maximize the number of pawns which can safely be placed. Find the largest number of pawns for $n \leq 5$.*

Exercise 5.4.4. *Write a computer program to expand your result in the previous problem to as large of an n as you can. Does the resulting sequence have any interesting problems? Try inputting it in the OEIS (see the comment after this problem).*

If you have not heard of the OEIS (the On-line Encyclopedia of Integer Sequences, https://oeis.org/), you are strongly encouraged to go to its website and explore. It's a wonderful resource, and for sequences in its database a lot of properties are listed, ranging from recurrence relations to generating functions to examples of where they arise; furthermore, there are often extensive references for additional reading.

Exercise 5.4.5. *How many ways are there to place n identical queens on an $n \times n$ board? For n large, approximate this number. Using the brute force approach, how many operations will it take to check all of these?*

Exercise 5.4.6. *Consider the problem of placing n queens on an $n \times n$ board with the goal of maximizing the number of pawns which may safely be placed. For each n, let that maximum number be $p(n)$. Find the best upper and lower bounds you can for $p(n)$. For example, trivially one has $0 \leq p(n) \leq n^2$; can you do better?*

Exercise 5.4.7. *For the 5×5 queen problem we saw we could safely place 3 pawns; is there a way to extend that argument to safely place 75 pawns on a 25×25 board? More generally, if you have a result for an $n \times n$ board, for what n, if any, can you extend to an $n^2 \times n^2$ board?*

Exercise 5.4.8. *Sometimes it's easier to do a simpler problem first and build intuition. Redo Exercise 5.4.6 where instead of queens one places rooks. If $p_{\text{rook}}(n)$ is the number of pawns that can safely be placed with n rooks on an $n \times n$ board, what can you say about $p_{\text{rook}}(n)$?*

Exercise 5.4.9. *Redo the previous problem with bishops instead of rooks, and discuss $p_{\text{bishop}}(n)$.*

Exercise 5.4.10. *Consider three-dimensional tic-tac-toe: the board is now 27 boxes in a $3 \times 3 \times 3$ space, and any three in a row wins. Does the first player have a winning strategy? If yes, what is it?*

Exercise 5.4.11. *Generalize the previous problem to a $4 \times 4 \times 4$ board, where now you need 4 in a row (if you're exceptionally brave, consider a $4 \times 4 \times 4 \times 4$ space, though it's better to treat that mathematically rather than pictorially!).*

Exercise 5.4.12. *Investigate the number of pawns that can safely be placed if you have m queens on an $n \times n$ board. Compare the answer for $m = f(n)$ for various f. Interesting choices could be $f(n) = \log n$, $f(n) = cn$ and $f(n) = cn^2$.*

Exercise 5.4.13. *How many distinct opening moves are there in Exercise 5.4.10? How many responses are there?*

Exercise 5.4.14. *Generalize the previous problem to an $n \times n \times n$ space.*

Duality and Linear Programming

Exercise 5.4.15. *Find the Dual Problem to the Diet Problem from §3.4.*

Exercise 5.4.16. *Find the Dual Problem of the Dual Problem in the previous exercise.*

Exercise 5.4.17. *Prove the Dual Problem of the Dual Problem is the original Linear Programming problem.*

One of the most common uses of duality is in probability, where it's called the **Law of Complementary Probability**. Given an event A we denote its probability by $\mathrm{Prob}(A)$, and the probability that A does not happen is A^c. Often it's easier to find $\mathrm{Prob}(A^c)$ and then use $\mathrm{Prob}(A) = 1 - \mathrm{Prob}(A^c)$.

Exercise 5.4.18. *Compute directly the probability that there is at least one shared birthday in a group of four, and then compute this by using the Law of Complementary Probability. For simplicity assume there are always 365 days in a year and each person is equally likely to be born on any day.*

Exercise 5.4.19. *Discuss the difficulty of generalizing the above problem to trying to compute directly the probability of at least one shared birthday among 20 people, or among n people; how would the computation look using complements?*

Exercise 5.4.20. *Assume a hand of five cards is chosen from a standard deck without replacement, and all hands are equally likely. What is the probability that at least two of the cards have the same 'number' (we consider the jacks, queens, kings and aces as four different numbers).*

Exercise 5.4.21. *A **derangement** of $\{1, 2, \ldots, n\}$ is a permutation of these numbers so that no number returns to where it began; thus $\{1, 3, 5, 4, 2\}$ is not a derangement (as 1 and 4 return to where they start), but $\{4, 3, 5, 1, 2\}$ is. Assume all $n!$ permutations of $\{1, 2, \ldots, n\}$ are equally likely. What is the probability of a derangement if $n = 5$?*

Exercise 5.4.22. *A k-strong derangement of $\{1, 2, \ldots, n\}$ is a permutation of these numbers so that no number returns to where it began, or to any of the $k-1$ neighbors above or below; for this problem we assume 1 and n are neighbors. Thus a standard derangement is a 1-strong derangement, and there are no 2-strong derangements of $\{1, 2, 3\}$. Investigate the probability of a k-strong derangement as $n \to \infty$; if you want, for simplicity assume $k = 2$.*

Exercise 5.4.23. *Generalize the previous problem and estimate the probability of a derangement as $n \to \infty$.*

Exercise 5.4.24. *Evaluate*

$$\int_{y=0}^{1} \int_{x=y}^{1} e^{-x^3} \, dx \, dy.$$

The next few problems involve the **Fourier Transform** $\widehat{f}$ of a function f. There are many normalizations; I prefer

$$\widehat{f}(y) = \int_{-\infty}^{\infty} f(x) e^{-2\pi i x y} \, dx,$$

where

$$e^{2\pi i x} = \cos(2\pi x) + i \sin(2\pi x), \quad i = \sqrt{-1}.$$

We say a function $f : \mathbb{R} \to \mathbb{R}$ is in $\mathbf{L}^p(\mathbb{R})$ if $\int_{-\infty}^{\infty} |f(x)|^p \, dx$ is finite; $\mathbf{L}^p([0,1])$ is defined similarly (with the integration now from 0 to 1).

The **Schwartz space** is the set of all functions f such that f and all of its derivatives decay faster than $1/(1 + |x|)^N$ for any N, though the transition point where $1/(1 + |x|)^N$ is larger than $|f^{(n)}(x)|$ can depend on x.

Exercise 5.4.25. *Prove that if $f \in L^1(\mathbb{R})$, then $\widehat{f}(y)$ exists for all y. Must $\widehat{f}$ be in $L^1(\mathbb{R})$?*

Exercise 5.4.26. *Prove or disprove: (1) if $f \in L^1(\mathbb{R})$, then $f \in L^2(\mathbb{R})$; (2) if $f \in L^2(\mathbb{R})$, then $f \in L^1(\mathbb{R})$.*

Exercise 5.4.27. *Prove or disprove: (1) if $f \in L^1([0,1])$, then $f \in L^2([0,1])$; (2) if $f \in L^2([0,1])$, then $f \in L^1([0,1])$.*

Exercise 5.4.28. *Prove that if $f \in L^1(\mathbb{R})$, then $\lim_{|x| \to \infty} |f(x)| = 0$.*

Exercise 5.4.29. *Prove the Gaussian density $e^{-x^2/2}/\sqrt{2\pi}$ is in the Schwartz space. Is the Cauchy density $\frac{1}{\pi} \frac{1}{1+x^2}$ as well?*

For functions f with suitable decay, we have the **Poisson Summation Formula**:

$$\sum_{n=-\infty}^{\infty} f(n) = \sum_{n=-\infty}^{\infty} \widehat{f}(n);$$

f being in the Schwartz space suffices. Poisson summation is often used to convert a slowly decaying sum, where we need to take many terms to obtain a good approximation, to another sum with rapid decay, where just a few terms (sometimes even just one term) provide an excellent approximation. The next problem gives an important example; that result plays a key role in a variety of subjects, such as the Riemann zeta function and the distribution of primes and Benford's law of digit bias.

Exercise 5.4.30. *Prove*

$$\frac{1}{\sqrt{N}} \sum_{n=-\infty}^{\infty} e^{-\pi n^2/N} = \sum_{n=-\infty}^{\infty} e^{-\pi n^2 N}.$$

As N tends to infinity, bound the error in replacing the sum on the right hand side with the zeroth term (i.e., taking just $n = 0$). Hint: the Fourier transform of a Gaussian is another Gaussian; if $f(x) = e^{-ax^2}$, then $\widehat{f}(y) = \sqrt{\pi/a}e^{-\pi^2 y^2/a}$.

Appendix: Fun Versions of Tic-Tac-Toe

Exercise 5.4.31. *Investigate bidding Tic-Tac-Toe, both when the losing bidder receives the winning bid and when that money is removed from the game. What is a good strategy?*

Exercise 5.4.32. *Investigate Gobble Tic-Tac-Toe. In particular, if you go first, do you think it's a good idea or a bad idea to place your largest piece in the middle square?*

Imagine Tic-Tac-Toe on an $n \times n$ board, where the first person to get m in a row wins; let's call this game $\mathcal{T}(n;m)$. Clearly it cannot be a disadvantage to go first, though it's not clear that Player One has a winning strategy.

Exercise 5.4.33. *Show the first person has a winning strategy in $\mathcal{T}(n;3)$ if $n \geq 4$.*

Exercise 5.4.34. *Does the first person have a winning strategy in $\mathcal{T}(n;4)$ for all n sufficiently large?*

Exercise 5.4.35. *To try and counterbalance the (perceived) advantage in going first in $\mathcal{T}(n;m)$, we select a fixed number of squares on the board to be filled with the second player's sign. Investigate how many and where one should place these to make the game seem fair for various m and n.*

Chapter 6

Basic Feasible and Basic Optimal Solutions

In the Diet Problem we saw that we only needed to check the vertices to find the optimal solution. While this is a tremendous improvement over checking all the points or all the boundary points, for most real-world problems there are far too many vertices to check by brute force in a reasonable amount of time. We need to find a good way to navigate through the astronomically large number of potential solutions. One successful approach is the Simplex Method, which we'll discuss later; first, though, we need to set some notation and prune the list of candidates. We do that by introducing basic feasible and basic optimal solutions below.

6.1. Review of Linear Independence

We quickly review some important concepts from linear algebra which play a key role in our discussion on solutions.

> **Linearly independent:** Non-zero vectors $\vec{v}_1, \ldots, \vec{v}_k$ are linearly independent if
> $$\alpha_1 \vec{v}_1 + \cdots + \alpha_k \vec{v}_k = \vec{0}$$
> implies that
> $$\alpha_1 = \alpha_2 = \cdots = \alpha_k = 0.$$
> If the vectors are not linearly independent, then they are said to be linearly dependent.

Linear independence is closely related to the notion of dimension. Consider the roads of Manhattan. They all go either North-South (the avenues) or East-West (the streets), two linearly independent vectors (or directions). It's thus easy to specify a location using these directions. Of course, one could use another basis, such as $(1,0)$ and $(1,1)$, but any choice would have exactly two directions. If there are more non-zero vectors than the dimension of a space, the vectors can't be linearly independent; there's always a redundancy somewhere.

Note that this doesn't mean that if there are fewer vectors than the dimension of the space, then there is linear independence. If all vectors are in the same direction, for example, they'll all be linearly dependent.

What does it mean to calculate $\mathbf{A}\vec{x}$? One interpretation is the one you're used to: it's a new vector which represents the effect of $\mathbf{A}$ on $\vec{x}$. We can rewrite this in a way that will highlight certain properties of the output. Let A_i be the ith column of $\mathbf{A}$ and x_i be the ith entry of $\vec{x}$. Then

$$
\begin{aligned}
\mathbf{A}\vec{x} &= \begin{pmatrix} \uparrow & & \uparrow \\ A_1 & \cdots & A_N \\ \downarrow & & \downarrow \end{pmatrix} \begin{pmatrix} x_1 \\ \vdots \\ x_N \end{pmatrix} \in \mathbb{R}^M \\
&= x_1 A_1 + x_2 A_2 + \cdots + x_N A_N.
\end{aligned}
$$

Notice the output of $\mathbf{A}\vec{x}$ is a linear combination of the columns of $\mathbf{A}$. In particular, if we want $\mathbf{A}\vec{x} = \vec{b}$ to have a solution, we need $\vec{b}$ to be in the **column space** of $\mathbf{A}$.

If $\mathbf{A}$ is an $M \times N$ matrix, the discussion above shows that the column space must be a subset of $\mathbb{R}^M$. While of course it can go either way, we almost always have more columns than rows, $N > M$, because if $N < M$ the system would be overdetermined and there would be good chance of no solution. Thus, throughout the book it's safe to assume that the column space is, in fact, all of $\mathbb{R}^M$. As a result of this, when we cover the Simplex Method one of our assumptions will be that the column space of $\mathbf{A}$ is all of $\mathbb{R}^M$ (or at the very least that $\vec{b}$ is in the column space).

6.2. Basic Feasible and Basic Optimal Solutions

We now begin pruning the set of feasible and optimal solutions which we must explore in solving our canonical Linear Programming problems. The subsets we introduce below have many nice properties: there are finitely many of them, there are easy ways to find them, and if a general solution exists an equivalent one exists from these spaces.

Definition 6.2.1 (Basic Feasible Solution). *Given a canonical Linear Programming problem with $\mathbf{A}, \vec{x}, \vec{b}, \vec{c}$, we say $\vec{x}$ is a basic feasible solution if*

(1) $\vec{x} \geq \vec{0}$,

(2) $\mathbf{A}\vec{x} = \vec{b}$, and

(3) if we have non-zero entries $x_{j_1}, \ldots, x_{j_k}$ of $\vec{x}$, then the columns $A_{j_1}, \ldots, A_{j_k}$ are linearly independent.

If $\vec{x}$ is an optimal solution and it's also basic feasible, then we say $\vec{x}$ is a **basic optimal solution**.

Why are basic feasible solutions good to study? If we take a look at the linear combination

$$\mathbf{A}\vec{x} = x_1 A_1 + x_2 A_2 + \cdots + x_N A_N,$$

we see that if an $x_i = 0$, it doesn't matter what the corresponding column is. The only columns in play are those for which the corresponding entry of $\vec{x}$ is non-zero. Thus, we can simplify our linear combination to only include those non-zero terms:

$$\mathbf{A}\vec{x} = x_{j_1} A_{j_1} + \cdots + x_{j_k} A_{j_k} \quad (x_{j_i} > 0).$$

We're getting closer to our condition for basic feasibility. All we need in addition is for the columns $A_{j_1}, \ldots, A_{j_k}$ to be linearly independent. The idea here is that if they're linearly dependent, there should be a way to replace some of the columns with some of the others, and get a few more of the $\vec{x}$ entries to be zero. By doing this, we should be able to winnow things down to a minimal solution.

6.3. Properties of Basic Feasible Solutions

We now explore some properties of basic feasible solutions. Perhaps the two most natural questions to ask are whether or not one exists, and if one exists, then how many are there? We start with the first question.

Theorem 6.3.1. *Consider a canonical Linear Programming problem. If a feasible solution exists, then a basic feasible solution exists.*

The idea of the proof is that, given a feasible solution, we can remove some of the columns and replace them with combinations of the other columns, eventually winnowing things down to a set of linearly independent columns. While doing this, we also need to make sure that the vector $\vec{x}$ stays non-negative. The importance of this theorem should be clear: if we just need a feasible solution, it suffices to search for it among the space of basic feasible solutions, which hopefully will greatly narrow our field.

Proof. The proof uses a common technique: the **method of infinite descent**. We find a special element among our candidate set which is minimal in some way, and show that if our desired conclusion fails, then our candidate could not have been minimal.

To make this idea work we need to choose a property for which our candidate will be minimal. We require the components of a feasible $\vec{x}$ to be non-negative, and for a basic feasible solution the columns of $\mathbf{A}$ associated to non-zero elements of $\vec{x}$ must be linearly independent. This suggests that, perhaps, a good property to consider would be the *number* of non-zero components. We try this, assuming we have a feasible $\vec{x}$ with the fewest non-zero entries. To obtain a contradiction we assume $\vec{x}$ is not basic feasible (if it were basic feasible there would be nothing to prove!). Note that such a minimal $\vec{x}$ exists, as we assumed that there was a feasible solution.

Thus

$$\mathbf{A}\vec{x} \ = \ x_{j_1} A_{j_1} + \cdots + x_{j_k} A_{j_k} \ = \ \vec{b},$$

where the A_{j_i}'s are not linearly independent. Therefore, we can find γ_{j_i}'s not all zero such that

$$\gamma_{j_1} A_{j_1} + \cdots + \gamma_{j_k} A_{j_k} \ = \ \vec{0}.$$

By relabeling and if necessary multiplying by -1, we may assume $\gamma_{j_1} > 0$. Next, we multiply throughout by a free constant λ, which we'll determine in a moment, to get

$$\lambda\gamma_{j_1} A_{j_1} + \cdots + \lambda\gamma_{j_k} A_{j_k} \ = \ \vec{0}.$$

We subtract this from our feasible solution $\vec{x}$:

$$\left(x_{j_1} - \lambda\gamma_{j_1}\right) A_{j_1} + \cdots + \left(x_{j_k} - \lambda\gamma_{j_k}\right) A_{j_k} \ = \ \vec{b}.$$

We now have a new vector $\vec{x}_\lambda$ such that $\mathbf{A}\vec{x}_\lambda = \vec{b}$. This solution is feasible as long as no components are negative: this is true for $\lambda = 0$ (by assumption), and when λ is small (the upper bound can be at most x_{j_1}/γ_{j_1}).

As λ increases, there must be a critical value λ_c such that $\vec{x}_{\lambda_c}$ has at least one more zero component than $\vec{x}$ without there yet being any negative components. This yields a new feasible solution with fewer non-zero columns than $\vec{x}$, a contradiction.

Thus, a feasible solution for which the number of non-zero columns is minimal must be basic feasible. A minimal case always exists as long as there exist feasible solutions, so a basic feasible solution always exists if a feasible solution exists; this completes the proof. $\square$

We now turn to our second question, showing that there are only finitely many basic feasible solutions (and obtaining a bound on their number).

Theorem 6.3.2. *Consider a canonical linear programming problem. There are only finitely many basic feasible solutions.*

Proof. Our matrix $\mathbf{A}$ has M rows and N columns. Thus, the dimension of the column space is at most M, and the number of linearly independent columns which we may choose is at most M. There are thus only finitely many possible sets of linearly independent columns; the number of candidate sets is at most

$$\sum_{m=0}^{M} \binom{N}{m} \ \leq \ \sum_{m=0}^{N} \binom{N}{m} \ = \ 2^N.$$

The proof is completed by showing that any choice of $m \leq M$ independent columns leads to a finite number of basic feasible solutions; we'll see it leads to at most 1.

We fix $A_{j_1}, \ldots, A_{j_k}$ linearly independent. We need to show that there are only finitely many feasible solutions. Let $\mathbf{A}' = (A_{j_1} \cdots A_{j_k})$ be the contracted matrix with these linearly independent columns. We want to solve $\mathbf{A}'\vec{x}' = \vec{b}$. Ideally, we'd multiply by the inverse of $\mathbf{A}'$, but since we're dealing with matrices we can't always do this; however, if $\mathbf{A}'^{-1}$ exists, then there's only 1 feasible solution, and we're done. If instead $\mathbf{A}'^{-1}$ doesn't exist, we just consider $\mathbf{A}'^T \mathbf{A}'$. This matrix is $k \times k$, and in Exercise 6.6.26 you are asked to prove it's always invertible. Thus, in this case we find

$$\vec{x}' = \left(\mathbf{A}'^T \mathbf{A}' \right)^{-1} \mathbf{A}'^T \vec{b}.$$

Although this vector $\vec{x}'$ may not be feasible, our arguments suffice to show that there are at most finitely many basic feasible solutions; we've just seen there is at most one basic feasible solution to any set of linearly independent columns. As there are only finitely many possible sets of linearly independent columns, each of which yields at most 1 basic feasible solution, the proof is completed. □

6.4. Optimal and Basic Optimal Solutions

We saw that restricting ourselves to basic feasible solutions allowed us to narrow down our search space to a finite number of cases to check; this is a tremendous improvement, because if there are two feasible solutions, then there are infinitely many (see Exercise 6.6.12). We also showed that if there is a feasible solution, then there is a basic feasible solution; a similar result holds for optimal solutions. The proof is similar to the one for feasible and basic feasible solutions, but requires some additional work to show optimality.

Theorem 6.4.1. *Consider a canonical Linear Programming problem. If there exists an optimal solution, then there exists a basic optimal solution.*

Proof. We again use the infinite descent/minimality approach. Assume $\vec{x}$ is an optimal solution with the fewest number of non-zero entries, and assume $\vec{x}$ is not basic optimal (if it were we would be done). We have

$$\mathbf{A}\vec{x} = x_{j_1} A_{j_1} + \cdots + x_{j_k} A_{j_k} = \vec{b},$$

where by assumption the columns A_{j_i} are not linearly independent. Therefore we can find γ_{j_i}'s not all zero such that

$$\gamma_{j_1} A_{j_1} + \cdots + \gamma_{j_k} A_{j_k} = \vec{0}.$$

By relabeling and if necessary multiplying by -1, we may assume $\gamma_{j_1} > 0$. We now multiply everything by some free constant λ (which we will fix soon) to get

$$\lambda \gamma_{j_1} A_{j_1} + \cdots + \lambda \gamma_{j_k} A_{j_k} = \vec{0}.$$

We subtract this from our solution $\vec{x}$ and find

$$\left(x_{j_1} - \lambda \gamma_{j_1} \right) A_{j_1} + \cdots + \left(x_{j_k} - \lambda \gamma_{j_k} \right) A_{j_k} = \vec{b}.$$

We now have a new vector $\vec{x}_\lambda$ such that $A\vec{x}_\lambda = \vec{b}$. This solution is feasible as long as no components are negative: this is true for $\lambda = 0$, and since all the coefficients are positive when $\lambda = 0$, we still have a feasible solution if λ is sufficiently small (we can take x_{j_1}/γ_{j_1} as a bound on how far we can vary $|\lambda|$).

As λ increases or decreases, there must be a critical value such that at least one of the components becomes 0, without there yet being any negative components. This yields a new feasible solution with fewer non-zero columns than $\vec{x}$.

To now show that this new solution is optimal, we need to look at its cost. Our old cost was $\vec{c}^{\,T}\vec{x}$. The new cost is

$$\vec{c}^{\,T}\vec{x}_\lambda \;=\; \vec{c}^{\,T}\vec{x} - \lambda\vec{c}^{\,T}\vec{\gamma}.$$

There are three cases to consider:

- $\vec{c}^{\,T}\vec{\gamma} > 0$. Taking a small, positive valued λ gives a feasible solution with lower cost than the optimal $\vec{x}$, a contradiction; thus this case cannot occur.

- $\vec{c}^{\,T}\vec{\gamma} < 0$. Taking a small, negative valued λ gives a feasible solution with lower cost than the optimal $\vec{x}$, a contradiction; thus this case also cannot occur.

- $\vec{c}^{\,T}\vec{\gamma} = 0$. By process of elimination, this is the only possible scenario, and it implies that our new solution $\vec{x}_\lambda$ always has optimal cost. The critical value of λ yields a new optimal solution with fewer non-zero columns than $\vec{x}$, contradicting our minimality assumption and completing the proof.

$\square$

Why is this process helpful to us in finding optimal solutions? As an example, let's say we worked in an admissions office and were tasked with selecting the best applicants for acceptance at our institution. We might be given a large amount of criteria and information on each applicant, too much to handle efficiently (does it really matter if the candidate is left-handed or, as one of my colleagues told me, has a letter of recommendation from a dentist about brushing habits). By winnowing out the excess criteria, we can streamline the process and avoid diluting the predictive power of each criterion. It can help to think of basic feasible and basic optimal solutions in this light: they are like the core criteria.

6.5. Efficiency and Euclid's Prime Theorem

One of the great aspects of these proofs for the existence of basic feasible and basic optimal solutions is that they are constructive: they provide us with a means of obtaining a basic feasible/optimal solution if we're given one that isn't basic. But one question we must always ask is whether this process is efficient.

Not all constructive processes are efficient. We have the following constructive theorem, with a proof due to Euclid, that there are infinitely many primes. This is one of the oldest results (and proofs!) in mathematics which is still taught, but we'll see it's horribly inefficient. Remember, a number $n \geq 2$ is **prime** if its only positive divisors are 1 and itself; the number 1 is a **unit**, and all other positive numbers are **composites**.

Theorem 6.5.1. *There are infinitely many primes.*

Proof. We proceed by contradiction: assume there are finitely many, say N, primes. We list them all exhaustively: $p_1 = 2$, $p_2 = 3$, $p_5 = 5$, ..., p_N. Let $x_N = p_1 p_2 \cdots p_N + 1$. There are two possibilities: either x_N is prime or it's composite.

- *Case 1: x_N is prime.* Thus, our list wasn't exhaustive, which is a contradiction and there are infinitely many primes.

- *Case 2: x_N is composite.* As x_N is composite it must be divisible by a prime. It can't be divisible by any of the primes listed (the remainder is always 1), so there must be a new prime not on the list which divides x_N. Thus, our list was not exhaustive.

$\square$

Our proof that there are infinitely many primes is constructive: it provides a method of obtaining new primes from a given list. But does it provide an efficient means of counting the primes?

Let
$$\pi(x) \;:=\; \#\{p \leq x : p \text{ prime}\}$$
count the number of primes at most x. The famous **Prime Number Theorem**, proven in 1896 independently by Jacques Hadamard and Charles Jean de la Vallée-Poussin, states that as $x \to \infty$,
$$\pi(x) \;\approx\; \frac{x}{\log x}.$$
Thus there are a lot of primes up to x; not quite as many as there are even numbers, but a lot more than the number of perfect squares (there are only on the order of $\sqrt{x}$ such integers).

What estimates does Euclid's argument give us on the growth rate of $\pi(x)$? If we always had the worst-case scenario, where the new number x_N is always prime, the list of primes generated would be

$$2$$
$$2 + 1 \;=\; 3$$
$$2 \cdot 3 + 1 \;=\; 7$$
$$2 \cdot 3 \cdot 7 + 1 \;=\; 42 \cdot 43 + 1 \;=\; 1807$$

and so on (though 1807 is composite as it equals $13 \cdot 139$). As the primes in our list nearly double in size at each step, our lower bound for $\pi(x)$ is

$$\pi(x) \;\geq\; c \log_2 \left(\log_2 x \right)$$

for some $c > 0$. While this does indeed go to infinity, it does so at an extremely slow growth rate, especially when compared to $x/\log x$.

Of course, this estimate assumes the worst case happens at each step, and we know that isn't always the case from 1807. The actual sequence is very strange: so far, we've got $2, 3, 7, 43, 13$. These are followed by $53, 5, 6221671, 38709183810571$, $139, 2801, 11, 17, 5471, \ldots$, and is called the **Euclid-Mullin sequence**. It goes all over the place. Interestingly, we don't know that many of its terms; currently we know only the first 50 or so.

This raises several questions. Is every prime number "hit" in the Euclid-Mullin sequence? Given a prime number p in the sequence, is there a way to find, in terms of p, how long it will take for p to arise in the sequence? These are just some of many open problems related to this sequence.

As this topic is not traditionally seen in an operations research course, it's reasonable to ask why we have devoted a section to it. The reason is that it becomes relevant when we speak of efficiency. Euclid's argument is often given as a simple proof of the infinitude of primes, but its efficiency is often not discussed. As we've seen, it does successfully generate infinitely many primes, but these grow incredibly fast, and the lower bound it yields for $\pi(x)$ is terrible. In operations research it must become second nature to inquire about efficiency and to always ask whether a task can be performed better; you are of course encouraged to ask these questions elsewhere as well.

6.6. Exercises

Review of Linear Independence

Exercise 6.6.1. *In the definition of linear independence we assumed the vectors are non-zero. Why is this a good condition to include?*

Exercise 6.6.2. *Consider two bases of $\mathbb{R}^3$, say $\vec{v}_1, \vec{v}_2, \vec{v}_3$ and $\vec{w}_1, \vec{w}_2, \vec{w}_3$. Thus, any $\vec{u}$ has expansions in terms of each:*

$$\vec{u} = a_1\vec{v}_1 + a_2\vec{v}_2 + a_3\vec{v}_3 = \alpha_1\vec{w}_1 + \alpha_2\vec{w}_2 + \alpha_3\vec{w}_3.$$

Express the α_i's in terms of the a_i's, $\vec{v}_i$'s, and $\vec{w}_i$'s.

Exercise 6.6.3. *Prove that any set of $n + 1$ non-zero vectors in $\mathbb{R}^n$ is linearly dependent.*

Exercise 6.6.4. *Let $\vec{v}_1, \ldots, \vec{v}_n$ be vectors in $\mathbb{R}^n$, and let $\boldsymbol{A}$ be the matrix whose i^{th} row is $\vec{v}_i$. Prove the vectors are linearly independent if and only if $\det(\boldsymbol{A}) \neq 0$ or, equivalently, the vectors are linearly dependent if and only if $\det(\boldsymbol{A}) = 0$.*

Exercise 6.6.5. *Let $\vec{v}_1 = (1, 2, 3)$ and $\vec{v}_2 = (2, 1, 0)$. Describe the set of all vectors $\vec{v}_3$ such that $\{\vec{v}_1, \vec{v}_2, \vec{v}_3\}$ is a linearly independent set.*

Exercise 6.6.6. *Given any positive integer $n \geq 2$, construct a $2 \times n$ matrix such that any two columns are linearly independent.*

Exercise 6.6.7. *Given any positive integer $n \geq 2$, construct an $n \times 2n$ matrix such that any set of n columns is linearly independent.*

Exercise 6.6.8. *Prove for any $n \geq 2$ that there exists an $n \times \infty$ matrix such that any set of n columns is linearly independent.*

Exercise 6.6.9. *Explicitly construct a matrix with the property from the previous problem.*

Basic Feasible and Basic Optimal Solutions

Exercise 6.6.10. *Consider*

$$
\begin{pmatrix} 1 & 2 & 4 & 1 & 9 \\ 3 & 0 & 7 & 1 & 2 \\ 2 & 3 & 3 & 5 & 4 \end{pmatrix} \begin{pmatrix} x_1 \\ x_2 \\ x_3 \end{pmatrix} = \begin{pmatrix} b_1 \\ b_2 \\ b_3 \end{pmatrix}.
$$

For what $\vec{b}$ does a solution exist? For what $\vec{b}$ does a basic feasible solution exist? If any exist find them.

Exercise 6.6.11. *Consider*

$$
\begin{pmatrix} 1 & 2 & 5 & 6 & -3 \\ 3 & 0 & 3 & 6 & -3 \\ 2 & 3 & 8 & 10 & -4 \end{pmatrix} \begin{pmatrix} x_1 \\ x_2 \\ x_3 \end{pmatrix} = \begin{pmatrix} b_1 \\ b_2 \\ b_3 \end{pmatrix}.
$$

For what $\vec{b}$ does a solution exist? For what $\vec{b}$ does a basic feasible solution exist? If any exist find them.

Exercise 6.6.12. *Prove there are only finitely many basic feasible solutions, but that if there are at least two feasible solutions, then there are infinitely many feasible solutions (thus the number of feasible solutions is either 0, 1 or infinity).*

Exercise 6.6.13. *For fixed M, find some lower bounds for the size of $\sum_{k=1}^{M} \binom{N}{k}$. If $M = N = 1000$ (which can easily happen for real world problems), how many basic feasible solutions could there be? There are less than 10^{90} sub-atomic objects in the universal (quarks, photons, et cetera). Assume each such object is a supercomputer capable of checking 10^{20} basic solutions a second (this is much faster than current technology!). How many years would be required to check all the basic solutions?*

Exercise 6.6.14. *Give an example of a 4×4 matrix such that each entry is positive and all four columns are linearly independent; if you cannot find such a matrix prove that one exists.*

Exercise 6.6.15. *Redo the previous problem, but for an arbitrary N (thus find an $N \times N$ matrix where all entries are positive and the N columns are linearly independent).*

Exercise 6.6.16. *For each positive integer N prove there exists a matrix with N rows and infinitely many columns so that all entries are positive and any set of N columns is linearly independent.*

Exercise 6.6.17. *For each positive integer N find a matrix with N rows and infinitely many columns so that all entries are positive and any set of N columns is linearly independent.*

Properties of Basic Feasible Solutions

The idea of starting with a minimal item and finding a contradiction by constructing an object with a smaller value under a false assumption is used in a variety of problems; below are a few examples.

Exercise 6.6.18. *Prove that $\sqrt{2}$ is irrational (i.e., we cannot write it as a ratio of two integers). Hint: assume it's rational, and let $\sqrt{2} = a/b$, where b is the smallest positive integer of any such rational expansion. Prove such a minimal representation exists and find a contradiction. See* [**MM**] *for some nice geometric proofs of contradiction of the irrationality of $\sqrt{2}$ and other select numbers.*

The next few problems show how the method of infinite descent can be used to express numbers as a fixed sum of squares.

Exercise 6.6.19. *Assume that -1 is non-zero square modulo p (this means there is an x such that $x^2 \equiv -1 \bmod p$, i.e., there is an x such that $x^2 + 1$ is a multiple of p). Then p can be written as a sum of two squares.*

Exercise 6.6.20. *Show that if n and m can be written as the sum of two squares, so too can $n \cdot m$. Hint: think about the norm of **complex numbers**; if $z = x + iy$ (with $x, y \in \mathbb{R}$ and $i = \sqrt{-1}$), then $|z|^2 = x^2 + y^2$.*

Exercise 6.6.21. *What numbers can be written as a sum of two squares?*

Exercise 6.6.22. *Show that if n and m can be written as the sum of four squares, so too can $n \cdot m$. Hint: think about the norm of **quaternions**, numbers q of the form $t + ix + jy + kz$ with $x, y, z \in \mathbb{R}$ and $i^2 = j^2 = k^2 = ijjk = -1$ ($|q|^2 = t^2 + x^2 + y^2 + z^2$).*

Exercise 6.6.23. *Show every prime can be written as a sum of four squares. Hint: use a descent argument and prove that a prime p where -1 is not congruent to a square modulo p is the sum of three squares, and thus conclude every integer is a sum of four squares.*

Another important application of infinite descent is in finding integer solutions to **Diophantine equations** (polynomial equations with integer coefficients).

Exercise 6.6.24. *Fermat's Last Theorem* *says that if $n \geq 3$ is a positive integer, then the only integer solutions to $x^n + y^n = z^n$ are trivial (i.e., have $xyz = 0$). While proving this in general is a challenge (see* [**Wi, TaWi**]*), special choices of n are not too difficult and have been known for a while. Prove that the only solutions of $r^4 + s^4 = t^2$ have $rst = 0$.*

The final set of problems from this section are on linear programming.

Exercise 6.6.25. *Consider the vector $\vec{x}_\lambda$ from our proof:*

$$\vec{x}_\lambda = (x_{j_1} - \lambda \gamma_{j_1}) A_{j_1} + \cdots + (x_{j_k} - \lambda \gamma_{j_k}) A_{j_k}.$$

Prove there is a critical value of λ, say λ_c, such that at least one additional component becomes zero and no components become negative. Give an example where at least two components become zero at the critical value.

Exercise 6.6.26. *Prove that if $\boldsymbol{A}'$ has M rows and k linearly independent columns, with $M \geq k$, then $\boldsymbol{A}'^T \boldsymbol{A}'$ is invertible. Verify this claim by using the matrix*

$$\boldsymbol{A} = \begin{pmatrix} 1 & 2 & 3 \\ 2 & 5 & 7 \\ 2 & 3 & 5 \end{pmatrix}$$

and any two columns.

Exercise 6.6.27. *We say $\vec{x}$ is an* ordered feasible solution *if its non-negative entries are ordered from smallest to largest; thus $(1, 0, 0, 4, 3, 0, 0, 5, 8)$ is not ordered (as 4 is less than 3) but $(1, 0, 0, 3, 3, 0, 0, 5, 8)$ is. Prove or disprove: if a canonical linear programming problem has a feasible solution, then it has an ordered feasible solution.*

Exercise 6.6.28. *With notation as in the previous exercise, prove or disprove: A canonical linear programming problem has at most finitely many ordered feasible solutions.*

Exercise 6.6.29. *Find all basic feasible solutions to*
$$\begin{pmatrix} 1 & 2 & 3 \\ 4 & 5 & 6 \\ 7 & 8 & 9 \end{pmatrix} \begin{pmatrix} x_1 \\ x_2 \\ x_3 \end{pmatrix} = \begin{pmatrix} 1 \\ 1 \\ 1 \end{pmatrix}.$$

Exercise 6.6.30. *Consider the following Linear Programming problem: $x_j \geq 0$,*
$$\begin{pmatrix} 1 & 4 & 5 & 8 & 1 \\ 2 & 2 & 3 & 8 & 0 \\ 3 & 2 & 1 & 6 & 0 \end{pmatrix} \begin{pmatrix} x_1 \\ x_2 \\ x_3 \\ x_4 \\ x_5 \end{pmatrix} = \begin{pmatrix} 311 \\ 389 \\ 989 \end{pmatrix},$$

where we want to minimize
$$5x_1 + 8x_2 + 9x_3 + 2x_4 + 11x_5.$$

Find all basic feasible solutions.

Exercise 6.6.31. *We obtained the following weak estimate:*
$$\sum_{m=0}^{M} \binom{N}{m} \leq \sum_{m=0}^{N} \binom{N}{m} = 2^N.$$

Find sharper bounds for various ranges of N and M. For example, if $M = N$ the inequality above is an equality, but if $M = \sqrt{N}$ we have greatly overestimated.

Optimal and Basic Optimal Solutions

Exercise 6.6.32. *Find the basic optimal solutions, if any, to the problem in Exercise 6.6.30.*

Exercise 6.6.33. *Prove there are only finitely many basic optimal solutions, but that if there are at least two optimal solutions, then there are infinitely many optimal solutions (thus the number of optimal solutions is either 0, 1, or infinity).*

Exercise 6.6.34. *We say $\vec{x}$ is an* ordered optimal solution *if its non-negative entries are ordered from smallest to largest; thus $(1, 0, 0, 4, 3, 0, 0, 5, 8)$ is not ordered (as 4 is less than 3) but $(1, 0, 0, 3, 3, 0, 0, 5, 8)$ is. Prove or disprove: if a canonical linear programming problem has an optimal solution, then it has an ordered optimal solution.*

Efficiency and Euclid's Prime Theorem

Exercise 6.6.35. *Prove there are infinitely many composite numbers.*

Exercise 6.6.36. *Dirichlet proved that if a and m are relatively prime, there are infinitely many primes congruent to a modulo n. For what pairs (a, m) are there infinitely many composites congruent to a modulo m?*

Exercise 6.6.37. *Prove there are infinitely many primes congruent to 3 modulo 4.*

Exercise 6.6.38. *Given any prime $q \geq 2$, prove there are infinitely many primes congruent to a non-square modulo q. For example, if $q = 11$ the numbers relatively prime to q are $\{1, 2, \ldots, 10\}$, with $\{1, 3, 4, 5, 9\}$ equivalent to a square and $\{2, 6, 7, 8, 10\}$ not equivalent to a square. See* [**MT-B, Mu**] *for more on these questions.*

Exercise 6.6.39. *The nth* **Fermat number** *is $F_n = 2^{2^n} + 1$. Show that any two Fermat numbers are relatively prime, and thus conclude there are infinitely many primes. See Chapter 1 of* [**AZ**] *for more proofs on the infinitude of primes.*

Exercise 6.6.40. *Find the analogue of the Euclid-Mullin sequence for the primes constructed from the Fermat numbers. How many terms can you compute?*

Exercise 6.6.41. *Prove that if a prime p divides a product ab, then $p|a$ or $p|b$; this result is surprisingly difficult to prove as it's not clear what you can and cannot use.*

Exercise 6.6.42. *A better approximation to $\pi(x)$ is given by the* **offset logarithmic integral***:*

$$\mathrm{Li}(x) := \int_2^x \frac{dt}{\log t}.$$

Bound $|\mathrm{Li}(x) - x/\log x|$ as $x \to \infty$.

Chapter 7

The Simplex Method

We've shown how to convert any linear programming problem to canonical form, and we've reduced the search for an optimal solution to checking the at most finitely many basic feasible solutions (remember it's possible that there are no feasible solutions). Unfortunately, for most problems of interest there are just too many candidates to check by brute force. We need an efficient way to search this space. One great approach is George Dantzig's Simplex Method. We give a thorough description of *how* it works but do not provide an analysis of *why* it works so efficiently, as that would take us too far afield for a first course. See [**Da1, Fr**] for more details.

7.1. The Simplex Method: Preliminary Assumptions

Before we get to the method itself, we first discuss the reasons behind the assumptions we'll make. As always, we start with a canonical Linear Programming problem:

$$\mathbf{A}\vec{x} = \vec{b}, \quad \vec{x} \geq \vec{0}, \quad \text{minimize } \vec{c}^{\,T}\vec{x}.$$

We assume the following:

(1) $\mathbf{A}$ is an $M \times N$ matrix (M is the number of rows, N the number of columns), with $M < N$.

(2) The rows of $\mathbf{A}$ are linearly independent.

(3) $\vec{b}$ is not a linear combination of fewer than M rows.

(4) $\vec{b} \geq \vec{0}$.

Why do we want more columns than rows? If we have more rows than columns the system is over-determined, meaning we have at most a single feasible solution, if any. If we only have one feasible solution, then that's also the optimal solution; there's no need to find an efficient way to search a space with one element (so long as you can find that one element). If, however, $M < N$, then we expect to have infinitely many solutions, and now we need an efficient way to traverse the candidates.

Why do we want linearly independent rows? If there are linearly dependent rows, some of the rows will be redundant or inconsistent. If there are redundant rows, we can remove them without loss of generality; if there are inconsistent rows, the problem is not solvable and there is no purpose trying in the first place.

Why do we want $\vec{b}$ not to be a linear combination of fewer than M rows? We'll see that this assumption is needed for some stability issues and other technicalities. But how bad, or how restricting, is it? Well, if $\vec{b}$ is a linear combination of k columns, where $k < M$, then $\vec{b}$ lies in one of $\binom{N}{k}$ possible hyperplanes; while this may seem like a lot, since we are looking at $\mathbb{R}^M$ the total M-dimensional volume of these hyperplanes still turns out to be 0. The proportion of cases excluded by this assumption is infinitesimal, and on the very off-chance that we might have a $\vec{b}$ that doesn't satisfy this, usually a very small tweak to the rows of $\vec{b}$ fixes the issue. Of course, there might be some problems with very small tweaks, as computers can have round-off errors and we may need to worry about numerical precision. In practice, however, this condition is almost always met, and we shall assume it holds.

Why do we insist that $\vec{b} \geq 0$ (i.e., that all of its entries are non-negative)? This assumption is more of a notational convention than an assumption, and it is not restrictive at all. If we have a $b_i < 0$, just multiply row i of A and of $\vec{b}$ by -1, and the assumption holds again. Thus, we can always assume the entries of $\vec{b}$ are non-negative; however, we cannot simultaneously insist that the entries of $\vec{b}$ and $\mathbf{A}$ are non-negative.

7.2. The Simplex Method: Statement

The Simplex Method has two steps, traditionally called "phases".

The Simplex Method:

Consider a canonical Linear Programming problem $\mathbf{A}\vec{x} = \vec{b}$, $\vec{x} \geq \vec{0}$, $\min \vec{c}^T \vec{x}$ which satisfies the assumptions from §7.1. The Simplex Method proceeds in two steps.

<u>Phase I</u>: Find a basic feasible solution to $\mathbf{A}\vec{x} = \vec{b}$, $\vec{x} \geq \vec{0}$, *or* prove that none exists.

<u>Phase II</u>: Given a basic feasible solution, find a basic optimal solution *or* produce a sequence of feasible solutions with cost $\vec{c}^T \vec{x} \to -\infty$. (Note that either way, this phase is fully constructive.)

We'll give the details of how to do the two steps later, but here is how it will (roughly) work in practice.

Step 1: Assume we can do Phase II; prove we can do Phase I.
Step 2: Assume we can do Phase I; prove we can do Phase II.

At first glance, this seems like a circular proof: we're assuming we can do Phase II to prove we can do Phase I, and then we assume we can do Phase I to do Phase II.

The reason this approach works is that Phase II and Phase I don't have to be applied to the *same* Linear Programming problem. Their problems need to be related, of course, since otherwise there's no way the Simplex Method will be useful, but they needn't be the same. This allows us to avoid any circular reasoning during our proof.

Essentially, we're exploiting the fact that there are many Linear Programming problems where it's trivial to write down a feasible solution. Sometimes it's easier to start with a problem not in canonical form, as the answer will be quite clear in that setting. For example, consider

$$\mathbf{A}\vec{x} \ \leq \ \vec{b}, \ \ \vec{x}, \vec{b} \geq 0.$$

We can use the Method of Divine Inspiration (see §3.2 for other examples) to find a feasible solution: **just take $\vec{x} = \vec{0}$!** The goal is to tweak any problem to something as easy as this.

7.3. Phase II implies Phase I

We start with $\mathbf{A}\vec{x} = \vec{b}$, with $\vec{x}, \vec{b} \geq \vec{0}$. Briefly, our game plan is the following. We first construct a related problem where we can immediately find a basic feasible solution. We then feed that into the Phase II machine, and try to crank out a basic optimal answer to our related problem. If such a solution exists, it turns out that it's a basic feasible solution to the original problem.

Now for the details. We consider the related constraints

$$(\mathbf{A} \quad \mathbf{I}) \begin{pmatrix} \vec{x} \\ \vec{z} \end{pmatrix} \ = \ \vec{b},$$

with $\vec{x} \geq \vec{0}$, $\vec{z} \geq \vec{0}$ (*warning: these are not necessarily the same zero vector*), and our goal is to minimize $z_1 + \cdots + z_M = \vec{1}^T \vec{z}$. We call $(\mathbf{A} \ \mathbf{I})$ the augmented constraints matrix.

The ith row in the augmented matrix leads to the equation

$$a_{i,1}x_1 + \cdots + a_{i,N}x_N + z_i \ = \ b_i, \quad x_j \geq 0, z_i \geq 0.$$

Fortunately it's very easy to find a solution to this by inspection; simply take $\vec{x} = \vec{0}, \vec{z} = \vec{b}$! This is a basic feasible solution, since any subset of the columns of the identity matrix $\mathbf{I}$ we appended (from choosing $z_i = b_i$) are linearly independent.

We now apply Phase II, as promised. We either get a basic optimal solution to our new, augmented problem, *or* we get a sequence of basic feasible solutions whose cost goes to $-\infty$. For this augmented problem the cost is $z_1 + \cdots + z_M$; this is always non-negative since all the entries of $\vec{z}$ are non-negative. Since this cost *cannot* go to $-\infty$, Phase II *must* yield a basic optimal solution to the augmented problem. Call this basic optimal solution $\vec{x}_{\mathrm{op}}, \vec{z}_{\mathrm{op}}$. We have 2 cases.

(1) *The optimal cost is zero.* Then $\vec{z}_{\mathrm{op}} = \vec{0}$, so $\vec{x}_{\mathrm{op}}$ is basic feasible for the original problem.

(2) *The optimal cost is positive.* Thus some $z_i > 0$. This means there is no basic feasible solution to the original problem, as if there were, taking that solution to be $\vec{x}$ and taking $\vec{z}$ to be $\vec{0}$ would have yielded a cost of zero, which is lower than the cost of this optimal solution.

Thus if we can do Phase II, then we can do Phase I.   □

7.4. Phase II of the Simplex Method

We now describe an algorithm for Phase II, namely how to pass from a basic feasible solution to a basic optimal solution (or prove one does not exist by constructing a sequence of basic feasible solutions with costs tending to minus infinity). We begin with a useful fact about basic feasible solutions.

Lemma 7.4.1. *If $\vec{x}$ is a basic feasible solution, then $\vec{x}$ must have exactly M non-zero entries.*

Proof. Remember we are assuming $\mathbf{A}$ is an $M \times N$ matrix with $M \leq N$ and $\vec{b}$ is a non-negative vector with M components which cannot be written as a linear combination of fewer than M columns of $\mathbf{A}$.

We first show that $\vec{x}$ has at most M positive entries. The rank of $\mathbf{A}$ is at most M, and the columns corresponding to the positive entries of the feasible solution $\vec{x}$ must be independent. Thus $\vec{x}$ cannot have more than M positive components.

Furthermore, $\vec{x}$ must have at least M positive components; if it didn't, then $\vec{b}$ could be written as the sum of fewer than M columns of $\mathbf{A}$. Thus, the only possibility is that $\vec{x}$ has exactly M positive components.   □

We now describe how to do Phase II. By the above we may assume that we have a basic feasible solution $\vec{x}$ with exactly M non-zero entries. Let $B = \{j : x_j > 0\}$; note $|B| = M$ by the above lemma. We call B the **basis**. Let $\vec{x}_B = (x_{j_1}, \ldots, x_{j_m})$ be the positive components of $\vec{x}$, and let $\mathbf{A}_B$ denote the matrix of columns A_j of $\mathbf{A}$ where $j \in B$; we call $\mathbf{A}_B$ the **basis matrix**. Thus we have

$$(7.1) \qquad\qquad \mathbf{A}_B \vec{x}_B = \vec{b}.$$

Furthermore, $\mathbf{A}_B$ is an invertible matrix (it's an $M \times M$ matrix with M linearly independent columns). Thus we can also study the linear system of equations

$$(7.2) \qquad\qquad \vec{y}^T \mathbf{A}_B = \vec{c}_B^T,$$

which has the unique solution

$$(7.3) \qquad\qquad \vec{y} = \vec{c}_B^T \mathbf{A}_B^{-1}.$$

There are two possibilities for this vector $\vec{y}$: either it's a feasible solution to the Dual Problem (the dual of the original Linear Programming problem, introduced in §5.2) or it isn't. Note that it's easy to see if $\vec{y}$ is feasible. We need only check that $\vec{y}^T A_j \leq c_j$ for all $j \in \{1, \dots, M\}$. By construction this holds for $j \in B$; thus we need only check these conditions for $j \notin B$.

Case 1: $\vec{y}$ is feasible for the Dual Problem.

If $\vec{y}$ is feasible for the Dual Problem, then from (7.1) and (7.2) we have
$$\vec{y}^T \vec{b} = \vec{y}^T \mathbf{A}_B \vec{x}_B = \vec{c}_B^T \vec{x}_B = \vec{c}^T \vec{x}.$$

By Lemma 5.2.1, the fact that $\vec{y}^T \vec{b} = \vec{c}^T \vec{x}$ means that $\vec{x}$ is an optimal solution of our Linear Programming problem. As $\vec{x}$ is a basic feasible solution, this means $\vec{x}$ is a basic optimal solution.

Case 2: $\vec{y}$ is not feasible for the Dual Problem.

As $\vec{y}$ is not a feasible solution for the Dual Problem, for some s we have $\vec{y}^T A_s > c_s$; by construction we know $s \notin B$. The idea is that we can lower the cost by bringing the column A_s into the basis matrix $\mathbf{A}_B$.

As the M columns of the $M \times M$ matrix $\mathbf{A}_B$ are linearly independent and A_s is a vector with M components, we have
$$A_s = \sum_{j \in B} t_j A_j$$
or

(7.4)
$$A_s = \mathbf{A}_B t, \quad t = \mathbf{A}_B^{-1} A_s.$$

From the relations
$$\sum_{j \in B} x_j A_j = \vec{b}, \quad A_s - \sum_{j \in B} t_j A_j = 0,$$
we find that

(7.5)
$$\lambda A_s + \sum_{j \in B} (x_j - \lambda t_j) A_j = \vec{b};$$

for λ sufficiently small and positive, we have a new feasible solution $\vec{x}'$ to the original Linear Programming problem, with
$$\vec{x}'_j = \begin{cases} \lambda & \text{if } j = s, \\ x_j - \lambda t_j & \text{if } j \in B, \\ 0 & \text{otherwise.} \end{cases}$$

The original cost (associated to the feasible solution $\vec{x}$) is
$$\sum_{j \in B} x_j c_j;$$

the new cost (associated to the feasible solution $\vec{x}'$) is

$$\lambda c_s + \sum_{j \in B}(x_j - \lambda t_j)c_j.$$

We now show that the new cost is less than the old cost. The new cost minus the old cost is

(7.6) $$\lambda\left(c_s - \sum_{j \in B}t_j c_j\right);$$

as $\lambda > 0$, we need only show that $c_s < \sum_{j \in B}t_j c_j$ to show the new cost is less than the old cost. From (7.2) we have $\vec{y}^T \mathbf{A}_B = \vec{c}_B^T$, and from (7.4) we have $A_s = \mathbf{A}_B t$. These relations imply

$$\sum_{j \in B}t_j c_j \;=\; \vec{y}^T \mathbf{A}_B t \;=\; \vec{y}^T A_s \;>\; c_s,$$

where the last inequality follows from our assumption that $\vec{y}$ is not feasible for the Dual Problem (the assumption that $\vec{y}$ is not feasible for the Dual Problem means that $\vec{y}^T A_s > c_s$).

There are two possibilities: either all $t_j \leq 0$ or at least one is positive.

(1) **Case 2: Subcase (i):** Assume all $t_j \leq 0$ in (7.4). Then we may take λ to be *any* positive number in (7.5), and each positive λ gives us another feasible solution. Therefore the cost in (7.6) tends to minus infinity as λ tends to infinity. This implies we can construct a sequence of feasible solutions to the original Linear Programming problem with costs tending to minus infinity, and therefore the original Linear Programming problem does not have an optimal solution.

(2) **Case 2: Subcase (ii):** Suppose now at least one $t_j > 0$. The largest positive λ we may take and still have a feasible solution is

(7.7) $$\lambda^* \;=\; \min_{j \in B}\left(\frac{x_j}{t_j} : t_j > 0\right).$$

We may assume the minimum for λ^* occurs for $j = p$. Note that p is unique; if not, we would find that $\vec{b}$ is a linear combination of fewer than M columns, contradicting our assumption on $\vec{b}$ (we are basically swapping the pth column for the sth column). We have thus found a new feasible basic solution (we leave it to the reader to check that our solution, in addition to being feasible, is also basic) with exactly M non-zero entries. We now restart Phase II with our new basic feasible solution as our basic feasible solution, and continue the search for a basic optimal solution.

We have described the algorithm for Phase II of the Simplex Method; we show that it must terminate either in an optimal solution or we can construct a sequence of basic feasible solutions with costs tending to minus infinity. When we start Phase

II, three things can happen: (1) we end up in Case 1: if this happens, then we have found an optimal solution; (2) we end up in Case 2, Subcase (i): if this happens, there is no optimal solution (and we have a sequence of basic feasible solutions with costs tending to minus infinity); (3) we end up in Case 2, Subcase (ii): if this happens, we restart Phase II with the new basic feasible solution.

The only way Phase II would not terminate is if we always end up in Case 2, Subcase (ii) each time we apply it. Fortunately, we can see this cannot occur. $\mathbf{A}$ is an $M \times N$ matrix $(M < N)$. There are only finitely many sets of M linearly independent columns of $\mathbf{A}$ (there are at most $\binom{N}{M}$). Each time we enter Case 2, Subcase (ii) we obtain a new feasible solution with cost *strictly* less than the previous cost. This implies, in particular, that all the solutions from Case 2, Subcase (ii) are distinct. As there are only finitely many possibilities, eventually Phase II must terminate with either a basic optimal solution to the original Linear Programming problem or with a sequence of feasible solutions with costs tending to minus infinity. This completes our analysis of the Simplex Method.

7.5. Run-time of the Simplex Method

While we've shown that the Simplex Method either terminates with a basic optimal solution or gives a sequence of feasible solutions with cost tending to minus infinity, we have *not* shown that it terminates in a reasonable amount of time! It is imperative we show this if we are interested in using the Simplex Method as a practical tool, and not merely in noting it as a theoretical curiosity.

Unfortunately, one of the reasons we haven't shown it runs in a reasonable amount of time is because some times it doesn't! Klee and Minty [**KM**] gave an example where it took exponentially long for the Simplex Method to run; however, such examples are 'rare'. This means that under reasonable assumptions on the possible values of the parameters, the average run-time is polynomial, not exponential.

We encourage the interested reader to peruse [**Fr**] and the references therein for more details, especially proofs of the efficiency of these methods.

7.6. Efficient Sorting

We end the chapter by discussing the problem of efficient sorting and some famous algorithms used to tackle this problem. The most basic version of the problem deals with a sequence of n integers, where the end goal is to place these numbers in ascending order in the most computationally efficient manner. Although sorting is itself trivial, sorting efficiently requires some thought. There are countless applications; just think of all the web searches performed where one wants the output displayed as a list ordered by price (or some other parameter).

We quickly describe some famous sorting algorithms, along with a few words about their run-times; the interested reader should consult the extensive literature for more information. We use **big-Oh** notation: we say a function $f(x)$ is big-Oh of $g(x)$, written $f(x) = O(g(x))$ or $f(x) \ll g(x)$, if there exist constants C, x_0 such that for all $x \geq x_0$ we have $|f(x)| \leq Cg(x)$. Thus $x^2 = O(x^3/2017)$ and $x^2 = O(x^2 \log x)$. Note this is discussing behavior as x grows to infinity; we could of course choose another point. Though we don't need it here, for completeness it's worth introducing **little-Oh** notation: $f(x) = o(g(x))$ means that as $x \to \infty$ we have $f(x)/g(x) \to 0$.

Bubblesort: Bubblesort operates by recursively stepping through lists, comparing each pair of adjacent items, and swapping them if they are in the wrong order (resulting in the larger of the two elements being in the higher index position). Each iteration reduces the effective size of the array, since after each iteration the largest element "bubbles" to the end of the list. This passing procedure is repeated until no swaps are required, indicating that the list is sorted. Bubblesort has a worst-case and average complexity of $O(n^2)$. The position of elements in bubblesort play an important role in determining performance. Large elements at the beginning are easily swapped, but small elements toward the end move to the beginning slowly.

Insertionsort: The idea of an insertionsort is to build a new array which we keep sorted at all times. While inserting the ith element (of n total elements) from the unsorted array into the new array, insert it such that the new array (now i elements long) is sorted. This is repeated until $i = n$. In some sense, it's like taking an element from the unsorted array and pushing it as far as it will go into the new, sorted array. We now investigate the worst-case run-time of insertionsort by calculating the total number of comparisons in an unsorted array of n elements. We need 0 comparisons to insert the first element, 1 comparison for the second element, 2 comparisons for the third element, and so on up to $n-1$ comparisons for the last element. Thus in aggregate this is: $1+2+3+\cdots+(n-1) = O(n^2)$. The best-case (already sorted) run-time complexity is $O(n)$, showing that the algorithm runs fast on nearly sorted data.

Selectionsort: In selectionsort, we recursively find the largest element in the array and move it to its final position in the sorted array (i.e., the end the of array). We start by finding the largest element and moving it to the highest index position. We can do this by swapping the end-position element and the largest element. We then reduce the effective size of the array by one element and repeat the process on the smaller subset, until we are left with a set of size 1. The worst-case run-time complexity (reverse-sorted array) is $O(n^2)$.

Quicksort: Quicksort is a divide-and-conquer algorithm that partitions an array into two parts on either side of a pivot, and then sorts the two parts independently. The crux of the method is the partitioning operation, which rearranges the array in the following way:

- A pivot is picked (usually at random) from the unsorted array A, call it p.

- All items smaller than p are moved to the left of p and all items larger than p are moved to the right of p. Note that now p is in its correct *ultimate* position in A.

- Apply this procedure recursively to the subarrays of A on the left and right of p.

The algorithm terminates when we reach subarrays of size 1 (already sorted). The time complexity of quicksort is $O(n \log n)$ on average, and $O(n^2)$ in the worst case (poor choice of pivot).

Mergesort: Mergesort is another divide-and-conquer algorithm. The input array is divided recursively into n lists of length 1. The crux of the sorting is achieved in the merge step which combines two *ordered* arrays to make one larger ordered array. It is possible to accomplish this merging in $O(n_1 + n_2)$ time where n_1 and n_2 are the lengths of the two sorted arrays. The idea then is to divide an unsorted array into two halves, sort the two halves (recursively), and then merge the results. The base case of the recursive merging is when we reach a subarray of size 1 (already sorted). Mergesort guarantees to sort an array of n items in time proportional to $n \log n$, regardless of the nature of the initial unsorted ordering.

7.7. Exercises

The Simplex Method: Preliminary Assumptions

Exercise 7.7.1. *Assume one row of $\boldsymbol{A}$ is a linear combination of the other rows. Prove that unless the corresponding element of $\vec{b}$ is the same linear combination of the corresponding entries there cannot be any solutions to $\boldsymbol{A}\vec{x} = \vec{b}$.*

Exercise 7.7.2. *Assume one row of $\boldsymbol{A}$ is a linear combination of the other rows, and the corresponding element of $\vec{b}$ is the same linear combination of the corresponding entries. Prove that there is a new problem, $\boldsymbol{A}'\vec{x}' = \vec{b}'$, where the solutions to this are in a one-to-one correspondence to the solutions of the original problem.*

Exercise 7.7.3. *Why can we not assume in general that the entries of $\boldsymbol{A}$ and $\vec{b}$ are both non-negative?*

Exercise 7.7.4. *Prove or disprove: given a square $n \times n$ matrix $\boldsymbol{A}$ (with $n \geq 4$) whose columns are linearly independent, and a vector $\vec{b} \in \mathbb{R}^4$ whose components are non-negative, there is always a solution $\vec{x}$ to $\boldsymbol{A}\vec{x} = \vec{b}$.*

Exercise 7.7.5. *Prove or disprove: given a square $n \times n$ matrix $\boldsymbol{A}$ (with $n \geq 4$) whose columns are linearly independent, and a vector $\vec{b} \in \mathbb{R}^4$, if the entries of $\boldsymbol{A}$ and $\vec{b}$ are non-negative, then there is always a solution $\vec{x} \geq \vec{0}$ to $\boldsymbol{A}\vec{x} = \vec{b}$.*

The Simplex Method: Statement

Exercise 7.7.6. *Consider the Diet Problem from §3.4. Find a feasible solution by inspection.*

Exercise 7.7.7. *Give an example of a canonical Linear Programming problem that does not have a feasible solution, though the augmented problem does.*

Exercise 7.7.8. *One of our assumptions was that we adjusted our matrix $\boldsymbol{A}$, if needed, so that $\vec{b} \geq \vec{0}$. What if $\vec{b}$ were not non-negative? Prove that in this case the augmented constraint matrix could lead to a situation where there is no basic feasible solution to the related problem, and give an explicit example where this happens.*

Phase II implies Phase I

Exercise 7.7.9. *Prove that A_B is invertible.*

Exercise 7.7.10. *In Case 2: Subcase (ii), we saw that we may assume the minimum for λ^* occurs for $j = p$, and then claimed p is unique. Prove this claim.*

Phase II of the Simplex Method

Exercise 7.7.11. *Consider the following Linear Programming problem: $x_j \geq 0$,*

$$(7.8) \qquad \begin{pmatrix} 1 & 4 & 5 & 8 & 1 \\ 2 & 2 & 3 & 8 & 0 \\ 3 & 2 & 1 & 6 & 0 \end{pmatrix} \begin{pmatrix} x_1 \\ x_2 \\ x_3 \\ x_4 \\ x_5 \end{pmatrix} = \begin{pmatrix} 311 \\ 389 \\ 989 \end{pmatrix},$$

where we want to minimize

$$(7.9) \qquad\qquad\qquad 5x_1 + 8x_2 + 9x_3 + 2x_4 + 11x_5.$$

Find an optimal solution (or prove one does not exist).

Exercise 7.7.12. *Assume a canonical linear programming problem has an optimal solution. As a worst-case scenario, how many times must we go through Phase II before we reach an optimal solution?*

Run-time of the Simplex Method

   Fermat's little Theorem states that if n is prime, then $a^n - a$ is a multiple of n. For example, if we take $a = 5$ and $n = 12$, then $a^n - 1 = 5^{1}2 - 5 = 244,140,620$, which is not a multiple of 12; thus we see 12 is not prime. If, however, we take $a = 3$ and $n = 11$, then $a^n - 1 = 3^{11} - 3 = 177,144 = 16104 \cdot 11$, showing it's a multiple of 11 as predicted.

   The consequence of all of this is that we have a way to detect if a number is prime; this is extremely important, since primes play a key role in many modern encryption and decryption schemes. Unfortunately, it's not a perfect test for primality. Clearly if $a^n - a$ is not a multiple of n, then n is not prime; however, some composites n have some a for which $a^n - a$ is a multiple of n. Some exceptionally bad composites have $a^n - a$ a multiple of n for all a; these are the **Carmichael numbers**, and their existence complicates using Fermat's little Theorem (which through repeated squaring is very fast to implement) in cryptography. Unfortunately there are infinitely many Carmichael numbers; in [**AGP**] the authors proved for large x that at least $x^{2/7}$ of the integers up to x are Carmichael numbers.

   For the next few problems we make the following assumptions:

- If n is a non-Carmichael composite number, there is an a such that $a^n - a$ is not a multiple of n; we call such an a a witness to n's compositeness. Assume the probability that a given a is a witness is $1/2$, and each a being a witness is independent of any other a being a witness.

- For simplicity, assume $x/\log x$ integers up to x are prime, and $x^{2/7}$ are Carmichael numbers.

Exercise 7.7.13. *As $x \to \infty$, if n is a non-Carmichael composite, what is the average run-time to find a witness?*

Exercise 7.7.14. *As $x \to \infty$, what is the expected worst-case run-time to find a witness for a non-Carmichael composite?*

Exercise 7.7.15. *What is the average run-time to discover that there is not a witness for compositeness if n is prime? If n is composite?*

Exercise 7.7.16. *For each $n \leq x$ search to see if it has a witness for compositeness. What is the average run-time?*

Efficient Sorting

Exercise 7.7.17. *Prove $x^n = O(e^x)$ for any $n > 0$.*

Exercise 7.7.18. *Prove $\log^a x = O(x^r)$ as $x \to \infty$ for any $a, r > 0$. What is the relation between these two functions as $x \to 0$?*

Exercise 7.7.19. *Apply the sorting methods to the set $\{5, 7, 3, 2, 6, 8, 1, 4\}$. Which is fastest?*

Exercise 7.7.20. *In insertionsort we are not taking advantage of the fact that our new array is sorted. We can use a binary search to speed up the comparisons (check the middle entry first against the new candidate; if that entry is smaller, then move forward a quarter, while if it's larger, move back a quarter). How long will this procedure take on average? Worst case?*

Exercise 7.7.21. *Use online resources to study the sorting algorithms in a bit more depth and note the advantages and disadvantages of each. These could relate to time/space complexity, restrictions on the data, difficulty of writing the code, et cetera.*

Exercise 7.7.22. *A big theme of this book is deciding what* metric *to consider when making decisions. What do you think is the right metric to use in comparing sorting algorithms? Is it worst-case run-time? Average runtime (and if so does this refer to mean or median)? Does the answer depend on the problem?*

Exercise 7.7.23. *Based on your metric from the previous question, pick a "good" sorting algorithm and a "bad" one. Write code to implement these two methods and report the run-time of each of the two algorithms for several randomly chosen permutations of $\{1, 2, \ldots, n\}$. For what n does a significant difference start to arise between the two methods? Does using the "bad" algorithm for small values of n confer any advantages?*

In comparing the different sorting algorithms, we had to decide what mattered most to us: the average run-time or the run-time in the worst-case situation. Below we give a famous problem that's frequently used in job interviews: the **egg drop problem**, which we now describe. Consider a building with n floors, labeled $1, 2, \ldots, n$. We have k identical eggs, and assume if an egg cracks when dropped from a floor it cracks from all floors higher, while if it does not crack when dropped it can be used again. We want to find the highest floor where we can safely drop eggs without their cracking.

If we only have one egg, there is only one possible strategy: start at floor 1, and keep dropping until the egg cracks. For the problems below try to minimize the number of drops in the worst case for the various strategies; if you are brave redo the problems where instead you minimize the average number of drops.

Exercise 7.7.24. *Assume you have two eggs. The binary approach is to drop an egg at the halfway point; if it cracks you then drop from the bottom incrementing by one floor, while if it does not crack you drop from the three-quarters point and then continue similarly. What is the number of drops needed in the worst case? Show that if we had $\log_2 n$ eggs we could find the floor in at most $\log_2 n$ drops.*

Exercise 7.7.25. *Assume you have two eggs. Keep moving up in increments of $\sqrt{n}$ until you have an egg that cracks, then move up by one floor at a time. What is the number of drops needed in the worst case? Show it's bounded by $2\sqrt{n}$.*

Exercise 7.7.26. *Redo the previous problem with k eggs, and show that the worst case requires at most $kn^{1/k}$ drops. Find k as a function of n that minimizes $kn^{1/k}$. How does this compare with the binary approach with $\log_2 n$ eggs?*

Exercise 7.7.27. *Return to the case of two eggs. One issue with the approach in Exercise 7.7.25 is that if the key floor is in the last set it takes significantly longer to find than if it's in the first set. We should thus be willing to drop the first egg a little higher than $\sqrt{n}$ and hence give us a larger window to potentially check one floor at a time, in order to cut down on the number of drops needed (in the worst case) before an egg cracks. One way to find the location of the first drop is to require that the number of drops needed to find the key floor be the same in all the worst cases. Thus, we drop the first egg at floor m, and worst case if it cracks there we need $m - 1$ more drops to find out the floor. If it doesn't break, then we drop the next egg at floor $m + m - 1$, and if it breaks here we need $m - 2$ more drops to find the floor in the worst case; thus both scenarios require m drops in the worst case. Continuing like this we find we want $m + (m - 1) + \cdots + 2 + 1 = n$; thus if n is a triangle number, then $n = m(m + 1)/2$ and the first drop is at approximately $\sqrt{2}n^{1/2}$. Generalize this argument to 3 eggs (or k if you are brave).*

Exercise 7.7.28. *Consider a two-dimensional version of the egg drop problem defined as follows. Before we had floors 1 through n and wanted to find the highest floor f from which a drop won't crack our eggs; now have pairs (i, j) with $1 \le i, j \le n$. We assume that if an egg cracks from (i, j) it will crack from all (i', j') where $i' \ge i$ and $j' \ge j$, and that the region of pairs where a drop causes a crack is a rectangle. We want to find the smallest pair (f_x, f_y) where it cracks. Generalize the previous problems and approaches to this problem. For the especially brave: what would the three-dimensional analogue be?*

Part 3

Advanced Linear Programming

Chapter 8

Integer Programming

It can be sometimes interesting to look at cases where constraints and variables are discrete. This is called *integer linear programming*, and it's a fascinating area of current study and applications. We won't spend too much time on the topic due to its complexity

So far all the variables in our Linear Programming problems are real, taking values from a continuum. For many problems this is reasonable, though at first glance it might not appear so. For example, we might worry about being able to ship any amount of oil (perhaps we can only send an integral number of gallons) or charge any amount (for most products the cost must be some number of cents); however, if the quantities are large, then for all practical purposes there's no harm in allowing these quantities to vary continuously. If we're looking at the cost of a flight, which is measured in hundreds of dollars, a hundredth of a cent is immaterial; if we're shipping oil on a tanker, which can transport millions of gallons, part of a gallon will not be noticeable.

Unfortunately, in other problems we cannot make such a simplifying assumption. For example, if we consider the Diet Problem and we only have a few items, you can't purchase just part of a product and leave the rest (as Kramer found out when we wanted to buy part of a can of Coke and part of an apple in the *Seinfeld* episode "The Seven", or as Steve Martin's character found out in the movie *Father of the Bride* with hot dog buns). In these settings, we need to restrict to integer amounts; not surprisingly, this leads to the subject of **integer programming**.

We'll see later that integer programming is useful for far more than just discrete variables. By using integer variables we can incorporate a lot of constraints and non-linear functions into linear programming problems, such as "and", "or", "if and only if", "max", "min", and "absolute value", to name just a few. The ability to use these greatly increases the types of problems we can study; unfortunately, it also greatly increases the difficulty of solving them. On top of this, the algorithms we have for solving integer programming problems are not as good as the ones we have for problems that use real variables. Fortunately, all is not lost. We do have powerful techniques that allow us to get close to an optimal solution quickly (recall Lemma 5.2.1, which provides a great test to see how close we are to optimal), and for many problems it suffices to rapidly find an approximate solution (especially since we are often estimating parameters, and thus the problem we are trying to solve is only an approximation of reality!).

As an introduction to integer linear programming, we'll look at a movie theater problem. The issues that arise here are representative of what happens in a variety of problems. We'll then explore some of the basics of integer programming and see how we can use it to attack a lot of situations.

We concentrate in this chapter on showing how we can use integer valued variables to bring a lot of non-linear relations into linear programming. Unfortunately, as with most things in life, *there ain't no such thing as a free lunch*. While we are able to incorporate a lot of non-linear constraints and functions, there is an enormous cost: we must handle integer variables and perform optimization on integer inputs. Earlier, we saw how easy it was to use calculus (see this discussion of Lagrange multipliers in §3.3) to find extrema. Such methods crucially use the fact that we have quantities which vary continuously; integer optimization can be significantly harder. In order to fully appreciate the power of integer variables, we postpone a discussion of the subtleties and difficulties that they produce until Chapter 9.

8.1. The Movie Theater Problem

Let's imagine we run a movie theater. Our goal, not surprisingly, is to maximize profits. To figure out our optimal strategy, we need to study our revenues and expenditures; below is a partial list which provides a great starting point.

Sources of revenue:

- selling tickets,
- selling food,
- selling advertisements,
- in-house arcades.

Expenditures:

- labor wages,
- rent/utilities,
- cost of movies,
- cost of food/arcades,
- taxes,
- advertisement budget.

It's important to remember that maximizing profits on the whole doesn't necessarily mean maximizing all of the revenue streams and minimizing all of the costs. Some theaters sell tickets for as low as \$1 in order to attract customers who spend money on food and concessions, which is a much more profitable source of income than the ticket sales. It can sometimes be worthwhile to take a loss on product A in order to sell Product B; you can also see this in a lot of stores, which frequently have leading items which aren't profitable for the store, but serve to get customers inside; once they arrive, they'll buy other items.

Our problem is to schedule the movies at our theater; this means we'll need constraints, and we'll have to find what our variables should be. There are a lot of similarities with the oil problem. There, one of our variables was *how much* oil we shipped from refinery i to city j; here, the key variable is *whether or not* at time t movie m is shown on screen s.

Parameters.

- screens $s \in \{1, \ldots, S\}$,
- capacity c_s (how many people can see a movie on screen s),
- demand $d_{t,m}$ (the number of people who wish to see movie m if it starts at time t),
- labor costs (different staff require different salaries),
- time $t \in \{1, \ldots, T\}$ (we break the day into a large number of discrete times),
- movies $m \in \{1, \ldots, M\}$ (the candidate movies),
- run-time r_m (how long movie m is),
- ticket prices (note these could be functions of the time of day and age of moviegoer).

We have to be careful with our parameters above. This is just meant to be a first pass on the problem so we'll ignore some issues in our formulation, but it's worth mentioning them. Let's examine the demands, $d_{t,m}$. The first difficulty is we need a way to estimate these values. Approximations can be done by comparing our movies to others that have been shown in the past. For example, we might have ticket sales from last week. Or we might have historical data on how action sequels do. Even if we know these values, however, there are other problems. Do movies compete with each other? If yes, does this mean the demands for movies at each moment must depend on what else has been shown during the day (and even potentially what might be in the queue for later)? For now we'll ignore issues such as these in order to introduce the main ideas.

Variables:

- decision variable

$$x_{t,m,s} = \begin{cases} 1 & \text{if at time } t \text{ movie } m \text{ starts on screen } s, \\ 0 & \text{otherwise.} \end{cases}$$

Variables such as these are called **binary variables**; they are either 0 or 1, and we'll see that they have a variety of nice properties.

While the above is an excellent choice for our decision variable, there are other options. We could have defined our decision variable as

$$y_{t,m,s} = \begin{cases} 1 & \text{if at time } t \text{ movie } m \text{ is running on screen } s, \\ 0 & \text{otherwise.} \end{cases}$$

The advantage of our first formulation is that it highlights the action of starting a movie. Since we know how long the movie is, we can then figure out when the screen is next available. In particular, we would have the following constraint:

$$\text{If } x_{t,m,s} = 1, \quad \text{then} \quad \sum_{\tau=t+1}^{t+r_m-1} \sum_{\sigma=1}^{S} x_{\tau,m,\sigma} = 0;$$

in other words, if at time t we start movie m on screen s, then the earliest that movie can be started on any screen is time $t + r_m$. We can encode constraints like this so concisely because our variables only take on the values 0 and 1. Thus, the only way the sum above can be zero is if each summand is zero.

Of course, the constraint above does not fit into our linear programming framework as written. We don't have a linear combination of our variables; instead, we have an if-then statement. We'll see below how to convert expressions such as this to our linear programming framework.

Objective function: For simplicity let's assume all ticket prices are P and that there is no concession revenue. Then we are trying to maximize

$$\sum_{t=1}^{T} \sum_{m=1}^{M} \sum_{s=1}^{S} P \min(c_s, d_{t,m}) x_{t,m,s}$$

(we need the minimum above as we cannot have more people watching a movie on screen s than can fit in that room). We could, and should, add terms related to food sales. The challenge is how to incorporate these. For example, which do you think generates more concession sales: a teen date movie, an action movie, a family movie, or a senior citizen favorite? Thus we might need to keep track of *who* is in the theater, not just the number of people. This suggests adding another parameter, say $\text{rev}_{t,m}$, which is the revenue generated by an average attendee who comes at time t to see movie m:

$$\sum_{t=1}^{T} \sum_{m=1}^{M} \sum_{s=1}^{S} \text{rev}_{t,m} \min(c_s, d_{t,m}) x_{t,m,s}.$$

8.2. Binary Indicator Variables

One of the attractive aspects of integer programming is its ability to deal with constraints that may not at first seem linear. **binary indicator variables** are a key ingredient in this success. As their name suggests, they take on only two values: 0 and 1. Their utility is that they can easily encode any binary alternative. They are particularly useful when considering whether or not something has happened, or making sure a condition that we want is enforced. Typically we use 1 for it has happened ("on") and 0 for it has not ("off").

For example, let's consider a constraint from the Diet Problem. Say we want to make sure we have a certain amount of nutrient j:

$$a_{j1}x_1 + a_{j2}x_2 + \cdots + a_{jn}x_n \geq b_j.$$

We might be worried, however, if we have *too* much of nutrient j (think of those poor laboratory mice who receive 10,000 times the recommended daily allowance, *for humans*, of various substances). Thus, we might want a penalty in our objective function if we consume too much of nutrient j. Here we will need a variable z which is 1 if some event E happens (which, here, is consuming too much of nutrient j) and 0 otherwise.

In mathematics, it's common practice to convert items to numerical values. This makes it easy to compare and allows us to apply various mathematical operations to them. For example, in cryptography (such as RSA) we don't transmit letters, but numbers; see for example the fast multiplication exercises from §1.7 and the last few from the Euclidean algorithm problems in §2.6. Thus, we'll often associate a numerical value with a threshold for the event E happening, and our variable exceeds that threshold if and only if E happens.

We now show how to code this using binary indicator variables. As we'll have a lot of variables throughout this chapter, we incorporate subscripts that highlight the items in play. We make the following two reasonable assumptions throughout this chapter:

Assumptions for Integer Programming: For problems involving integer variables, we assume the following always hold:

(1) Any quantity A under consideration is **bounded**; this means there is some number N such that $|A| \leq N$. We frequently include a subscript and write N_A for the bound to highlight its dependence on A.

(2) We assume our quantities are discrete with a fixed smallest unit (where everything else is an integral multiple). We often denote this small quantity by δ.

These assumptions are harmless in practice. In the real world, we can never have infinitely much of something. Furthermore, if δ is sufficiently small (say $10^{-100!}$, for example!), then for all practical purposes a continuum of possibilities is available. This means we cannot ship an arbitrary amount of oil on our tanker, but since the answers are in millions of gallons there's no harm in saying we cannot measure more accurately than a drop! Or, equivalently, we can only determine the recommended daily allowance of a nutrient to one-trillionth of a gram, and we cannot consume an unbounded amount of food.

Theorem 8.2.1. *Consider a quantity A such that $-N_A \leq A \leq N_A$ (i.e., $|A|$ is at most N_A) and A is discrete (i.e., $A \in \{0, \pm\delta, \pm 2\delta, \dots\}$). The following constraints ensure that z_A is 1 if $A \geq 0$ and z_A is 0 otherwise:*

(1) $z_A \in \{0, 1\}$.

(2) $\frac{A}{N_A} + \frac{\delta}{2N_A} \leq z_A$.

(3) $z_A \leq 1 + \frac{A}{N_A}$.

Proof. The first condition ensures that z_A is a binary indicator variable, taking on only the values 0 or 1.

The second condition implies that if $A \geq 0$, then $z_A = 1$. To see this, if A is positive, then the left hand side is positive, and the only way the inequality can hold is if z_A is 1. If, however, $A \leq 0$ this condition provides no information. This is because if $A < 0$, then $A \leq -\delta$, and thus the left hand side is negative; as this is negative, any value of z_A satisfies the inequality.

Thus the first two conditions ensure that z_A is 1 if $A \geq 0$, but are silent on whether or not it's 0 or 1 if $A < 0$. We now turn to the third condition. If $A < 0$, then it forces $z_A = 0$; to see this, note that if $A < 0$, then the right hand side is a non-negative number strictly less than 1, and hence the inequality can only hold if $z_A = 0$. If $A \geq 0$ this condition provides no information on z_A, and either value is permissible; this is excellent, as we do not want to destroy the results from the second condition.   □

8.3. Logical Statements

Building on our ability to encode indicator relations, we now show how to handle a variety of logical statements ("or", "and", "if-then") within the confines of integer linear programming. It's important to emphasize the fact that we are only able to incorporate these at the cost of introducing integer random variables. There are advantages and disadvantages. The advantage is clear: we greatly increase what we can handle so we have a far richer system to explore. The disadvantage is that there is no known fast way to exactly solve integer linear programming problems; the simplex method works for real inputs but not for discrete ones. It can unfortunately be very difficult to find integer maxima; we save the discussion of these challenges for Chapter 9 and savor for now the success we have in expanding what we can encode.

Inclusive and Exclusive Or

Often we want one of two constraints to hold. There is the **exclusive or** (*exactly* one of two holds) and the **inclusive or** (both may hold, but *at least* one holds). For example, if you're asked if you want coffee, tea, or milk, it's expected that you would choose at most one of the three (an exclusive or), while if you're asked if you want hamburgers or hotdogs at a barbecue it's absolutely socially acceptable to have both (an inclusive or).

As always, we assume below that all quantities are bounded by $N = \max(N_a, N_b)$ and are discrete multiples of some fixed δ. We let A and B be two expressions involving quantities of interest in the problem. The goal is to create a series of constraints that force exactly one of these two to be non-negative and the other to be negative.

Theorem 8.3.1 (Exclusive Or). *The following constraints ensure that either $A \geq 0$ or $B \geq 0$ but not both:*

(1) $z_A \in \{0, 1\}$.

(2) $\frac{A}{N} + \frac{\delta}{2N} \leq z_A$.

(3) $z_A \leq 1 + \frac{A}{N}$.

(4) $z_B \in \{0, 1\}$.

(5) $\frac{B}{N} + \frac{\delta}{2N} \leq z_B$.

(6) $z_B \leq 1 + \frac{B}{N}$.

(7) $z_A + z_B = 1$.

Proof. We start by building on our previous work, where we showed how to construct constraints to detect if a quantity was non-negative.

The first three conditions ensure z_A is 1 if $A \geq 0$ and 0 otherwise; the next three ensure z_B is 1 if $B \geq 0$ and 0 otherwise.

The last condition ensures that exactly one of z_A and z_B is 1 (and the other is 0). For example, if $z_A = 0$, then condition 2 implies that $A < 0$, and if $z_B = 1$, then condition 6 implies that $B \geq 0$.   □

Theorem 8.3.2 (Inclusive Or). *The following constraints ensure that $z_A = 1$ if $A \geq 0$ or $B \geq 0$ (and possibly both are greater than or equal to zero), and $z_A = 0$ otherwise:*

(1) $z_A \in \{0, 1\}$.

(2) $\frac{A}{N} + \frac{\delta}{2N} \leq z_A$.

(3) $z_A \leq 1 + \frac{A}{N}$.

(4) $z_B \in \{0, 1\}$.

(5) $\frac{B}{N} + \frac{\delta}{2N} \leq z_B$.

(6) $z_B \leq 1 + \frac{B}{N}$.

(7) $z_A + z_B \geq 1$.

The proof is similar to the exclusive or case and is left as Exercise 8.8.10.

If-Then Statements

Theorem 8.3.3 (If-Then). *The following constraints allow us to program the statement: If* $(A < 0)$, *then* $(B \geq 0)$.

(1) $z \in \{0, 1\}$.

(2) $Nz \geq A$.

(3) $A + (1 - z)N \geq 0$.

(4) $B \geq -zN$.

Proof. If $A < 0$, the third constraint is satisfied only when $z = 0$. The fourth constraint now becomes $B \geq 0$.

If $A > 0$, the second constraint makes $z = 1$. The third constraint is trivially satisfied (as $A > 0$), and the fourth constraint becomes $B \geq -N$, which is trivially satisfied (we are assuming $|B| \leq N$).

If $A = 0$, the second and third constraints are always satisfied. Taking $z = 1$ we see the fourth is satisfied. $\qquad\qquad\qquad\qquad\qquad\qquad\qquad\qquad\qquad\qquad\square$

It's often valuable to see just how expensive it is to introduce all these new constraints. For example, we can encode if-then statements by adding one binary variable and three constraints (four if you count the binary declaration). We could also do "If $(A < A_0)$ Then $(B \geq 0)$" just as easily.

8.4. Truncation, Extrema and Absolute Values

We now show how to recast some very important non-linear functions as linear functions, though at a cost of increasing the number of constraints and introducing integer variables.

Truncation

Given an integer variable or expression X, we show how to add constraints so a variable Y equals X if $X \geq X_0$, and $Y = 0$ otherwise. We constantly use the if-then constraints.

Theorem 8.4.1 (Truncation). *Let* $z \in \{0, 1\}$ *and let* X *be a non-negative random variable which is at most* N_X. *The following encodes truncation by creating a random variable* Y *which equals* X *if* $X \geq X_0$ *and is zero otherwise:*

(1) IF $(X < X_0)$, THEN $z = 0$.

(2) IF $(X > X_0 - \frac{\delta}{2})$, THEN $z = 1$.

(3) $(Y - X) + N_X(1 - z) \geq 0$.

(4) $Y - N_X z \leq 0$.

(5) $0 \leq Y \leq X$.

Proof. The first two steps involve `IF-THEN`. We are able to do these by taking $B = z - \frac{1}{2}$ in the `IF-THEN` section. Note that the $-\frac{\delta}{2}$ allows us to re-write the If condition as $X \geq X_0$.

We now turn to the final three conditions. We look at their effect on Y when $z = 0$ and then when $z = 1$.

If $z = 0$ (i.e., $X < X_0$), then constraint (3) holds as constraint (5) forces Y to be non-negative and at most X. As Y is non-negative, constraint (4) forces $Y = 0$.

If $z = 1$ (i.e., $X \geq X_0$), then (3) forces $Y \geq X$ which, combined with the final condition, forces $Y = X$. As $Y \leq X \leq N_X$, the fourth constraint holds. $\square$

Minima and Maxima

Theorem 8.4.2 (Minima)**.** *The following constraints ensure that* $Y = \min(A, B)$:

(1) $Y \leq A$,

(2) $Y \leq B$,

(3) $Y = A$ OR $Y = B$.

Proof. The first condition forces $Y \leq A$ and the second $Y \leq B$. Without loss of generality, assume $A \leq B$. The third condition says either $Y = A$ or $Y = B$ (or both). If $Y = B$ this contradicts the first condition. Thus $Y \leq A$ is improved to $Y = A$. $\square$

Theorem 8.4.3 (Maxima)**.** *The following constraints ensure that* $Y = \max(A, B)$:

(1) $Y \geq A$,

(2) $Y \geq B$,

(3) $Y = A$ OR $Y = B$.

Absolute Values

Theorem 8.4.4. *The following constraints ensure that* $X = |A|$:

(1) $A \leq X$,

(2) $-A \leq X$.

(3) $X \leq A$ OR $X \leq -A$.

Proof. The first two constraints force X to be a non-negative number, of size at least $|A|$. We just need to make sure $X \not\geq |A|$.

For the third constraint, if $A = 0$, the two OR clauses are the same, and $X = 0$. If $A \neq 0$, as X is non-negative it can only be less than whichever of A and $-A$ is non-negative. $\square$

One application is the following.

Lemma 8.4.5. *Assume* $|\vec{u}^T \vec{x} + \vec{v}| \leq M$ *for all* $\vec{x}$ *that are potential solutions of the integer programming problem, and assume* $\vec{u}$, $\vec{x}$, *and* $\vec{v}$ *are integral. We may* *replace a term $|\vec{u}^T \vec{x} + \vec{v}|$ with y by introducing three new variables (y, z_1 and z_2) and the following constraints:*

(1) $z_1, z_2 \in \{0, 1\}$,

(2) $\vec{u}^T \vec{x} + \vec{v} \leq y$,

(3) $-(\vec{u}^T \vec{x} + \vec{v}) \leq y$,

(4) $0 \leq y$, y is an integer,

(5) $y \leq M$,

(6) $y - (\vec{u}^T \vec{x} + \vec{v}) \leq 2z_1 M$,

(7) $y + (\vec{u}^T \vec{x} + \vec{v}) \leq 2z_2 M$,

(8) $z_1 + z_2 = 1$.

8.5. Linearizing Quadratic Expressions

Now that we've assembled an extensive toolbox on how to incorporate logical connectives and linearize certain functions, we can use these to perform some impressive feats! One great example is linearizing quadratic expressions. First, though, it's worth discussing why this is so important. Why do we want to examine cases that have product terms? It turns out that product-like interaction terms occur frequently in Linear Programming problems. For instance, if a movie theater is showing two hot new action movies at the same time, such as both a new *X-Men* and a new *Spiderman* movie, it would have to calculate the demand by adding the movies' individual demands, but would also need to subtract their shared demand, representing viewers who would be interested in both movies, but could only see one!

We'll limit ourselves to the case where we only have binary variables and leave the more general case to the exercises. Let's assume that $x, y \in \{0, 1\}$ and that we have an xy term.

What do we know about xy? We know that:

$$xy = \begin{cases} 1, & \text{if } x + y = 2, \\ 0, & \text{if } x + y \leq 1. \end{cases}$$

This is a truncation! We have:

$$\text{IF } (x + y \geq 1.5) \text{ THEN } (z = 1) \text{ ELSE } (z = 0).$$

Now replace xy with z, and we have removed a quadratic (i.e., degree 2) term! Add more variables, and do the same to get polynomials of whatever desired degree.

This concludes our coverage of linearization techniques, which allows us to convert problems into a linear programming framework. Unfortunately, while this framework allows us to obtain a good real solution, the real solution is not always anywhere near the integer solution, which is the goal of the much more complicated integer linear programming. We will examine such a case now before moving on to a new topic.

8.6. The Law of the Hammer and Sudoku

We end this chapter by showing how we can solve a variety of problems *if* we can solve an integer programming problem quickly. It's important to note that we are *not* saying that this is the best way to solve these problems; we are saying it is *a* way to do it. This is an important distinction, and worth dwelling on, as it provides a wonderful example of the **Law of the Hammer**.

Principle (or Law) of the Hammer: There are many versions, including:

◇ Abraham Kaplan: I call it the law of the instrument, and it may be formulated as follows: Give a small boy a hammer, and he will find that everything he encounters needs pounding.

◇ Abraham Maslow: I suppose it is tempting, if the only tool you have is a hammer, to treat everything as if it were a nail.

◇ Bernard Baruch: If all you have is a hammer, everything looks like a nail.

There are several lessons we can glean from this. If you have something you do well, try to convert whatever problem you have to a situation where you can use your skills. This is great advice, but it's also passive and only half the picture. The other half is to seek out new problems and challenges. Remember, if you work in a field and attack the same problems in the same ways as your colleagues who are similarly trained, it will be hard to shine and distinguish yourself. There are a lot of smart people in the world,[1] and this is a tough battle to win. If, however, you take your tool set and travel to another area where people don't know these methods, you have a much greater chance of success. For example, if there is a land of only screwdrivers, your hammer will solve problems they cannot, and you will be hailed as a conquering hero for doing what their brightest minds could not. The following quote from Richard Feynman's book *Surely you're joking, Mr. Feynman*, illustrates this well:

> *One thing I never did learn was contour integration. I had learned to do integrals by various methods shown in a book that my high school physics teacher Mr. Bader had given me.*
>
> *The book also showed how to differentiate parameters under the integral sign. It's a certain operation. It turns out that's not taught very much in the universities; they don't emphasize it. But I caught on how to use that method, and I used that one damn tool again and again. So because I was self-taught using that book, I had peculiar methods of doing integrals.*
>
> *The result was that, when guys at MIT or Princeton had trouble doing a certain integral, it was because they couldn't do it with the standard*

---

[1] If you are one in a million, then there are a thousand like you in China.

methods they had learned in school. If it was contour integration, they would have found it; if it was a simple series expansion, they would have found it. Then I come along and try differentiating under the integral sign, and often it worked. So I got a great reputation for doing integrals, only because my box of tools was different from everybody else's, and they had tried all their tools on it before giving the problem to me.

Let's see this in action by showing how we can use integer programming to solve Sudoku. The standard version of the puzzle has a 9×9 grid, where each entry is chosen from the set $\{1, 2, \ldots, 9\}$ subject to the following constraints:

- Each row has each number from 1 to 9 once and only once.
- Each column has each number from 1 to 9 once and only once.
- The 9×9 board is split into nine 3×3 subgrids, and each of these subgrids has each number from 1 to 9 once and only once.

For a proper Sudoku puzzle, certain squares are filled in as a start, and from those values and the rules there is a unique way to fill in the remaining entries (without having to guess); see Figure 1 for an example.

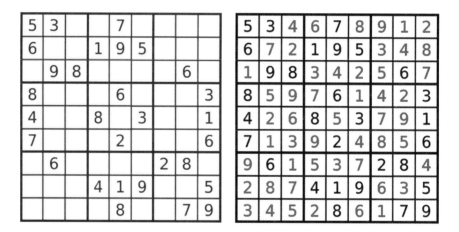

Figure 1. An example of a Sudoku problem (left) and its solution (right). Images from Wikimedia Commons (left from Lawrence Leonard Gilbert, right from Colin M. L. Burnett; licensed under the Creative Commons Attribution-Share Alike 3.0 Unported (https://creativecommons.org/licenses/by-sa/3.0/deed.en) license).

We could solve this as an integer programming problem or as a binary integer programming problem. If we do it as a binary integer programming problem, one option is to have variables

$$x_{ijk} = \begin{cases} 1 & \text{if in cell } (i,j) \text{ have a } k, \\ 0 & \text{otherwise.} \end{cases}$$

We then incorporate the given values (so $x_{115} = 1$ and $x_{11k} = 0$ for $k \neq 5$). Next we encode the three constraints. For example, to make sure each row contains each number from 1 through 9 we can do:

$$\forall i \in \{1, \ldots, 9\}, \ \forall k \in \{1, \ldots, 9\} : \sum_{j=1}^{9} x_{ijk} = 1.$$

In other words: for each fixed row (our i) and for each fixed value (our k) it occurs once and only once. We leave the other constraints as exercises.

All that is left is determining an objective function. As a good Sudoku problem has a unique answer, if we find a feasible solution then we are done![2] Thus we have enormous freedom in choosing the objective function; for example, we could take our objective function to be x_{111}.

The above provides a solution to the Sudoku problem using binary variables. Note that we have $9^3 = 729$ variables with this approach. We could solve it using fewer variables by letting

$$y_{ij} \in \{1, 2, \ldots, 9\}.$$

The advantage is that this has only $9^2 = 81$, but the variables can take on many more values, and the constraints become harder to state. For example, how would we encode that each number occurs once and only once in a column? We could do this by a massive use of OR statements, such as

$$y_{11} = 1 \ \texttt{OR} \ y_{12} = 1 \ \texttt{OR} \ \cdots \ \texttt{OR} \ y_{19} = 1,$$

but this is extremely expensive and requires the introduction of many binary indicator variables.

Is there a better way? Amazingly, there is. Notice that we don't really care that the values are 1 through 9; all that matters is that we have nine *distinct* objects. A clever choice makes formulating the constraints significantly easier:

$$w_{ij} \in \{1, 10, 100, 1000, \ldots, 10^8\}.$$

Why is this better? Notice the only way to have nine numbers from this list sum to 111,111,111 is to have one of each. Thus, instead of a massive use of OR statements we could do

$$\forall i \in \{1, \ldots, 9\} : \sum_{j=1}^{9} x_{ij} = 111, 111, 111.$$

This requires fewer constraints than our original approach with binary indicator variables, but now the variables take on more values. Which of the two approaches is best?

There are several metrics we can use to determine which approach is best. A natural one, of course, is to find the formulation that leads to the shortest run-time. Another possibility is to minimize the time we spend coding. Typically we don't care about the difference in a program which runs in .1 seconds versus 1 second; if the slower code is much faster to write and we only need to use it a few times, it's probably a worthwhile trade to sacrifice efficiency in run-time for ease in coding.

---

[2] Even if it isn't a well-posed problem and there are multiple solutions, so long as we only care about finding *a* solution this approach still works.

8.7. Bus Route Example

One very important application of integer programming is to assign schedules. These are ideally suited to the techniques of this chapter, as they involve binary options: we start a given option at a given time, or we don't. Below we discuss how to set up the program to create a bus route for a district.

For simplicity, let's assume we know where the bus stops are located and how many buses are to be used. Of course, both of these could be free and for the model to determine. Of the two, it's much easier to vary the number of buses than the location and number of stops. We can just run the program again and again with a different number of buses and compare the results. If instead we vary the stops we have to decide where to place them; it's much easier if each kid is picked up at their house, but perhaps we have several houses that send their kids to a common spot (which of course opens up the need to figure out where they go).

What are the issues in constructing the Linear Programming problem? We must make sure all kids are picked up and delivered to the school, and that no bus ever exceeds its capacity. We'll assume we have a sufficient number of buses such that each bus only goes to the school once (if we had 100 people and one bus with capacity 20, we would have to make at least 5 trips to the school) and that a bus either picks up everyone from a stop, or no one. While this assumption can fail if we have a very large number of people at one stop, it's quite reasonable. If the program returns no answer, we just run it again with a big stop split into two smaller stops at the same location.

Let's begin with the parameters; we choose names for the variables to make it easy to look down at the equations and interpret. We label the bus stops by s (for stop), with $s \in \{1, \ldots, S\}$, and we assume there are p_s people at stop s. Let there be $b \in \{1, \ldots, B\}$ buses with capacity c_b (which we might take to be constant, but let's allow some freedom as not all buses are the same size). Let d_{ij} be the distance from stop i to stop j; distance and time should be closely related.

Our main variable is x_{bij}, which equals 1 if bus b goes from stop i to stop j (picking up the people from stop j), and 0 otherwise (for convenience we set $x_{bii} = 0$). We could instead have introduced a time subscript and looked at y_{bni} if bus b's n^{th} stop is at i; you are encouraged to redo the problem using these variables and compare the formulations. Let's also assume that all the buses start at the same place, though it's easy to remove this. We make stop 1 this common start, and we must make sure all buses end at the same place (say stop S, the school). Thus $p_1 = p_S = 0$.

We now turn to the constraints and the objective function. We need to make sure everyone makes it to school, but what should we minimize? We could minimize total time of students on the bus (so people aren't sitting too long) or the total distance the buses travel. We choose the latter, though again the reader is encouraged to explore the other alternative, or additional possibilities (such as the previously mentioned number of buses).

Let's find the constraints. These ensure everyone reaches the school.

- For all b: $\sum_{i \neq S} x_{biS} = 1$ and $\sum_{j \neq S} x_{bSj} = 0$. This means each bus ends at the school and doesn't go further.

- For all b: $\sum_{j \neq 1} x_{b1j} = 1$ and $\sum_{i \neq 1} x_{bi1} = 0$. This means each bus starts at the the depot, and doesn't go anywhere before. Notice the bus might just go straight to the school without getting anyone! We could also have the depot be the same location as the school, but it's good to have added flexibility.

- For all b: $\sum_{i,j=1}^{S} p_j x_{bij} \leq c_b$. This means no bus exceeds its capacity.

- For all $j \in \{2, \ldots, S-1\}$: $\sum_{b=1}^{B} \sum_{i \neq j} x_{bij} = 1$. This means exactly one bus picks up people at stop j. In other words, we get all the kids.

- For all $j \in \{2, \ldots, S-1\}$: if bus b comes to j then it must leave j: IF $(\sum_{i \neq j} x_{bij} = 1)$ THEN $(\sum_{k \neq j} x_{bjk} = 1)$. Notice bus b starts at 1, goes to a few stops, and then ends at S. Also observe that we're using an IF-THEN; we've shown earlier how these can be coded linearly, but it's not worth explicitly writing the details here. Also, many computer systems that solve linear programming problems are designed to be easy to use and allow the user to directly use such statements.

Now that we have the constraints, we turn our attention to optimization. We'll minimize the total distance traveled by the buses; this is a good but not perfect proxy for the total time the buses are driving.

$$\text{Minimize} \sum_b \sum_i \sum_j d_{ij} x_{bij}.$$

8.8. Exercises

The Movie Theater Problem

Exercise 8.8.1. *Write a constraint that prevents two movies from being on the same screen at the same time.*

Exercise 8.8.2. *Write a constraint that prevents any movie from running after the theater closes.*

Exercise 8.8.3. *Write a constraint that in any 8 consecutive time blocks a movie must start somewhere in the theater.*

Exercise 8.8.4. *The constraint from Exercise 8.8.3 turns out to cause enormous trouble; prove if you do not have constraints such as this that a feasible movie schedule exists. Must a feasible schedule exist with this constraint?*

Exercise 8.8.5. *Assume two movies can compete with each other for demand. Create a set of variables that allows you to handle this situation linearly, as well as non-linearly.*

Binary Indicator Variables

Exercise 8.8.6. *Let X and Y represent two binary indicator random variables. Is there a binary indicator random variable Z that represents XY? How would you interpret Z?*

Exercise 8.8.7. *Let X and Y represent two binary indicator random variables. Is there a binary indicator random variable Z that represents $X + Y$? How would you interpret Z?*

Exercise 8.8.8. *Let X be a random variable taking on integer values in $\{0, 1, \ldots, N\}$. Is it possible to replace this one random variable with many binary indicator random variables? If yes, how?*

Exercise 8.8.9. *Let X be a random variable taking on integer values in $\{0, 1, \ldots, N\}$ and let $Y = X^n$. Is it possible to replace this one random variable with many binary indicator random variables? If yes, how? What if $Y = e^X$?*

Logical Statements

Exercise 8.8.10. *Prove Theorem 8.3.2.*

Exercise 8.8.11. *We showed how to code the inclusive and exclusive OR; show how one can code AND. For example, we want $z_{A\&B}$ to be 1 if A and B are non-negative, and 0 otherwise.*

Exercise 8.8.12. *Encode IF AND ONLY IF.*

Exercise 8.8.13. *Encode A or B or C or D, assuming each is an inclusive OR. Do so using as few constraints as you can.*

Truncation, Extrema and Absolute Values

Exercise 8.8.14. *Modify the construction from the chapter so that, given a non-negative random variable X, we find a new variable Y such that $Y = X$ whenever $X \leq X_0$.*

Exercise 8.8.15. *Expand the IF-THEN statements in the truncation construction and write all conditions out explicitly.*

Exercise 8.8.16. *Prove Theorem 8.4.3.*

Exercise 8.8.17. *Prove the Lemma 8.4.5.*

Exercise 8.8.18. *Frequently in problems we desire two distinct tuples, say points $(a_1, \ldots, a_k) \neq (\alpha_1, \ldots, \alpha_k)$. Find a way to incorporate such a condition within the confines of integer linear programming.*

Exercise 8.8.19. *Extending the previous problem, sometimes we need a stronger condition than the points being different; we need that no reordering of one equals another. For example, $(1, 3, 4)$ is not the same as $(4, 3, 1)$, but we can reorder the second triple to $(1, 3, 4)$. How would you program that $(a_1, \ldots, a_k)$ is not equal to any reordering of $(\alpha_1, \ldots, \alpha_k)$?* Hint: as this problem is in the section on absolute values, perhaps those will be of use.

Linearizing Quadratic Expressions

Exercise 8.8.20. *Generalize the methods of this section to linearize xy, where now x and y are integer random variables taking on values in $\{0, 1, \ldots, N\}$.*

Exercise 8.8.21. *Show how to replace $x_1 x_2 \cdots x_n$, where each x_i is a binary integer variable, with sums of binary integer variables.*

The next few problems involve population dynamics and lead to product terms.

Let $x(t)$ represent the population of some species at time t. A very simple growth model is that

$$x'(t) = ax(t)$$

for some real a. This is an idealized situation, where there are no predators and an unlimited food supply (so no crowding out of resources).

Exercise 8.8.22. *Solve the differential equation above, $x'(t) = ax(t)$, in terms of the parameter a and the initial population $x(0)$. Interpret what it means for the solution for $a > 0$, $a = 0$, and $a < 0$.*

The previous model, of course, is too simple to be realistic in many situations (though it does a great job for modeling bacteria or radioactive decay). We consider a more involved model with two species, the prey (whose population is $x(t)$ at time t) and the predators (whose population is $y(t)$ at time t). The **Lotka-Volterra equations** (also called the **predator–prey equations**) are often used to model such situations:

$$\begin{aligned} x'(t) &= \alpha x(t) - \beta x(t) y(t), \\ y'(t) &= \delta x(t) y(t) - \gamma y(t), \end{aligned}$$

where α, β, γ, and δ are positive parameters.

Exercise 8.8.23. *Give an interpretation to the terms in the Lotka-Volterra equations. For example, what would be the story if $\alpha = \beta = \delta = 0$?*

Exercise 8.8.24. *Solve the Lotka-Volterra equations; how does your answer depend on the values of the parameters?*

The Law of the Hammer and Sudoku

The first few problems are examples of differentiating under the integral sign; you may assume in all these problems that you may interchange the integral and the derivative, though sometimes interchanging operations is not permissible (see Exercise 8.8.31).

Exercise 8.8.25. *If*

$$F(t) = \int_0^\infty e^{-tx} dx,$$

show that $F'(t) = -1/t^2$, and more generally $F^{(n)}(t) = (-1)^n n!/t^{n+1}$. Show this implies

$$\int_0^\infty x^n e^{-tx} dx = n!.$$

Note: I heard of this example (and the next) from my colleague Leo Goldmakher who learned of this method from Noam Elkies' website who reports it was used by his student Inna Zakharevich, a former student of mine, on a problem set.

Exercise 8.8.26. *If $F(t) = \int_0^1 \frac{x^t - 1}{\log x} dx$, show F is well defined for $t > 0$ and that $F'(t) = \frac{1}{t+1}$. Use this to compute $F(3)$ and $F(1701)$.*

Exercise 8.8.27. *Let $F(t) = \int_0^1 x^t \log^n(x) dx$. Prove $F(t) = (-1)^n n!/(t+1)^{n+1}$.*

The next exercise requires some results from analysis, as we need to justify differentiating under an integral sign and then exchanging an integral and a limit. Assume all those operations are justified here (or, even better, justify them!).

Exercise 8.8.28. *The goal is to evaluate $\int_0^\infty \frac{\sin x}{x} dx$. Let*

$$G(t) = \int_0^\infty \frac{\sin x}{x} e^{-tx} dx$$

(so our desired integral is $G(0)$) and note that

$$G'(t) = -\int_0^\infty e^{-tx} \sin x dx.$$

Prove

$$G'(t) = -\frac{1}{1+t^2},$$

and thus

$$G(t) = C - \arctan t$$

for some C; by looking at $\lim_{t \to \infty} G(t)$ show $C = \pi/2$ and thus determine $G(0)$.

Instead of differentiating under the integral sign one could differentiate under the summation sign; both involve interchanging the order of differentiation and another operation (a sum or an integral). This is a powerful technique in probability theory and is often used to determine the mean, variance, or other moments of a distribution. The two problems below are related to such investigations; for more see [**M**].

Exercise 8.8.29. *Recall the geometric series formula from §13.3. Thus if $|x| < 1$ we have*

$$\sum_{n=0}^\infty x^n = \frac{1}{1-x}.$$

Prove the derivative of the sum is the sum of the derivatives, and thus

$$\sum_{n=0}^\infty n x^n = \frac{x}{(1-x)^2}$$

(note we multiplied both derivatives by x). Hint: the geometric series has the wonderful property that it's tail is a geometric series. Thus break the sum into $n < N$ and $n \geq N$, evaluate the sum over $n \geq N$ by the geometric series formula, and notice that we now have a finite sum equaling $1/(1-x)$.

Exercise 8.8.30. *Recall the Binomial Theorem from Exercise 2.6.16. Given*

$$\sum_{k=0}^n \binom{n}{k} x^k y^{n-k} = (x+y)^n,$$

prove that

$$\sum_{k=0}^{n} k \binom{n}{k} \frac{1}{2^n} = \frac{n}{2}.$$

Exercise 8.8.31. *Let $f_n(x)$ be the triangular function which rises linearly from 0 at $1/n$ to n at $2/n$ and then falls linearly to 0 at $3/n$, and is zero elsewhere. Prove for any x that $\lim_{n\to\infty} f_n(x) = 0$, but*

$$\lim_{n\to\infty} \int_0^1 f_n(x)dx \neq \int_0^1 \lim_{n\to\infty} f_n(x)dx.$$

Thus the limit of an integral is not always the integral of a limit.

Exercise 8.8.32. *Write down the other constraints for the Sudoku problem using integer variables x_{ijk}.*

Exercise 8.8.33. *Find a "smallest" set of distinct positive integers $\{a_1, \ldots, a_9\}$ such that the only way to have nine numbers from the list sum to $a_1 + \cdots + a_9$ is to use each number once and only once. For example, if our set were $\{1, 2, \ldots, 9\}$, then $1 + \cdots + 9$ and $2 + 2 + 3 \cdots + 7 + 8 + 8$ both sum to 45, and thus our coding trick would not work. There are many notions one can have for "smallest"; a natural choice is $a_9 - a_1$. Is there more than one choice which works?*

Exercise 8.8.34. *Generalize the previous problem. For each $n \geq 3$ find the smallest set of distinct positive integers $\{a_1, \ldots, a_n\}$ such that the only way to have n numbers from the list sum to $a_1 + \cdots + a_9$ is to use each number once and only once.*

Exercise 8.8.35. *Investigate the different approaches we discussed for solving the Sudoku problem (as well as any others you can think of and wish to study). Compare the number and type of variables, the number of constraints, and the run-time to find a solution. Which approach is "best"?*

The next few problems involve the **Battleship puzzle**, a one-player variant of the popular Battleship game (see Table 1). In the most common version you're given an $n \times n$ checkerboard and k ships of dimension $1 \times d_1$, $1 \times d_2$, $\ldots$, $1 \times d_k$. The ships are placed parallel to the x- and y-axes, and each square of the board is occupied by at most one ship. After each row is a number indicating the number of ship pieces in that row, and similarly for each column. The ships are then removed and the board, with the $2n$ numbers, is given. Your job is to determine the location of the ships.

Exercise 8.8.36. *Set up the Battleship puzzle as a Linear Programming problem.*

Exercise 8.8.37. *Solve the Battleship puzzle for the configuration in Table1.*

Exercise 8.8.38. *Generalize the Battleship puzzle to a hexagonal grid. What should the constraints be on ship placement?*

Exercise 8.8.39. *In the Sudoku and Battleship problems we were concerned with finding a solution. What if we want all solutions? Is it possible to do this using integer programming? If yes, is it possible in an efficient way?*

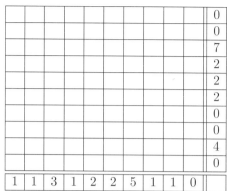

										0
										0
										7
										2
										2
										2
										0
										0
										4
										0
1	1	3	1	2	2	5	1	1	0	

Table 1. Example of the Battleship puzzle. The grid is the same as the standard two-player game, as are the ships: one destroyer (1×2), one cruiser (1×3), one submarine (1×4), one battleship (1×4), and one carrier (1×5).

Figure 2. Three versions of the Flow Free game from Big Duck Games LLC, bigduckgames.com, Classic, Bridges, and Hexes.

Exercise 8.8.40. *Figure 2 shows three versions of the game Flow Free. One is given a board and n pairs of colored disk on the board, where no two pairs share a color. The object is to connect each pair by a continuous path through cells so that no lines cross (unless there is a bridge, which one can use) and also ensuring each cell on the board is occupied by either a colored disk or part of a path. Set up finding a valid solution as an integer Linear Programming problem.*

Exercise 8.8.41. *Consider the game from the previous problem on an $n \times n$ square board with m pairs. Is there a choice of m and n, and a placement of the pairs, such that there are at least two distinct valid solutions?*

Bus Route Example

Exercise 8.8.42. *Add a constraint to the bus route problem to ensure that the buses reach the school by a given time.*

Exercise 8.8.43. *Redo the bus route problem where now the objective is to minimize the number of buses used.*

Exercise 8.8.44. *Redo the bus route problem where now we wish to minimize the time students spend on the bus (or perhaps to make sure that no one spends more than a certain amount of time on a bus).*

Exercise 8.8.45. *Try to generalize the bus route problem to airlines. What constraints do you want for a simple model? What data/parameters will you need?*

Exercise 8.8.46. *Many airlines keep some planes in a strategic reserve, so that if a plane goes down in a route (or perhaps only in a key route), then there is a replacement that can be brought into service relatively quickly. The problem, of course, is that this requires planes to not fly, it could require flight crews to be ready, and depending on where the plane is stored it could take a significant amount of time to get the plane to where it is needed. Discuss the constraints arising from this insurance policy, as well as the expected revenue (the airline must gain something from the reserve or they would not do it, unless of course they are mandated to).*

Exercise 8.8.47. *Implement the approach to the bus route problem by writing a linear programming problem to solve it.*

Chapter 9

Integer Optimization

We've seen how powerful integer random variables are – many non-linear functions (absolute values, products, truncations) and logical operations can be easily encoded. Unfortunately, the cost is that we must now optimize with the restriction that certain variables take on only integer values. This means, in particular, that the tools from calculus are unavailable because our quantities are not continuously varying. While it's beyond the scope of this book to analyze these issues completely, we can be productive and look at some related problems.

For some problems we can find the real extrema and are fortunate that the integer extrema is very close; for other problems, the optimal value with integer inputs is nowhere near the optimal value with real inputs. We explore an example of each of these cases below to highlight the possibilities and pitfalls that can happen.

9.1. Maximizing a Product

Imagine we have a large integer, S, and we want to write it as a sum of real numbers such that the product of the summands is as large as possible. If we play with decompositions we quickly realize we can make the product arbitrarily large through some trivialities. For example, if $S > 1$ we can write

$$S = S + S + S + (-S) + (-S),$$

which has product S^5. If $S < 1$ this example doesn't work, and a little more inspection gives us an answer that will work for all S:

$$S = S + 10^m + 10^m + (-10^m) + (-10^m);$$

the product is now $S \cdot 10^{4m}$, which we can make arbitrarily large by sending m to infinity.

Thus, in order for the problem to be interesting, we need some restrictions on the summands. From our experiments above we see it's natural to require everything to be non-negative (which also immediately implies each summand is at most S). We'll first solve the problem under the assumption that the summands are integers, and then generalize to the summands being real. The integral optimization difficulty is *not* coming from the restrictions on the summands; you're encouraged to ponder for a bit what causes the integrality issues.

Case 1: Integral Summands

As we require the summands to be integers, there must be a number n and integers $a_1, \ldots, a_n$ such that

$$S = a_1 + \cdots + a_n.$$

Our goal is to choose n and the a_i to make

$$a_1 a_2 \cdots a_n$$

as large as possible.

We can resolve this very quickly by looking at some cases. First, we clearly don't want any a_i to be zero. Furthermore, if one of them were 1 we could increase the product by adding the 1 to another summand (so long as $S \neq 1$, of course). Thus, we can assume each $a_i \geq 2$.

How large can an a_i be? If we had $a_i = 10$ we could replace that with $8 + 2$, which has a larger product. This works not just for 10 but for any integer at least 5 (as $3 \cdot 2$ has a larger product than 5). If we had 4 we *could* replace that with $2+2$ and not decrease the product. Thus, we can assume the summands are at most 3.

We now need to figure out how many of the summands are 2 and how many of the summands are 3. Playing around with products, we see that three 2's have the same sum as two 3's, but a smaller product ($2 \cdot 2 \cdot 2 = 8 < 9 = 3 \cdot 3$); thus whenever we have three 2's it's better to replace with two 3's. We leave the final step as Exercise 9.4.1.

Case 2: Real Summands

We now consider the more interesting case where the summands are allowed to be real. Much of our analysis from before holds. We still have $1 < a_i < 4$, but we can no longer say $a_i \in \{2, 3\}$; now we have $a_i \in [1, 4]$ (it's technically easier to allow a_i to equal 1 or 4, as now we have compact intervals). The difficulty is not that the a_i vary continuously; this actually helps and suggests we can use calculus! The difficulty is that we must have an integral number of summands.

We attack the problem in stages. We first fix a positive integer n and find the optimal decomposition, and then try and vary n. Consider a decomposition

$$a_1 + \cdots + a_n = S, \quad a_i \geq 0.$$

While we know each $a_i \in [1, 4]$, it's a bit easier right now to assume $a_i \in [0, S]$. We know from real analysis that a continuous function on a compact set attains its maximum and minimum (see Exercise 4.9.6), and thus there must be some choice, say $\widetilde{a}_1, \ldots, \widetilde{a}_n$, with maximal product. We claim that all these values are the same. If not, without loss of generality assume $\widetilde{a}_1 < \widetilde{a}_2$. Then by Exercise 9.4.2 we can increase the product by replacing these two summands with two copies of $(\widetilde{a}_1 + \widetilde{a}_2)/2$, which clearly does not affect the sum. This is a contradiction. (You may be familiar with this as the Farmer Brown problem from calculus, where he wants the rectangular pen with maximal area for a given perimeter; this can be solved either through calculus or by plotting a parabola.) A common mistake in problems like this is to *assume* there is a maximum; we have one in this particular case due to compactness and continuity.

We have shown that for each n the maximum product is attained when

$$a_1 = a_2 \cdots = a_n = S/n,$$

giving a product of $(S/n)^n$. We've reduced our problem to finding the *integer* n where $g(n) := (S/n)^n$ attains its maximal value.

We first consider a far simpler problem: finding the *real* x such that $g(x) = (S/x)^x$ attains its maximum; the hope is that we can use calculus here, easily find the solution, and then somehow pass on to the integral max. It's convenient to re-write $g(x)$ to facilitate differentiation:

$$g(x) = e^{x \log(S/x)} \quad \text{so} \quad g'(x) = e^{x \log(S/x)} \left(\log(S/x) - 1 \right).$$

The candidates for the maximum are the boundary points and the critical points. The only critical point (i.e., where $g'(x) = 0$) is $\log(S/x) = 1$ or $x = S/e$; the boundary points ($x \to 0$ and $x \to \infty$) cannot yield the maximum (see Exercise 9.4.3).

We now must prove that $x = S/e$ gives a maximum and not a minimum. One way to see this is to do the second derivative test (see Exercise 9.4.4); we prefer to use the first derivative test as this allows us to review curve sketching (justifying another topic from Calc I) *and* it will allow us to find the integer maximum! If the first derivative is positive the function is increasing, while if it's negative the function is decreasing. Looking at $g'(x)$, we see that the sign of $g'(x)$ is the same as the sign of $\log(S/x) - 1 = \log(\frac{S/e}{x})$. If $x < S/e$ this is positive (and the function is increasing), and if $x > S/e$ this is negative and the function is decreasing. Thus $x = S/e$ is the maximum.

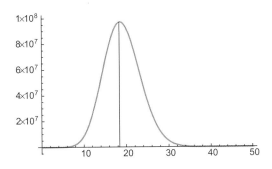

Figure 1. Plot of $g(x) = (S/x)^x$ for $S = 50$; the vertical line is at $x = S/e$.

Figure 1 plots $g(x)$ when $S = 50$, though similar shapes hold for other S. Our first derivative analysis now pays enormous dividends. Not only do we know $x = S/e$ is the real maximum, but we know $g(x) = (S/x)^x$ is increasing for $x < S/e$ and decreasing for $x > S/e$. Thus the *integer* maximum must occur either for the integer n just before S/x or just after S/x.

Again, we're remarkably lucky here. As the problem in the next section shows, sometimes the integer maximum is nowhere near the real maximum; in this situation the sign of the first derivative allowed us to narrow our search to just two points, the floor and the ceiling of S/e.

Remark 9.1.1. *We conclude this section by extracting a useful observation from our argument above. Sometimes the best way to understand a function on the integers is to extend it to a function on the reals, as then the tools of calculus and real analysis are available. See Exercises 9.4.10 and 9.4.11 for other examples.*

9.2. The Knapsack Problem

Much of the difficulty in integer programming stems from the fact that a problem may have optimal real solutions and optimal integer solutions, but the optimal integer solutions are not close to the optimal real solutions. To see this, we consider the famous **knapsack problem**, following the presentation in [**Fr**], pages 127–128 (which introduced me, many many years ago, to the beauty and difficulty of the subject).

Imagine we have a knapsack that can hold at most 100 kilograms. There are three items we can pack. The first weighs 51 kilograms and is worth \$150 per unit; the second weights 50 kilograms and is worth \$100 per unit; the third weighs 50 kilograms and is worth \$99 per unit. The goal is to pack as much as we can in the knapsack and maximize the value. Thus if x_j is the amount of the jth item, we have the constraint

$$51x_1 + 50x_2 + 50x_3 \ \leq \ 100, \quad x_j \geq 0,$$

where we want to maximize

(9.1) $$150x_1 + 100x_2 + 99x_3.$$

If we allow the x_j's to be real numbers, the optimal answer is easily found; we look and see which item is worth the most per kilogram, and fill up on that. With these numbers, we find $x_1 = 100/51$ and $x_2 = x_3 = 0$; the value of the knapsack is about \$294.12.

We can focus on the best value per kilogram *only if* we are free to take a non-integral amount. If we require the x_j to be integers, the optimal solution is $x_2 = 2$ and $x_1 = x_3 = 0$; the value of the knapsack is \$200. There are many ways to find this; since the quantities are sufficiently small we can do a brute force exhaustive search.

Note how different both the revenue and the solution itself are between the real case and the integer case here. Not only is the optimal value significantly different when the x_j's are integers, but the answer itself is very different. Rather than almost 2 units of the first item and none of the second, we have 2 units of the second and none of the others. This just goes to show how far the real optimal configurations can be from the integer optimal configurations, and how challenging integer programming can be.

9.3. Solving Integer Programs: Branch and Bound

We can quantify how difficult integer programming is; it belongs to a class of problems that are **NP-Hard**. We briefly describe what this means, along with a method to find solutions. This section is independent from most of the rest of the book, and it is meant to *briefly* introduce the reader to the literature and terminology. See [**Cl**] and the references therein for additional details and reading.

We'll discuss an NP-Complete problem in Chapter 11, the Traveling Salesman Problem (TSP). NP-Complete problems are difficult problems for which there is no known algorithm capable of solving it in polynomial time; however, they have the special property that if you can solve one such problem, then you can solve them all! An NP-Hard problem is one that is at least as difficult as all NP-Complete problems – given a polynomial-time solution to an NP-Hard problem, we can solve all NP-Complete problems. NP-Hard and NP-Complete are subtly different classes. NP-Complete is a subset of NP-Hard. NP-Complete problems are problems that are both NP-Hard *and* easy enough to solve on a non-deterministic Turing machine in polynomial time. This is where the NP (non-deterministic polynomial) comes from.

Proving that integer programming is NP-Complete is beyond the scope (and interest!) of this book. However, given the versatility of integer programming, it's not inconceivable that if we can solve integer programming problems quickly, then we can solve many additional difficult problems quickly as well (just by putting them into integer programming problem form). We will see an example of this in Chapter 11 when we solve the TSP with integer programming. Since the TSP is known to be NP-Complete, this is almost a proof that integer programming is NP-Hard. We just need to show that this "reduction" from the TSP to integer programming takes polynomial time. Then, if we can find a polynomial-time algorithm to solve integer programming, we've found one for the TSP, and by extension, all NP-Complete problems! So, although it's clearly a difficult task, let's try to find the fastest algorithm we can to solve integer programming. While we cannot find (yet?) a polynomial algorithm, in many cases **Branch and Bound** (B&B) comes quite close.

Branch and Bound is, by a wide margin, the most well-used algorithm for solving NP-Hard problems. Although the algorithm has worst-case exponential run-time, this is often avoidable if it's set up with the correct set of parameters and components. Specifying the best version of the algorithm for a given problem is the challenge. We'll explore the various components of the Branch and Bound algorithm. The choices we make in setting up the algorithm greatly affect the run-time.

The algorithm starts with a solution space. A B&B algorithm dynamically constructs and searches through a solution space for optimal solutions. Unfortunately, due to the nature of NP-Hard problems, the solution space is usually exponential in size. Thus, it would take exponential time to search through it. The utility of a B&B algorithm is that it eliminates large chunks of the solution space without exploring them.

In order to describe the algorithm, we'll need a couple of new terms. Given an integer programming problem in the canonical form, let S be the set of feasible solutions. We let P be a set of *potential solutions* such that $S \subseteq P$, where the objective function f is still well defined on P. Next we let $g(x)$ be some function such that $g(x) \leq f(x)$ for all $x \in S$. We focus on P and g in our description of the Branch and Bound algorithm.

A *subproblem* is a new problem derived from the original via the addition of new constraints. The solution of a problem with a B&B algorithm is best described as a search through a tree. The root corresponds to the original problem and each child node is a subproblem of its parent. The leaves of the tree are feasible solutions, or subproblems that have been discarded.

We are now equipped to describe the Branch and Bound framework; the discussion below is influenced by online notes by Jens Clausen [**Cl**].

We begin by selecting a live node N to be processed. Of course, we start with the root (the original problem). Next we *branch* on this node N via some branch strategy, which involves adding constraints in order to generate subproblems (children). For each new subproblem, use the bounding function g to generate a lower bound (remember we are attempting to minimize the objective function). If, for some new subproblem, a feasible solution exists which when evaluated with the objective function is strictly less than the evaluated incumbent solution, then this feasible solution becomes the new incumbent solution. Otherwise, if the lower bound *exceeds* the incumbent, we *prune* the new node; this means we no longer branch on this node and thus no longer explore its subproblems. It is here that the exponential savings arise. Unfortunately, oftentimes we cannot eliminate a node since the lower bound is greater than or equal to the incumbent. In this case, the node must be added to the pool of live nodes to be further explored.

This completes the first iteration; now we must determine which nodes to explore next. There are many ways to answer this question. We describe some (depth-first, breadth-first, and best-first searches) later.

We have described the three main components of the algorithm, which we explore in greater detail in the following subsections.

- **Bounding**: A bounding function yields a lower bound on the optimal solution to a given subproblem. Importantly, this bound applies to all its subproblems as well.

- **Branching**: If a subproblem cannot be discarded, we create subproblems to explore by introducing new constraints. These subproblems are investigated in subsequent iterations.

- **Node Selection**: Method of selecting which subproblem to investigate in the next iteration.

9.3.1. Bounding. Usually, the bounding function g must satisfy the following three properties:

(1) $g(P_i) \leq f(P_i)$ for all nodes P_i in the tree,

(2) $g(P_i) = f(P_i)$ for all leaves in the tree, and

(3) $g(P_j) \leq g(P_i)$ if P_j is the parent of P_i.

In other words:

(1) g provides a lower bound on the objective function,

(2) every leaf agrees with the objective function, and

(3) as more constraints are added, the lower bound $(g(P_i))$ increases.

There is usually a trade-off between quality and run-time when we work with bounding functions. Frequently the more time we devote to determining the bound, the better the value obtained. As we want to keep the size of the search tree as small as we can, it's often a good idea to take as strong of a bounding function as one can work with. While we would love its value for a subproblem to equal the value of the best feasible solution to the problem, obtaining that level is often NP-Hard. Thus, instead we try to come as close as we can in a reasonable (i.e., polynomial) amount of time.

How do we find a bounding function that is computable in polynomial time but also provides a tight lower bound? The most standard method is *relaxation*. Relaxation means to leave out some constraints, which does result in additional feasible solutions. The same objective function is used ($g = f$). Thus, remembering P, the set of potential feasible solutions, we are minimizing f over P. If a solution to the relaxed subproblem happens to satisfy all constraints of the original, it's a feasible solution and can therefore be considered as a replacement for the incumbent. Otherwise, the value yielded by the original objective function is a lower bound because the minimization is performed over a superset P of the feasible solutions S. A common form of relaxation is *linear relaxation*, which removes the requirement that the variables be integers.

As an example, consider using the *fractional* knapsack problem as a relaxation of the binary knapsack problem. If we compute the benefit per weight metric for every item and sort greatest to least, and then greedily choose fractional (integral or non-integral) quantities, we get an upper bound on the benefit of a given subproblem.

Another possibility is to modify the objective function so that it's more efficiently computable but still satisfies the three properties outlined earlier.

Finally, we can combine both methods, which corresponds to minimizing g over P. At first glance this would seem to produce weaker bounds than each of these. The bounds calculated by *Lagrangian relaxation* (based on this method) are generally extremely tight yet computationally demanding. They can be used to (more) efficiently solve the TSP.

9.3.2. Branching. The branching rules lead to a subdivision of the search space by incorporating additional constraints. One can often show that the B&B algorithm converges if the generated subproblems are smaller than the original problem and if there are only finitely many solutions to the original problem.

In solving the knapsack problem described in this chapter, a subproblem is created by making the decision to *include or not include* an item. If we choose to *include* the item, we gain it's benefit (by adding it to some accumulated benefit) but subtract it's weight from the capacity of the backpack. Notice that now we have fewer items to choose from. The problem has become smaller. If we choose to *not include* an item, we simply remove it from the item list and leave the backpack capacity and accumulated benefit untouched.

9.3.3. Node Selection. Three popular ways of node selection are best-first (BeFS), breadth-first (BFS), and depth-first (DFS) search. BFS involves visiting all nodes of each level before moving on to the next. DFS means moving all the way down to a leaf before the next node at the current level. BeFS means selecting the current live node with the least lower bound.

9.3.4. Initial Incumbent Solution. A key for a great B&B algorithm is a low initial incumbent solution. It should be clear by now that the better the incumbent, the more nodes we can prune. Thus, if we start with a low incumbent, we can prune nodes much earlier. For example, if we are doing binary branching and we prune the left child of the root, we've most likely halved our run-time. We can use a heuristic to select the initial incumbent solution.

9.4. Exercises

Maximizing a Product

Exercise 9.4.1. *For S a positive integer, determine the optimal solution to its decomposition as a sum of positive integers with maximal product.*

Exercise 9.4.2. *If $x, y \geq 0$ prove the maximum product xy, given $x + y = S$, occurs when $x = y = S/2$.*

Exercise 9.4.3. *For $g(x) = (S/x)^x$, $x \geq 0$, show the maximum cannot occur as $x \to 0$ or $x \to \infty$.*

Exercise 9.4.4. *Prove $g(x) = (S/x)^x$ has a maximum at $x = S/e$ by using the second derivative test.*

Exercise 9.4.5. *Modify the decomposition problem so that we write S as a sum of non-negative summands, but now we want to maximize the product of the squares of the summands. What is the answer?*

Exercise 9.4.6. *Modify the decomposition problem so that we write P as a product of real numbers and we want to maximize the sum of the factors. Is there a finite maximum? Why or why not?*

Exercise 9.4.7. *Modify the previous problem so that each factor is at least 1. What choice maximizes the product?*

The final exercises in this section highlight the power of extending a function from the integers to the reals. Recall the **factorial function** is defined by $n! = n(n-1)\cdots 3 \cdot 2 \cdot 1$ for n a positive integer, and $0! = 1$; we can interpret $n!$ as the number of ways of ordering n items when order matters. The **Gamma function** is defined for $\mathrm{Re}(s) > 0$ by

$$\Gamma(s) := \int_0^\infty x^{s-1} e^{-x} dx$$

(though the definition makes sense for s complex, so long as the real part is positive). It extends the factorial function, and the famous **Stirling's formula** states

$$\Gamma(x) \sim \sqrt{2\pi} x^{x-1/2} e^{-x}$$

as $x \to \infty$; in other words,

$$\lim_{n \to \infty} \frac{n!}{n^n e^{-n} \sqrt{2\pi n}} = 1.$$

Exercise 9.4.8. *Prove the Gamma function converges for* $\mathrm{Re}(s) > 0$.

Exercise 9.4.9. *By integrating by parts, deduce* $\Gamma(s+1) = s\Gamma(s)$ *if* $\mathrm{Re}(s) > 0$, *and use this to deduce* $\Gamma(n+1) = n!$ *for* n *a non-negative integer.*

Exercise 9.4.10. *Find an approximation to Stirling's formula by using the integral test applied to the sum* $\log 1 + \log 2 + \cdots + \log n$.

Laplace's Method is a powerful way to estimate an integral where the integrand is large in a very small region. Given a twice differentiable function f which has a unique global maximum at $x_0 \in (a,b)$, most of the contribution to $\int_a^b e^{sf(x)} dx$, as $s \to \infty$, comes from x near x_0. One finds

$$\int_a^b e^{sf(x)} dx \approx \sqrt{\frac{2\pi}{s|f''(x_0)|}} e^{Mf(x_0)}$$

as $x \to \infty$; the presence of the factor $2\sqrt{\pi}$ is related to an integral of a Gaussian density that arises from expanding $f(x)$ near x_0.

Exercise 9.4.11. *From the Gamma function we have*

$$n! = \int_0^\infty e^{n \log x - x} dx = e^{n \log n} n \int_0^1 e^{n \log y - y} dy.$$

Apply Laplace's Method to finish the estimation of $n!$.

The Knapsack Problem

Exercise 9.4.12. *Consider the integer knapsack problem from the chapter. List all allowable choices of the products and the value of the knapsack so filled.*

Exercise 9.4.13. *Generalize the knapsack problem so that in addition to needing the total weight to be below a critical threshold, there is also a volume constraint. Set this up as a Linear Programming problem.*

Exercise 9.4.14. *Choose some reasonable values for the volume parameters and solve the resulting knapsack problems, both when we are allowed arbitrary amounts of the three objects and when we are only allowed integral amounts.*

Solving Integer Programs: Branch and Bound

Exercise 9.4.15. *Since $S \subseteq P$ and $g(x) \le f(x)$ on P, show that*

$$\min_{x \in P} g(x) \ \le \ \left\{ \min_{x \in P} f(x), min_{x \in S} g(x) \right\} \ \le \ \min_{x \in S} f(x).$$

Why is this significant in a Branch and Bound algorithm?

Exercise 9.4.16. *Describe an instance of the knapsack problem in which DFS outperforms BeFS. Describe an instance where BeFS outperforms DFS.* Hint: The expensive operation in this case is the heuristic evaluation.

Exercise 9.4.17. *Solve the knapsack problem described in the chapter using a B&B algorithm. Use the heuristic mentioned in the chapter and use BeFS. How many times did you evaluate the heuristic? Was it worth it?*

Chapter 10

Multi-Objective and Quadratic Programming

There are many ways to complicate the canonical linear programming problem. Our first example is **multi-objective linear programming**, where we have multiple items we wish to simultaneously maximize. After that we turn to **quadratic programming**.

While we have seen that some non-linearities can be brought in to linear programming, others cannot and require new methods. The bad news is that there is no method that works as well as the Simplex Method for general problems. The good news is that in many real-world situations small errors would not cause any great harm. There are several reasons for this. First, we often have to estimate parameters for our model, and thus as the model is only an estimate, perhaps it's not too bad that we cannot exactly solve an approximation to reality. Second, we can often get *very* close to the true optimal answer in a reasonable amount of time. For example, if your company measures revenues in the billions and you can quickly find a decision that is within $50 of the optimal, for all practical purposes you have found the optimal decision.

10.1. Multi-Objective Linear Programming

We now turn to multi-objective linear programming. One of the first problems we examined was the Diet Problem, where we tried to find the cheapest possible diet that satisfied the single objective of staying alive. In practice, however, there are many other objectives that could and should be involved. We might wish for the food to be palatable, for example. Or perhaps we want variety, or we need to minimize preparation time, or we care whether or not the food is organic or locally grown.

Importantly, the only objective that is not a luxury is that of staying alive. All the others are secondary. This means that we have to assign weights to the objectives, and prioritize. The advantage of having everything be a number is that it facilitates comparisons; we're comparing apples and apples, not apples and oranges.

The need to make such hard choices is not limited to the diet problem. Perhaps we're shopping for a new car and have to weigh how much price and miles per gallon are worth to us; of course, we might also care about items such as where the car is made, how it looks, ... For one last example, imagine you own a major sports franchise and are faced with allocating your salaries among players with different skills (in baseball maybe balancing pitching versus hitting versus fielding needs).

The general framework in all of these problems is the following, where we start with a canonical Linear Programming problem but modify the objective function.

Multi-Objective Linear Programming: Consider a constraint matrix $\mathbf{A}$, leading to $\mathbf{A}\vec{x} = \vec{b}$ with $\vec{x} \geq \vec{0}$. Choose non-negative weight $w_1, \ldots, w_n$ such that $w_1 + \cdots + w_n = 1$ (which immediately implies $0 \leq w_i \leq 1$) and objective vectors $\vec{c}_1, \ldots, \vec{c}_n$. The goal is to minimize

$$w_1 \vec{c}_1^{\,T} \vec{x} + \cdots + w_n \vec{c}_n^{\,T} \vec{x}.$$

In some sense, our phrasing above is misleading as this is, in fact, *a canonical Linear Programming problem!* If we let

$$\vec{c}_w = w_1 \vec{c}_1 + \cdots + w_n \vec{c}_n,$$

then our objective is to minimize $\vec{c}_w^T \vec{x}$, exactly as it should be. The challenge in multi-objective linear programming is determining the relative weights. Returning to the diet example: how much more do we care about cost than the taste of food? For purchasing a car: how much is fuel efficiency worth to us compared to convenience? We need to be able to make these relative evaluations.

Further, we need to keep everything linear. While linearity is reasonable in certain ranges, at some point such an assumption probably breaks down. For example, before streaming audio and video people used to listen to CDs, cassette tapes, records, ... (take your pick depending on your age). One could record at various quality levels, but at some point the human ear is unable to detect the improvements, and thus it would make no sense to continue giving credit for further gains.

10.2. Quadratic Programming

As we just saw, we can handle multi-objective linear programming with linear weights; this means we have a set of functions to weigh and minimize the weighted sum. There are, however, situations where we don't want linear weights; the statistical Method of Least Squares is a major example of this, which plays a large role in problems of regression in statistics.

Thus, we may not always want linear terms. We examine the simplest such case, in which the terms are quadratic.

Recall the canonical form for linear programming problems:

- $\mathbf{A}\vec{x} = \vec{b}$,
- $\vec{x} \geq \vec{0}$,
- minimize $\vec{c}^{\,T}\vec{x}$.

How could we insert quadratic elements into this canonical form? We would need to have certain elements depend on $\vec{x}$. There are only two places for this dependence.

(1) We could have the constraint matrix depend on $\vec{x}$: $\mathbf{A} = \mathbf{A}(\vec{x})$ or

(2) we could have the objective function depend on $\vec{x}$: $\vec{c} = \vec{c}(\vec{x})$.

Which of these two options seems more sensible? While both are easily realized in a variety of problems (as shown in the next section), the first is significantly harder to solve. In Exercise 10.6.5 you will prove that if we could incorporate and handle quadratic constraints, then we could solve *any* binary integer linear programming problem!

However, it turns out that we can handle quadratic elements in the objective function. Consider the following generalization of the canonical Linear Programming problem:

- $\mathbf{A}\vec{x} = \vec{b}$,
- $\vec{x} \geq \vec{0}$,
- minimize $(\vec{c}^{\,T}\vec{x} + \vec{x}^T\mathbf{C}\vec{x})$ where $\mathbf{C}$ is a **positive definite matrix** (see Exercise 10.6.6).

The Simplex Method can be generalized in this case. Recall that when we discussed the Simplex Method, we never used the linearity of the cost function; we simply needed it to be decreasing. (We also never proved that it runs as fast as we claimed it does, so perhaps the linearity is used there.) It turns out that the Simplex Method can be generalized to the case where the matrix $\mathbf{C}$ is positive definite, which means that $\vec{x}^T\mathbf{C}\vec{x} \geq \vec{0}$, with $\vec{x}^T\mathbf{C}\vec{x} = \vec{0}$ if and only if $\vec{x} = \vec{0}$. While a proof that the Simplex Method generalizes is beyond the scope of this book, the knowledge that it could should serve to give a glimpse of the broader applications of the Simplex Method.

10.3. Example: Quadratic Objective Function

In this section we introduce how quadratic and higher order functions could be introduced in our programming problem. In the next section we attempt to solve this problem using linear programming, with the hope that it will demonstrate how we could generalize the process to a large class of quadratic programming problems.

Consider the following example. We own a movie theater, and our goal is to schedule movies appropriately (and of course to make money). Assume the time slots are in 10 minute increments (labeled $0, 1, \ldots, T$), there are M movies (labeled $1, \ldots, M$), and there are S screens (labeled $1, \ldots, S$). We have the following decision variables:

$$y_{tm} = \begin{cases} 1 & \text{if movie } m \text{ is playing at time } t, \\ 0 & \text{otherwise} \end{cases}$$

and

$$d_{tm} \;=\; \text{demand for movie } m \text{ at time } t.$$

Our objective function would be to maximize the revenue of the theater. To do that, we need to know d_{tm}. However, often d_{tm} is not fixed but rather depends on the other movies that are playing. For example, it's reasonable to think that there are moviegoers who do not decide on what they want to see beforehand, and they may simply be in the mood of seeing an action movie. Assume that in our movie set we have two action blockbusters that are played at the same time. Since the two movies are competing for customers' attention, the total demand for these two movies would be smaller than the sum of the individual demand for each had they been played at different time slot.

This problem calls for some correcting factor. Here's an attempt at one. For a fixed time t, we consider the polynomial:

$$
\begin{aligned}
p(x) \;&=\; \prod_{m=1}^{m_1} y_{tm} \prod_{m=m_1+1}^{m_1+m_2} (1 - y_{tm}) \\[2mm]
&=\; \begin{cases} 1 & \begin{array}{l}\text{if at time } t \text{ movies 1 through } m_1 \text{ are being shown} \\ \text{and movies } m_1 + 1 \text{ through } m_1 + m_2 \text{ are not being shown;}\end{array} \\ 0 & \text{otherwise,} \end{cases}
\end{aligned}
$$

in which $m = m_1 + m_2$. Here we are dividing our set of movies into two subsets of size m_1 and m_2 respectively. By considering all possible polynomials of this form, we can handle any m_1-tuple of movies playing and m_2-tuple of movies not playing.

This correction would bring polynomial non-linearities in the objective function. In the next section we will investigate on how to linearize them. Although we are concentrating on removing non-linearities in the objective function (the schedule to maximize revenue), the method is identical for removing such non-linearities from the constraints.

Note that every variable occurs to either the zeroth or first power: as $y_{tm} \in \{0, 1\}$, $y_{tm}^n = y_{tm}$ for any integer $n \geq 1$. **This is an extremely useful consequence of having binary variables!**

10.4. Removing Quadratic (and Higher Order) Terms in Constraints

As mentioned in §10.2, there are some generalizations to linear programming methods such as the Simplex Method which work on quadratic constraints and objective functions. Describing these techniques and proving their correctness, however, is too much of a detour. We thus take a different approach and show how we can linearize these constraints, taking the problem from the above section as an example. Of course, just because we can linearize these quantities does not mean this will be the faster way to solve them; perhaps there are special techniques which could exploit the polynomial structure.

Our goal is to replace terms in the objective function of the form

$$-\text{Const} \cdot p(x)$$

with linear terms, possibly at the cost of additional variables and constraints. In our example, these $m_1 + m_2$ movies compete with each other for demand, and we must adjust the demand of each movie based on the competition concurrently screened.

How could we do that? One approach would be tò look at the simplest case and try to generalize. Let $m_1 = 1, m_2 = 1$. We need to generalize

$$p_t = y_{t1}(1 - y_{t2}).$$

Notice that $p_t = 1$ if and only if $y_{t1} = 1$ and $y_{t2} = 0$, or $1 - y_{t2} = 1$. Thus we could say $p_t = y_{t1}$ **AND** $(1 - y_{t2})$. This condition could be expressed through the two constraints:

(1) $y_{t1} + (1 - y_{t2}) - 2p_t \geq 0$,

(2) $y_{t1} + (1 - y_{t2}) - p_t \leq 1$.

If $y_{t1} = 1$ and $y_{t2} = 0$, then the second constraint forces $p_t = 1$ while the first constraint does not impose any limitation. Otherwise, the second condition does not constraint p_t, but the first forces $p_t = 0$.

Now we generalize the above approach for the general case

$$p_t = \prod_{m=1}^{m_1} y_{tm} \prod_{m=m_1+1}^{m_1+m_2} (1 - y_{tm}).$$

Using similar reasoning, we replace the product with the following two constraints:

(1) $\sum_{m=1}^{m_1} y_{tm} + \sum_{m=m_1+1}^{m_1+m_2}(1 - y_{tm}) - (m_1 + m_2)p_t \geq 0$,

(2) $\sum_{m=1}^{m_1} y_{tm} + \sum_{m=m_1+1}^{m_1+m_2}(1 - y_{tm}) - p_t \leq m_1 + m_2 - 1$.

The detailed reasoning on why these constraints are equivalent to our product is left for the reader as a small practice.

10.5. Summary

As remarked, we have only touched the beginning of a very important generalization of linear programming. It is important to analyze the *cost* of linearizing our problem specifically for real world problems. Can the linearized problems be solved (or approximately solved) in a reasonable amount of time?

We are reminded again of the quote of Abraham Maslow, who remarked that if all one has is a hammer, pretty soon all problems look like nails. Once we know how to do and solve Linear Programming problems, it's tempting to convert other problems to Linear Programming problems. While this yields a reasonable solution in many situations, there are additional techniques that are better able to handle many of these problems.

10.6. Exercises

Multi-Objective Linear Programming

Exercise 10.6.1. *Revisit the diet problem. Add at least one additional objective to the objective function, choose some reasonable numbers, and determine.*

Exercise 10.6.2. *Consider the Diet Problem from §3.5. Assume we are also concerned with our fat intake and wish to minimize that. If one unit of the first food has 3 grams of fat and the second has 2, we now want a cheap diet with little fat. Solve this problem as a function of the weights $w_1, w_2 = 1 - w_1$.*

Exercise 10.6.3. *Consider a multi-objective Linear Programming problem. Must the solution vary continuously with the weights?*

Exercise 10.6.4. *Consider a multi-objective Linear Programming problem. Must the solution vary differentiably with the weights?*

Quadratic Programming

Exercise 10.6.5. *Show that if we had the ability to incorporate quadratic constraints and solve the resulting problem, then we would immediately be able to solve any binary integer Linear Programming problem (or, more generally, any integer Linear Programming problem!).*

The next few problems involve positive definite matrices. Recall a square matrix $\mathbf{C}$ is **positive definite** if for any non-zero vector $\vec{x}$ we have $\vec{x}^T \mathbf{C} \vec{x} > 0$.

Exercise 10.6.6. *Prove that if $\boldsymbol{C}$ is a real-symmetric matrix whose eigenvalues are all positive, then $\boldsymbol{C}$ is positive definite.*

Exercise 10.6.7. *Give an example of a 10×10 positive definite matrix such that every element of the matrix is non-zero.*

Exercise 10.6.8. *If a matrix has all of its eigenvalues distinct and positive, is it positive definite?*

Exercise 10.6.9. *One place where positive definite matrices arise is in the multivariable generalization of the second derivative test. Consider a twice continuously differentiable function $f(x_1, \ldots, x_n)$ with a critical point at $(a_1, \ldots, a_n)$ (so $(\nabla f)(a_1, \ldots, a_n) = \vec{0}$). The Hessian matrix $\boldsymbol{H}_f(a_1, \ldots, a_n)$ has a $(i,j)^{th}$ entry of $\frac{\partial^2 f}{\partial x_i \partial x_j}(a_1, \ldots, a_n)$. Prove the Hessian matrix is real-symmetric and that if all of its eigenvalues are positive, then f has a minimum at $(a_1, \ldots, a_n)$.*

Note: if you have never seen this language before, the above problem hopefully explains the involved second derivative tests from multivariable calculus, where we have a minimum or a maximum depending on values of partial derivatives and their products.

Removing Quadratic (and Higher Order) Terms

Exercise 10.6.10. *In the chapter we looked at terms arising from m_1 movies playing and m_2 movies not. What if now all we care about is that at least n_1 of the first m_1 movies are playing and at least n_2 of the next m_2 are not. Is it still possible to introduce binary indicator variables and constraints to encode this linearly?*

Exercise 10.6.11. *Now we return to the safe pawns problem we introduced in Exercise 5.4.6. Here are some suggestions on how to formalize the problem as a linear programming problem.*

Let x_{ij} represent whether position (i,j) is occupied by any queen and let y_{ij} represent whether position (i,j) is safe (i.e, not occupied by any queen and not attacked by any queen).

What is our objective function? How could we write y_{ij} as a function of x_{ij}? Could we replace the function with some linear constraints? Formalize the problem with linear constraints and objective function.

Chapter 11

The Traveling Salesman Problem

The **Traveling Salesman Problem** (often abbreviated **TSP**) is one of the most famous and important problems in operations research. It's an example of an **NP-hard** problem, which is a group of problems which are extremely hard to solve, but for which potential solutions are easily verifiable. We've already seen an example of this: factorization. Given a large product N of two primes p and q, we currently do not have any known, fast way to factor N and find p and q; however, if someone gives us a claimed factorization of N we can quickly and trivially check and see if it works. This is the defining characteristic of NP-Hard problems: they are hard to solve, but proposed solutions are easy to check.

11.1. Integer Linear Programming Version of the TSP

As the name suggests, the problem involves the route a salesman takes as he goes from city to city, trying to sell his wares.

Formulation of the Traveling Salesman Problem

Given
  • a collection of n cities $\mathcal{C} = \{c_1, c_2, \ldots, c_n\}$,
  • from city i can travel to any city in $\mathcal{C}_i = \{c_{i_1}, \ldots, c_{i_{r_i}}\}$,
  • each city must be visited exactly once,
find a valid route which minimizes total distance traveled.

If each city is accessible from every other city, forming a complete graph, the number of possible paths is on the order of $n! \approx \left(\frac{n}{e}\right)^n \sqrt{2\pi n}$; this gets astronomically large as n grows. (See the Stirling formula problems, Exercises 9.4.10 and 9.4.11, for a sketch of the proof of this bound.) Note that there are far too many paths to make a brute force approach feasible. Also, there are several possible objectives that can be given for the TSP; while we concentrate on minimizing distance we could also minimize cost.

Many problems can be reduced to the TSP, beyond just the question of a salesman's most efficient route. The TSP is equivalent to many scheduling problems involving networks of tasks, to problems of efficiency in the laser-manufacture of data chips, and many others. Our goal here is to present a broad overview of the topic. To describe the problem in a linear programming format, we need to find parameters, variables, constraints, and an objective function.

We start by representing our network of connected cities as a graph, with vertices representing the cities and edges representing their connections to each other. Notably, there are several possible layouts for which there are no solutions to the original TSP given the restriction that cities cannot be revisited: take a layout of cities in an "H" for example (see Figure 1).

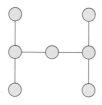

Figure 1. A configuration where there is no solution to the TSP.

One way to formulate the problem is to use the following variables:

$$x_{tij} = \begin{cases} 1 \text{ if at time } t \text{ leave city } i \text{ for city } j, \\ 0 \text{ otherwise}, \end{cases}$$

$$S_j = \text{the set of cities accessible from city } j,$$

$$d_{ij} = \text{the distance from city } i \text{ to city } j.$$

Note that, except for the first and last cities, each city must be entered exactly once and left exactly once; however, just having this hold is not sufficient to ensure a valid path, as we also need that the city entered at time t is the one left at time $t + 1$.

To take care of the problem posed by the first and last cities in our route, we add two extra, special cities: the *starter* city (labeled 0) and the *ending* city (labeled $n + 1$). We require exactly one departure from the starter city and no arrivals there; similarly we insist that there's exactly one arrival and no departures from the final city. This way we can treat the n cities on our route equally; each now has exactly one arrival and one departure.

We identify our salesman's path by listing the cities in the order they're visited, with the extra starter city as 0 and the ending city as $n + 1$. This way at time $t = 0$ we leave the starter city and arrive at our first city at time $t = 1$. We then continue, eventually leaving the last of our n cities at time n and arriving at the ending city at time $n + 1$. See Figure 2 for a depiction.

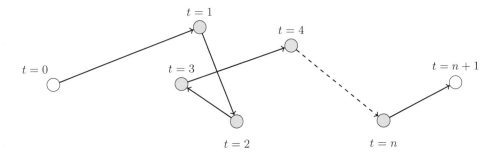

Figure 2. In this example the start and end cities are labeled by empty circles; $t \in \{-1, 1, \ldots n+1\}$ is such that at $t = 0$ the salesman travels from the start city and at time $t = n + 1$ the salesman arrives at his final city.

We now set up our constraints.

- $\forall j \in \{1, \ldots, n\} : \sum_{t=0}^{n-1} \sum_{i=0}^{n} x_{tij} = 1$ (each of our n cities is entered exactly once, and not from the ending city).

- $\sum_{t=0}^{n} \sum_{i=0}^{n+1} x_{ti0} = 0$ (we never travel to the starter city).

- $\sum_{j=1}^{n} x_{00j} = 1$ (at time 0 we leave the starter city for one of our n cities).

- $\forall i \in \{1, \ldots, n\} : \sum_{t=1}^{n} \sum_{j=1}^{n+1} x_{tij} = 1$ (each of our n cities is left exactly once, and not going to the starter city).

- $\sum_{t=0}^{n} \sum_{j=0}^{n+1} x_{t,n+1,j} = 0$ (we never travel from the ending city).

- $\sum_{j=1}^{n} x_{n,i,n+1}$ (at time n we leave for the ending city from one of our n cities).

- $\forall t \in \{0, \ldots, n\} : \sum_{i=0}^{n+1} \sum_{j=0}^{n+1} x_{tij} = 1$ (only one motion happens at a given time).

- $\forall j \in \{1, \ldots, n\}$ and $\forall t \in \{0, \ldots, n-1\}$:
  IF $x_{tij} = 1$ THEN $\sum_{k \in S_j} x_{t+1,j,k} = 1$ (a city which is entered on one turn is left on the following turn).

With these variables and constraints we can easily choose an objective function to minimize:

$$\min \sum_{t=0}^{n} \sum_{i=0}^{n+1} \sum_{j=0}^{n+1} d_{ij} x_{tij}.$$

11.2. Greedy Algorithm to the TSP

We've seen how we can formulate the TSP as a Linear Programming problem. Unfortunately, as we mentioned earlier, the TSP is an *extremely* hard problem, and we don't have a means to solve it efficiently with linear programming methods. We'll now discuss some alternative ways of approaching the TSP. While these may not always yield an optimal solution, at the very least they do terminate.

Assuming that all cities are always accessible, the greedy algorithm (also known as the nearest neighbor algorithm) goes as follows:

- Start at a city.

- Look at the remaining cities.

- Take the lowest cost of the remaining cities (i.e., whichever city is closest).

- Repeat.

What are the good and bad aspects of this algorithm? The main advantage of the greedy algorithm is that it always chooses what is locally best. Unfortunately, it can often miss the big picture (see Figure 3). This is another example where doing at each moment what is *locally* best does not always lead to what is *globally* best (see §3.3 for more on these issues).

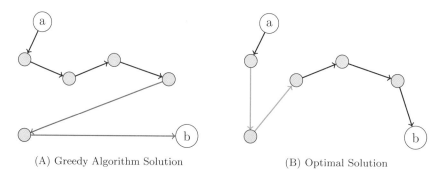

(A) Greedy Algorithm Solution   (B) Optimal Solution

Figure 3. For this arrangement of cities, the greedy algorithm finds a sub-optimal path from city a to city b.

While in some cases the greedy algorithm can miss the optimal result by a very large amount, it has the great advantage of being incredibly easy to code, and as such it's one of the more frequently used approaches to the TSP. See §3.3 for more on locally best choices not leading to a global optimal solution.

11.3. The Insertion Algorithm

We describe yet another approach to the Traveling Salesman Problem, the Insertion Algorithm.

- Start with some number of connected cities.

- Choose an unused city.

- Insert this city so that the new path has the lowest cost of all paths where we insert this, doing the local best with this city.

- Repeat.

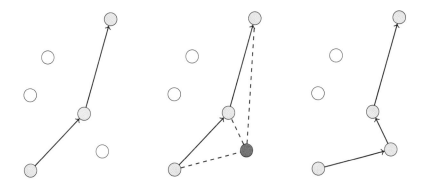

Figure 4. Given a starting arrangement with 3 connected cities (left), a random unused city is chosen. If the start and end cities are not fixed, there are four possible ways to insert the new city (middle). The city is added to minimize total path length (right).

Using this algorithm, you can choose to keep the endpoints of the trip fixed or allow cities to be added to the start or to the end of a trip. As such, it seems like the Insertion Algorithm should do better than the greedy algorithm, although it's more involved to code. We could also try to come up with an improved Insertion Algorithm, but this requires us to think about the question of which city we should choose to add.

One option would be to add the city with the lowest new cost. But this now raises questions of run-time. Say we have n cities. In the zeroth round we choose a city from n cities. In the first round we have a single city, and thus we have $n - 1$ cities to check and 2 directions in which to create a path between two cities. In the second round we have two connected cities and 3 insertion points, so we actually need to check $3(n - 2)$ possibilities. The next step has $4(n - 3)$ possibilities to check, then $5(n - 4)$, and so on. Adding up all of these possibilities that must be considered yields

$$S = n + 2(n - 1) + 3(n - 2) + \cdots + (n - 1)(2) + n.$$

This sum should be a reasonable approximation to the run-time of the insertion algorithm:

$$S \;=\; \sum_{k=1}^{n} k(n-(k-1)) \;=\; \sum_{k=1}^{n} kn - \sum_{k=1}^{n} k^2 + \sum_{k=1}^{n} k.$$

The sum turns out to be roughly on the order of n^3, which we can find with a rough bounding argument on the terms in the sum. So this algorithm is covering on the order of n^3 out of the $n!$ cases: it's clearly not examining every single path. A good question, which we leave to the reader to think about, is whether this algorithm is close to the optimal path in spite of missing so many values.

11.4. Sub-problems Method

We look at one last approach to the TSP. The idea is to break the TSP up into smaller subproblems by breaking the graph into smaller subsections which can each be examined separately.

The sub-problems method generally goes as follows.

- Divide the cities into groups (preferably by location).
- Solve the TSP for each group.
- Connect the solutions of all the groups.

This approach produces optimal or close to optimal solutions when points are grouped in obvious clusters, and it produces less optimal solutions for evenly distributed points. The advantage is that if the clusters are small, we can exhaustively search among the cities there.

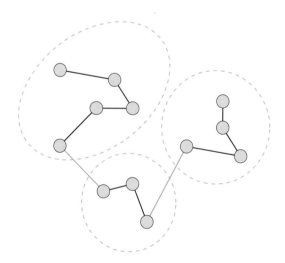

Figure 5. An example of the sub-problems method being implemented.

11.5. Exercises

Integer Linear Programming Version of the TSP

Exercise 11.5.1. *Factor 1,542,957,170,702,437,211,422,391.*

Exercise 11.5.2. *Explicitly give an example of a Traveling Salesman Problem (other than ones in this chapter!) where there is no solution.*

Exercise 11.5.3. *Assume that every city is accessible from every other city, but now assume we are trying to minimize the cost of a valid route. Come up with a set of travel costs among the n cities so that it's clear what the cheapest route is.*

Exercise 11.5.4. *Assume that every city is accessible from every other city. Show there is a valid route that satisfies the constraints.*

Exercise 11.5.5. *Verify that 1,948,138,430,863 divides the number from Exercise 11.5.1: 1,542,957,170,702,437,211,422,391. Note these two problems demonstrate how much easier it is to verify a factorization than it is to find it.*

Greedy Algorithm to the TSP

Exercise 11.5.6. *Choose a collection of cities, a starting city, and distances between them so that the greedy algorithm does not give a route of minimal distance.*

Exercise 11.5.7. *Write a simple program to code the greedy algorithm approach to the Traveling Salesman Problem, and try it on the state capitals in the continental United States.*

The Insertion Algorithm

Exercise 11.5.8. *Without doing out the algebra, find upper and lower bounds for the sum approximating the Insertion Algorithm's run-time: $\sum_{k=1}^{n} k(n-(k-1))$. How close can you get to the true value?*

Exercise 11.5.9. *Approximate all the sums with integrals in order to estimate $\sum_{k=1}^{n} k(n-(k-1))$.*

Exercise 11.5.10. *By expanding, we see that $\sum_{k=1}^{n} k(n-(k-1))$ can be determined if we can find $\sum_{k=1}^{n} k$ and $\sum_{k=1}^{n} k^2$. The first is just $n(n+1)/2$ (see Exercise 1.7.13); find the sum of the squares.*

Exercise 11.5.11. *Choose a collection of cities, a starting city, and distances between them so that the insertion algorithm does not give a route of minimal distance.*

Exercise 11.5.12. *Write a simple program to code the Insertion Algorithm approach to the Traveling Salesman Problem, and try it on the state capitals in the continental United States.*

Sub-problems Method

Exercise 11.5.13. *Choose a collection of cities, a starting city, and distances between them so that the sub-problem method does not give a route of minimal distance.*

Exercise 11.5.14. *Write a simple program to code the sub-problem method approach to the Traveling Salesman Problem, and try it on the state capitals in the continental United States.*

Chapter 12

Introduction to Stochastic Linear Programming

We've made good progress on the canonical Linear Programming problem, namely trying to minimize $\vec{c}^T\vec{x}$ subject to $\mathbf{A}\vec{x} = \vec{b}$ with $\vec{x} \geq \vec{0}$ (sometimes for convenience we allow inequalities in the matrix of constraints). We described the Simplex Method in Chapter 7, which frequently (though not always) finds a solution quickly. One important extension is integer programming (Chapter 8), where we now require the entries of $\vec{x}$ to be integers. There are tremendous advantages to considering such problems. We are able to linearly encode a variety of important operations such as AND, OR, IF--THEN, MAX, MIN, to name a few. However, there is a trade-off: we cannot find the solution quickly, though we are frequently able to rapidly find a very good approximation.

We now turn to another important generalization: stochastic linear programming. The idea is that frequently we do not know the values of the parameters in our model but rather must rely on extrapolations or estimates. For example, what is the true demand for a particular movie next week? How much oil will a city demand? How many passengers desire to fly from Albany to Tucson? The Simplex Method requires us to have known values of the entries of $\mathbf{A}$ and the vectors $\vec{b}, \vec{c}$; however, in many cases some or all of these parameters are either unknown at the time when we make our decisions for $\vec{x}$ or will be random and follow some joint distribution.

In this chapter we give a brief introduction to *some* of the approaches to dealing with these uncertainties. For definiteness we have chosen a specific problem, the Oil Problem from §4.8, as our focus. We'll explore various refinements[1] of the problem and see how different techniques handle the resulting issues.

---

[1] Yes, the choice was made for the sake of this pun.

12.1. Deterministic and Stochastic Oil Problems

Oil Problem: We have R refineries, each with a given maximum capacity s_i, $i \in \{1, 2, \ldots, R\}$, and C cities, each with a given level of demand for oil d_j, $j \in \{1, 2, \ldots, C\}$. We are given costs $c_{i,j}$ of shipping oil from each refinery i to each city j, which we express as a vector $\vec{c}$. Our goal is to determine $x_{i,j}$, how much oil to ship from refinery i to city j, in order to minimize the total shipping cost $\vec{c}^T \vec{x}$ while meeting demand in all of the cities and not exceeding capacity in any of the refineries.

For example, let's consider the case with 2 refineries and 3 cities. One set of parameters and variables would be

$$
A = \begin{pmatrix} -1 & -1 & -1 & 0 & 0 & 0 \\ 0 & 0 & 0 & -1 & -1 & -1 \\ 1 & 0 & 0 & 1 & 0 & 0 \\ 0 & 1 & 0 & 0 & 1 & 0 \\ 0 & 0 & 1 & 0 & 0 & 1 \end{pmatrix},
$$

$$
(12.1) \qquad \vec{x} = \begin{pmatrix} x_{1,1} \\ x_{1,2} \\ x_{1,3} \\ x_{2,1} \\ x_{2,2} \\ x_{2,3} \end{pmatrix}, \quad \vec{b} = \begin{pmatrix} -s_1 \\ -s_2 \\ d_1 \\ d_2 \\ d_3 \end{pmatrix}, \quad \vec{c} = \begin{pmatrix} c_{1,1} \\ c_{1,2} \\ c_{1,3} \\ c_{2,1} \\ c_{2,2} \\ c_{2,3} \end{pmatrix},
$$

and we can solve this using standard methods. We refer to this as the **deterministic problem**, since all parameter values are known (i.e., determined) in advance. If instead only some of the parameters are unknown, taking on random values, we have a **stochastic problem**.

For example, let's suppose that the demand is not known when we must make our decision about quantities $\vec{x}$ to ship, but that the demand levels (d_1, d_2, d_3) follow a known joint distribution. For simplicity, we assume that the vector of demands can only take on some finite number of values. From (12.1), we see that only the vector $\vec{b}$ is stochastic, since some of its elements are random, while the other parameters of the problem, $\mathbf{A}$ and $\vec{c}$, are known in advance. By assumption, we know that $\vec{b}$ takes on some finite number of values $\vec{b}_s$, $s \in \{1, 2, \ldots, S\}$, with each $\vec{b}_s$ occurring with a known probability p_s such that $\sum_{s=1}^{S} p_s = 1$.

We note that it is of course possible that the other parameters of the problem, $\mathbf{A}$ and $\vec{c}$, may also be stochastic in nature. In that case, we can also assume that $\mathbf{A}, \vec{b}, \vec{c}$ follow some joint distribution, taking on some finite number of values $\mathbf{A}_s, \vec{b}_s, \vec{c}_s$ for $s \in \{1, 2, \ldots, S\}$.

We end with a simple example, which we'll revisit later. Suppose we have two possible scenarios. In each we have $\vec{c}^T = (20, 20, 10, 50, 10, 15)$, and $\mathbf{A}$ is as shown in (12.1); however, suppose there are two possible values for $\vec{b}$.

Scenario 1: $\vec{b}_1 = (-300, -200, 300, 100, 100)^T$; $p(\vec{b}_1) = \dfrac{2}{3}$,

(12.2)   Scenario 2: $\vec{b}_2 = (-300, -200, 200, 250, 50)^T$; $p(\vec{b}_2) = \dfrac{1}{3}$

(where $p(\vec{b}_i)$ is the probability that $\vec{b}$ takes on the value $\vec{b}_i$).

If the probabilities are such that one of the two options happens with probability 1 and the other with probability 0, then our problem reduces to the canonical linear programming problem. In the next sections we explore how to handle the more general case.

12.2. Expected Value approach

In the **expected value approach**, we compute the following:

(12.3) $$\mathbf{A}_E = \sum_{s=1}^{S} p_s A_s, \quad \vec{b}_E = \sum_{s=1}^{S} p_s \vec{b}_s \quad \vec{c}_E = \sum_{s=1}^{S} p_s \vec{c}_s,$$

and then solve an associated canonical Linear Programming problem:

(12.4) $$\min \vec{c}_E^T \vec{x} \text{ subject to } \mathbf{A}_E \vec{x} \geq \vec{b}_E \text{ and } \vec{x} \geq \vec{0}.$$

There is a lot to commend about this approach. The quantities we need to compute are readily found if we have the probabilities (if instead of taking on a discrete set of values we had continuous distributions, all that would change is replacing sums with integrals). The name of the method comes from the fact that the associated problem uses the expected values of the parameters. Unfortunately, there are serious issues with this problem. Note the optimal $\vec{x}$ may not only be sub-optimal in some (or all) scenarios, but may in fact not be feasible in some (or any) of the individual scenarios!

For example, the two scenarios described as in (12.2) yield

$$\vec{b}_E = \left(-300, -200, 266\frac{2}{3}, 150, 83\frac{1}{3}\right)^T,$$

while $\mathbf{A}_E$ is the same as in (12.1) and the expected cost vector

$$\vec{c}_E^T = (20, 20, 10, 50, 10, 15)$$

similarly remains unchanged since the costs were determined in advance. Solving the associated canonical Linear Programming problem from (12.4), we obtain the solution

$$\vec{x}^* = \left(266\frac{2}{3}, 0, 33\frac{1}{3}, 0, 150, 50\right).$$

We note that $A\vec{x}^* = \vec{b}_E$. Now, we see that if Scenario 1 were to occur, then we would have failed to meet the demand in city 1, while if Scenario 2 happened, then we would have failed to meet the demand in city 2.

The important takeaway is that the expected value method can yield solutions which are always infeasible!

12.3. Recourse Approach

The **recourse approach** is a more popular approach than the expected value approach and has many applications (see for example [**SH**]). We first describe the method in the context of the same Oil Problem as above, and show how we can generalize it further to situations where all the parameters of the problem might be random.

Suppose that demand is unknown and we have the opportunity to ship quantities of oil $\vec{x}$ now, at costs $\vec{c}$. Then, after demand becomes known, we may need to ship additional quantities of oil $\vec{y}_s$, at scenario-specific "recourse" costs $\vec{r}_s$, in order to meet demand in all the cities. In many situations we expect $\vec{r}_s \geq \vec{c}$ for all $s \in \{1, 2, \ldots, S\}$ (maybe we're shipping smaller amounts and don't have a volume savings per unit, or we need to add an additional ship or truck or train). We can already see the advantages of this approach; specifically, the second shipments allow us to meet demand (something which we could fail to do in the expected value approach).

The constraints for the recourse problem, as in the original Oil Problem, can be divided into supply and demand constraints:

- In any scenario s the total amount shipped out of a refinery over the initial and recourse stages must not exceed the capacity of the refinery.

- In any scenario s the total amount shipped into a city over the initial and recourse stages must not fall short of the demand in that scenario.

Using the notation in (12.1), we can express this as

$$\mathbf{A}\vec{x} + \mathbf{A}\vec{y}_s \geq \vec{b}_s$$

for each scenario $s \in \{1, 2, \ldots, S\}$.

We have some freedom in our choice of the objective function. A common decision is to minimize the expected total cost. For any scenario s, the total cost we incur is $\vec{c}^T \vec{x} + \vec{r}_s^T \vec{y}_s$. Therefore, the expected total cost is

$$\sum_{s=1}^{S} p_s(\vec{c}^T \vec{x} + \vec{r}_s^T \vec{y}_s) = \vec{c}^T \vec{x} + \sum_{s=1}^{S} p_s \vec{r}_s^T \vec{y}_s$$

$$= \begin{pmatrix} \vec{c}^T & p_1 \vec{r}_1^T & \cdots & p_S \vec{r}_S^T \end{pmatrix} \begin{pmatrix} \vec{x} \\ \vec{y}_1 \\ \vdots \\ \vec{y}_S \end{pmatrix}.$$

Notice this is just a canonical Linear Programming problem. The cost vector is now

$$\begin{pmatrix} \vec{c}^T & p_1 \vec{r}_1^T & \cdots & p_S \vec{r}_S^T \end{pmatrix};$$

what is the matrix?

Let's do a concrete example with numbers, say the recourse costs are $\vec{r}_1^T = 2\vec{c}^T$ and $\vec{r}_2^T = 3\vec{c}^T$. We can summarize the recourse problem in block matrix form as

$$\min \begin{pmatrix} \vec{c}^T & p_1\vec{r}_1^T & p_2\vec{r}_2{}^T \end{pmatrix} \begin{pmatrix} \vec{x} \\ \vec{y}_1 \\ \vec{y}_2 \end{pmatrix}$$

(12.5)   subject to $\begin{pmatrix} \mathbf{A} & \mathbf{A} & \mathbf{0} \\ \mathbf{A} & \mathbf{0} & \mathbf{A} \end{pmatrix} \begin{pmatrix} \vec{x} \\ \vec{y}_1 \\ \vec{y}_2 \end{pmatrix} \geq \begin{pmatrix} \vec{b}_1 \\ \vec{b}_2 \end{pmatrix}, \quad \vec{x}, \vec{y}_1, \vec{y}_2 \geq \vec{0},$

where $\mathbf{0}$ is a matrix of zeros of the same dimensions as $\mathbf{A}$. This problem is of the same form as the canonical problem, and we obtain the solution

$$\vec{x} = (200,\ 0,\ 0,\ 0,\ 100,\ 50)^T,$$
$$\vec{y}_1 = (100,\ 0,\ 0,\ 0,\ 0,\ 50)^T, \quad \vec{y}_2 = (0,\ 100,\ 0,\ 0,\ 50,\ 0)^T.$$

We could of course generalize the problem further by incorporating storage costs in the objective function (see Exercise 12.5.5). Another generalization is to consider more time frames, and hence more opportunities for decisions. This leads to **multistage recourse**.

In many applications we have to make decisions at multiple stages in the future, as new information is revealed to us. For example, we might extend the Oil Problem to supplying a set of cities each month, subject to unknown levels of demand, and possibly also of supply.

Let's explore a simple case where we have three stages. Figure 1 illustrates the possible sequences of scenarios.

At each stage, the decision vector we need to pick depends on the decisions we made leading up to that stage. Each possible scenario at each stage will, therefore, comprise a set of parameters that define the relationship between that decision and the decisions preceding it, as well as some set of costs associated with that decision. Thus we can express this three-stage problem in the following form:

$$\min \begin{pmatrix} \vec{c}_1^T & (p_1 + p_2 + p_3)\vec{c}_2^T & p_4\vec{c}_3^T & p_1\vec{c}_4^T & p_2\vec{c}_5^T & p_3\vec{c}_6^T & p_4\vec{c}_7^T \end{pmatrix} \begin{pmatrix} \vec{y}_1 \\ \vec{y}_2 \\ \vec{y}_3 \\ \vec{y}_4 \\ \vec{y}_5 \\ \vec{y}_6 \\ \vec{y}_7 \end{pmatrix}$$

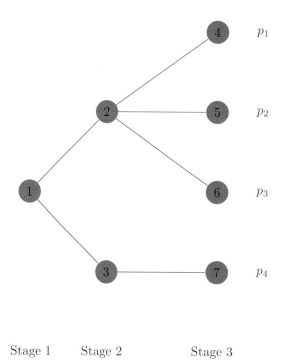

Stage 1   Stage 2   Stage 3

Figure 1. A simple scenario tree with three stages. Each p_i refers to the probability of that sequence occurring.

subject to

$$\begin{pmatrix} \mathbf{A}_1 & 0 & 0 & 0 & 0 & 0 & 0 \\ \mathbf{A}_2 & \mathbf{B}_2 & 0 & 0 & 0 & 0 & 0 \\ \mathbf{A}_3 & 0 & \mathbf{B}_3 & 0 & 0 & 0 & 0 \\ \mathbf{A}_4 & \mathbf{B}_4 & 0 & \mathbf{C}_4 & 0 & 0 & 0 \\ \mathbf{A}_5 & \mathbf{B}_5 & 0 & 0 & \mathbf{C}_5 & 0 & 0 \\ \mathbf{A}_6 & \mathbf{B}_6 & 0 & 0 & 0 & \mathbf{C}_6 & 0 \\ \mathbf{A}_7 & 0 & \mathbf{B}_7 & 0 & 0 & 0 & \mathbf{C}_7 \end{pmatrix} \begin{pmatrix} \vec{y}_1 \\ \vec{y}_2 \\ \vec{y}_3 \\ \vec{y}_4 \\ \vec{y}_5 \\ \vec{y}_6 \\ \vec{y}_7 \end{pmatrix} \geq \begin{pmatrix} \vec{b}_1 \\ \vec{b}_2 \\ \vec{b}_3 \\ \vec{b}_4 \\ \vec{b}_5 \\ \vec{b}_6 \\ \vec{b}_7 \end{pmatrix}.$$

12.4. Probabilistic Constraints

While a great advantage of the recourse approach is that all constraints are met, it can be excessive. Imagine that 99.99% of the time each city requires 1 million barrels, but .01% of the time one city needs an additional trillion barrels. Going through the algebra of the recourse method, we see that this low probability case has a significant impact on the solution, as we must make sure that in every situation the demands are met. What if, instead, we are willing to risk occasionally having some constraints fail?

Suppose that some of our constraints *must* hold true in any scenario. For example, in the Oil Problem we must always satisfy the supply constraints, as we cannot get more oil from one place than is available. We include these constraints as we did in (12.1). On the other hand, we might be willing to have a demand shortfall perhaps 5% of the time in each city, and would satisfy demand in the remaining 95% of scenarios. We call this the **probabilistic constraint** method, and describe how we handle this change.

We create binary indicator variables $t_{j,s}$ for each city j and scenario s such that

$$t_{j,s} = \begin{cases} 1 & \text{if demand is satisfied in city } j \text{ in scenario } s, \\ 0 & \text{otherwise.} \end{cases}$$

Then, for each city j, we would want

$$\sum_{s=1}^{S} t_{j,s} \geq 0.95S;$$

this ensures that demand is met in at least 95% of the cities. Subject to these constraints, we then minimize the standard objective function $\vec{c}^T \vec{x}$, the total shipping cost.

This approach is often more appealing than the recourse approach, particularly if we are in reality willing to violate some of the constraints a small percentage of the time. However, we should bear in mind that these constraints do not take into account the *extent* of the failure to meet any constraint. In the context of the Oil Problem, we would be indifferent between a small shortfall in demand in some scenario, and an extremely large shortfall. The recourse approach, in contrast, *does* account for the extent of shortfalls through the recourse costs, since a large shortfall requires us to ship a greater recourse quantity at a higher cost. In this case, we can think of the recourse costs $\vec{r}_s$ as being the penalty per unit shortfall in demand, rather than the recourse cost of shipping. In addition to this, using probabilistic constraints means that we now must solve an integer linear programming problem, which, when combined with a large number of scenarios to consider, can become computationally difficult to solve.

12.5. Exercises

Deterministic and Stochastic Oil Problems

Exercise 12.5.1. *Set up constraints for the movie theater problem where there are three options for the demands of the movies, occurring with probabilities p_1, p_2, and p_3.*

Exercise 12.5.2. *Set up constraints for an airline where there are n options for the demands of flying from city i to j, occurring with probabilities $p_1, p_2, \ldots, p_n$,*

Expected Value approach

For the next two problems, choose reasonable values for the parameters and solve.

Exercise 12.5.3. *Solve the problem from Exercise 12.5.1 using the Expected Value approach.*

Exercise 12.5.4. *Solve the problem from Exercise 12.5.2 using the Expected Value approach.*

Recourse Approach

Exercise 12.5.5. *Formulate the recourse approach to the Oil Problem when there are costs for storage.*

Exercise 12.5.6 (Alternative Objective Functions)**.** *While the standard objective function in recourse problems is the expected total cost, other choices might be more relevant to specific applications. One possibility is to minimize the maximum total cost that could occur in any scenario:*

$$T_{\max} = \max_s \{\vec{c}^T \vec{x} + \vec{r}_s^T \vec{y}_s\}.$$

Thus we would seek to minimize $T_{\max}$ in the Linear Programming problem. Investigate the difference in answers between this and the original approach for the Oil Problem with plausible numbers.

Exercise 12.5.7 (**Network flows**)**.** *Another possible refinement of the Oil Problem would be to allow oil to be shipped between cities. One way to code this is not to explicitly differentiate between refineries and cities, and consider these to be N nodes in a network, rather than R refineries and C cities. We would require the $N(N-1)/2$ costs of shipping between any two nodes, and would need to solve for the $N(N-1)/2$ quantities to ship between any two nodes.*

Probabilistic Constraints

Exercise 12.5.8. *Solve the Oil Problem from this chapter under the additional constraint that the demand is met in at least 95% of the cities.*

Part 4

Fixed Point Theorems

Chapter 13

Introduction to Fixed Point Theorems

The study of fixed points plays an important role in a variety of problems because a large number of questions can be reformulated to be about fixed points. After introducing the terminology we give some examples, and then briefly review some needed results from real analysis; those results will help us in both this and future chapters. In later chapters we'll explore applications of fixed point theorems to optimization.

13.1. Definitions and Uses

Definitions.

We begin with some notation. Briefly, given a map $f : S \to S$, a point $x \in S$ is a **fixed point** for f if $f(x) = x$. It turns out that fixed points are extremely useful in attacking a variety of problems, with both theoretical and implementation applications.

To put this concept in context with material you've previously seen, think back to matrices acting on vectors. In general $\mathbf{A}\vec{v}$ has a different magnitude and direction than $\vec{v}$. If the direction is the same and only the magnitude changes (so $\mathbf{A}\vec{v} = \lambda\vec{v}$ for some λ), we say $\vec{v}$ is an **eigenvector** of $\mathbf{A}$ with **eigenvalue** λ. We've already seen how useful eigenvectors and eigenvalues are, ranging from §1.6 with an application to Fibonacci numbers to §10.2 (and the subsequent exercises on positive definite matrices) to generalizations of the second derivative test to find extrema. Thus, just as eigenvectors have many uses, it's reasonable to suspect fixed points will as well.

Like much of mathematics, results here come in two flavors: constructive and non-constructive. What this means is that some Fixed Point Theorems come with not just a statement that a fixed point exists but also a way to find one; other results merely assert the existence of such a point but provide no way to find it. Sometimes, of course, it may not be immediately clear which type of result we have.

For example, think back to Euclid's proof of the infinitude of primes from §6.5. At first it might appear to be a second flavor type of result, but if we inspect the proof we find it either generates a new prime or a number divisible by a prime not on our list. This led to the fascinating Euclid-Mullin sequence. The exercises for this section sketch some beautiful non-constructive proofs of the infinitude of primes though a quick introduction to the **Riemann zeta function**:

$$\zeta(s) := \sum_{n=1}^{\infty} \frac{1}{n^s} \quad \text{for} \quad \mathrm{Re}(s) > 1.$$

Uses of fixed points.

So why are fixed point problems important? We give two uses. The first is that frequently we have some function f, and iterating it yields the dynamics of a system. Think of this as a deterministic process, where f takes us from one state to another. Thus, if we start at some point x_0, then $x_1 = f(x_0)$, $x_2 = f(x_1)$, and in general $x_{n+1} = f(x_n)$. When there is no cause for confusion we often write $f^2(x)$ for $f(f(x))$; when there is a danger of misinterpretation we write $f^{\circ 2}(x) = f(f(x))$ with of course f^n and $f^{\circ n}$ defined similarly. We discuss two important examples in the next section, the $3x + 1$ map and the Mandelbrot set.

For the second use, we consider one of the most general and important problems in mathematics: given a function f and a value a, find all inputs x sent to a under f; equivalently, solve $f(x) = a$. While there are no dearth of applications to this problem, it will be very hard, if not impossible, to say much about this due to how generally it's phrased. Our hope is that if we restrict to nice f and a, then we can make significant progress.

Of course, while the above is an important problem, as stated it's not a fixed point problem. Fortunately, it's trivial to convert $f(x) = a$ to a fixed point problem. We want a function $g(x)$, where g depends on f and a, such that if $g(x) = x$, then $f(x) = a$. If $f(x) = a$, then by adding x to both sides we get $f(x) + x = a + x$. Subtracting a from both sides gives

$$\text{if } f(x) = a \quad \text{then} \quad f(x) + x - a = x.$$

Thus if

$$g(x) := f(x) + x - a,$$

then

$$g(x) = x \quad \text{if and only if} \quad f(x) = a,$$

and we have recast our original problem as a fixed point one. In other words, *if* we can find fixed points, *then* we can solve equations! This is an example of one of the most important techniques in mathematics: *adding zero*. While it may seem as if we've done nothing, we've recast the algebra to a more useful way and now a large class of problems can be attacked similarly. Notice how similar this is to our work on the canonical Linear Programming problem of Chapter 4.

Unfortunately, several difficult questions remain:

- Given f, does it have a fixed point?

- How many fixed points does f have?
- How do we find the fixed points of f?

As remarked earlier, these questions lead us to constructive versus non-constructive methods. Constructive methods are of course more desirable, but they are often more difficult to come by. It turns out, however, that for fixed points it's often possible to obtain constructive methods.

13.2. Examples

The $3x + 1$ Problem

Consider the $3x + 1$ **map** given by

$$T(x) = \begin{cases} 3x + 1 & \text{if } x \text{ is odd,} \\ x/2 & \text{if } x \text{ is even.} \end{cases}$$

The famous $3x + 1$ **Conjecture** (sometimes called the **Collatz problem** or the **Syracuse problem**) states that no matter what our starting value may be, eventually it iterates to 1. For example,

$$11 \rightarrow 34 \rightarrow 17 \rightarrow 52 \rightarrow 26 \rightarrow 13 \rightarrow 40 \rightarrow 20 \rightarrow 10 \rightarrow 5 \rightarrow$$
$$16 \rightarrow 8 \rightarrow 4 \rightarrow 2 \rightarrow 1,$$

while more interestingly

$$27 \rightarrow 82 \rightarrow 41 \rightarrow 124 \rightarrow 62 \rightarrow 31 \rightarrow 94 \rightarrow 47 \rightarrow 142 \rightarrow$$
$$71 \rightarrow 214 \rightarrow 107 \rightarrow 322 \rightarrow 161 \rightarrow 484 \rightarrow 242 \rightarrow 121 \rightarrow 364 \rightarrow$$
$$182 \rightarrow 91 \rightarrow 274 \rightarrow 137 \rightarrow 412 \rightarrow 206 \rightarrow 103 \rightarrow 310 \rightarrow 155 \rightarrow$$
$$466 \rightarrow 233 \rightarrow 700 \rightarrow 350 \rightarrow 175 \rightarrow 526 \rightarrow 263 \rightarrow 790 \rightarrow 395 \rightarrow$$
$$1186 \rightarrow 593 \rightarrow 1780 \rightarrow 890 \rightarrow 445 \rightarrow 1336 \rightarrow 668 \rightarrow 334 \rightarrow$$
$$167 \rightarrow 502 \rightarrow 251 \rightarrow 754 \rightarrow 377 \rightarrow 1132 \rightarrow 566 \rightarrow 283 \rightarrow$$
$$850 \rightarrow 425 \rightarrow 1276 \rightarrow 638 \rightarrow 319 \rightarrow 958 \rightarrow 479 \rightarrow 1438 \rightarrow$$
$$719 \rightarrow 2158 \rightarrow 1079 \rightarrow 3238 \rightarrow 1619 \rightarrow 4858 \rightarrow 2429 \rightarrow 7288 \rightarrow$$
$$3644 \rightarrow 1822 \rightarrow 911 \rightarrow 2734 \rightarrow 1367 \rightarrow 4102 \rightarrow 2051 \rightarrow 6154 \rightarrow$$
$$3077 \rightarrow 9232 \rightarrow 4616 \rightarrow 2308 \rightarrow 1154 \rightarrow 577 \rightarrow 1732 \rightarrow 866 \rightarrow$$
$$433 \rightarrow 1300 \rightarrow 650 \rightarrow 325 \rightarrow 976 \rightarrow 488 \rightarrow 244 \rightarrow 122 \rightarrow$$
$$61 \rightarrow 184 \rightarrow 92 \rightarrow 46 \rightarrow 23 \rightarrow 70 \rightarrow 35 \rightarrow 106 \rightarrow 53 \rightarrow$$
$$160 \rightarrow 80 \rightarrow 40 \rightarrow 20 \rightarrow 10 \rightarrow 5 \rightarrow 16 \rightarrow 8 \rightarrow 4 \rightarrow 2 \rightarrow 1.$$

The problem has resisted numerous attempts at proofs. Kakutani described it as a Soviet a conspiracy to slow down American mathematical research because of all the time people spent on it; Erdös went further and said mathematics is not yet ready for such problems! See [**Lag1, Lag2**] for more on this problem, and [**KonMi, LagSo**] for connections between this function and Benford's law of digit bias.

The idea of a Fixed Point Theorem is, given f, find x such that $f(x) = x$. For a dynamical system (such as the $3x + 1$ map), functions are used to describe the evolution of a system:

$$x \rightarrow f(x) \rightarrow f(f(x)) \rightarrow f(f(f(x))) \cdots.$$

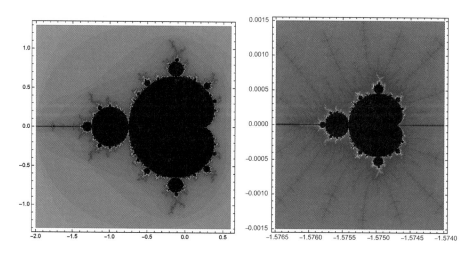

Figure 1. Plot of the Mandelbrot set (from Mathematica). The points in black are in the set, and the color of the points indicates how rapidly the absolute value of the iterates grow. The right is the canonical box; the left is zoomed in to center -1.57525.

If x is a fixed point, then the system remains static, as each iteration of the function yields the same output. One interpretation of the $3x + 1$ conjecture is that the unique fixed point 1 is strongly attractive; in other words, all other points iterate to it.

The Mandlebrot Set

For each complex number c we define a function $f_c : \mathbb{C} \to \mathbb{C}$ by $f_c(z) = z^2 + c$. Starting with 0 as our seed, we keep iterating:

$$0 \ \to \ f(c) = c \ \to \ f(f(c)) = c^2 + c \ \to \ f(f(f(c))) = (c^2 + c)^2 + c^2 + c \ \to \ \cdots.$$

The **Mandelbrot set** is defined to be the complex numbers c for which the iterates of 0 under f_c remain bounded. We plot the set in Figure 1. Entire courses are devoted to this and related sets; what follows is meant to whet the appetite, and will be explored a bit further when we discuss Newton's Method in §13.5.

For a fixed value of c, what are the fixed points of f_c? We want $f_c(z) = z$, or

$$z^2 + c \ = \ z.$$

This is equivalent to solving the quadratic

$$z^2 - z + c \ = \ 0,$$

which has roots

$$z \ = \ \frac{1 \pm \sqrt{1 - 4c}}{2}.$$

13.3. Real Analysis Preliminaries

We reviewed many of the results and terminology we need from real analysis in §4.1; below we quickly discuss two additional items.

The first is the definition of a limit. Informally speaking, we want to say that $x_n \to x$ as $n \to \infty$ if $|x_n - x| \to 0$. We're following standard convention and using the absolute value to denote magnitude. Thus, if x_n is a vector, say $x_n = (a_n, b_n, c_n)$, and we expect it to converge to $x = (a, b, c)$, then a common choice is to take

$$|x_n - x| = \sqrt{(a_n - a)^2 + (b_n - b)^2 + (c_n - c)^2}.$$

We now give the formal definition of limits of sequences.

Definition 13.3.1. *We say $x_n \to x$ as $n \to \infty$ if for all $\epsilon > 0$, there exists N such that for $n \geq N$, we have $|x_n - x| < \epsilon$. We call x the **limit of the sequence** x_n, and say that this sequence **converges** to x.*

Of note is the fact that this limit is unique. This is relatively intuitive: as a sequence gets arbitrarily close to one point, it necessarily moves a certain distance away from all others.

It turns out that we can show that a sequence converges even if we don't know its limit. This is an exceptionally useful result, as in many problems it's difficult to write down the limit exactly.

Theorem 13.3.2 (Cauchy Criterion). *If for all $\epsilon > 0$ there exists N such that whenever $n, m > N$ we have $|x_n - x_m| < \epsilon$, then the sequence x_n converges.*

While this is a useful result, note that it does not tell us what the limit point is. In practice, however, we can use this to home in on the fixed point in many cases.

Before we do this, there is one final result from real analysis we need to look at: the convergence of geometric series. As this is essential for our discussion of contraction maps, we give the standard proof here, and sketch an alternative one in the exercises which illustrates a powerful concept: the notion of a memoryless process.

Theorem 13.3.3 (Geometric Series Formula). *Let $a_n = ar^n$. Then if $r < 1$ we have*

$$\sum_{n=0}^{\infty} a_n = \frac{a}{1-r}.$$

Proof. Let $S_n = \sum_{k=0}^{n} a_k$. Then

$$S_n = a + ar + ar^2 + \cdots + ar^n,$$
$$rS_n = ra + ar^2 + ar^3 + \cdots + ar^{n+1},$$
$$(1-r)S_n = a - ar^{n+1}$$
$$\implies S_n = \frac{a}{1-r} - a\frac{r^{n+1}}{1-r}.$$

If $r < 1$, then $\lim_{n\to\infty} r^{n+1} = 0$, and thus

$$\sum_{n=0}^{\infty} a_n = \lim_{n\to\infty} S_n = \frac{a}{1-r},$$

completing the proof. $\square$

13.4. One-Dimensional Fixed Point Theorem

If $f : [0,1] \to [0,1]$ is "nice", the existence of a fixed point is remarkably easy. The heavy lifting is done by the following result, which you should have seen in calculus or real analysis.

Theorem 13.4.1 (Intermediate Value Theorem (IVT)). *If f is continuous on $[a,b]$ and C is between $f(a)$ and $f(b)$, then there exists a $c \in [a,b]$ such that $f(c) = C$.*

The Intermediate Value Theorem states that if we vary continuously, we cannot get from $f(a)$ to $f(b)$ without passing through all intermediate values. In other words, if we're walking from e to π we must pass through 3; there's no way to continuously 'jump'. While the result is intuitively clear, for completeness we sketch a proof.

Proof. Without loss of generality we assume $f(a) < C < f(b)$. Let x_1 be the midpoint of $[a,b]$ and look at $f(x_1)$. If that equals C we are done, else it's either less than or greater than C. If $f(x_1) < C$ we look at the interval $[x_1, b]$, while if $f(x_1) > C$ we look at the interval $[a, x_1]$.

Notice in both cases we get a new interval, call it $[a_1, b_1]$, which is half the size of the previous and $f(a_1) < C < f(b_1)$. We continue in this manner, repeatedly taking the midpoint and looking at the appropriate half-interval. This yields sequence of points $\{a_n\}$, $\{x_n\}$, and $\{b_n\}$.

If there is an n with $f(x_n) = C$, we are done. If not, we consider our sequence of sub-intervals $[a_n, b_n]$. Notice each is contained in the previous and of half the length. At this point we require some results from real analysis to finish the proof. As $\{a_n\}$ is non-decreasing, $\{b_n\}$ is non-increasing, and $\lim |b_n - a_n| = 0$, we have a_n and b_n converge to a common value, which we denote by c. As f is continuous, since $f(a_n) < C < f(b_n)$ we must have

$$f(c) \;=\; f(\lim_{n \to \infty} a_n) \;=\; \lim_{n \to \infty} f(a_n) \;=\; \lim_{n \to \infty} f(b_n) \;=\; C.$$

$\square$

We give an example in Figure 2 of a function $f : [0,1] \to [0,1]$ with $f(0) = 0$, $f(1) \approx 0.994604$, and $C = \pi/6 \approx 0.523599$. As predicted, there is a choice of c such that $f(c) = C$; however, it's not easy to find an exact value for c (though we can easily determine how many such c there are).

This illustrates the greatest difficulty with using the Intermediate Value Theorem. It's a wonderful theoretical tool and asserts the *existence* of a point with desired properties, but it does not use an explicit *construction* for that point. Note, however, that our method of proof *can* be used to create a sequence of points converging to our point.

Armed with this result, we prove our first Fixed Point Theorem.

Theorem 13.4.2. *Let f be a continuous function with $f : [0,1] \to [0,1]$. Then there exists an x such that $f(x) = x$; in other words, f has a fixed point.*

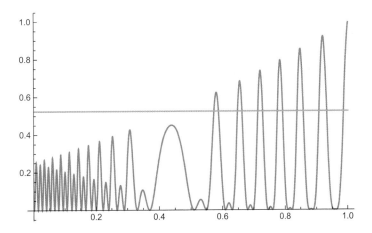

Figure 2. Plot of the function $f(x) = \left(x + \sin\left(x^2 + x\exp(5.5\cos x)\right)\right)^2/4$ with $a = 0$, $b = 1$ and $C = \pi/6$.

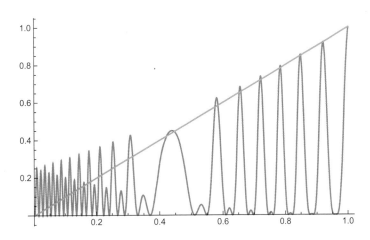

Figure 3. Fixed points for the function $f(x) = \left(x + \sin\left(x^2 + x\exp(5.5\cos x)\right)\right)^2/4$ on $[0,1]$.

Proof. Without loss of generality we may assume $f(0) > 0$ and $f(1) < 1$, as if either of these conditions failed to hold the result would be immediate.

Set $g(x) = f(x) - x$. We see that $g(0) > 0$ and $g(1) < 0$, so by the Intermediate Value Theorem there exists a c such that $g(c) = 0$. Thus $f(c) = c$, and c is our fixed point. $\qquad\square$

Our function from Figure 2 satisfies the conditions of Theorem 13.4.2, as it maps $[0,1]$ to $[0,1]$. It's clearly positive as the numerator is a square. To see that it's at most 1, notice that the sine term is between -1 and 1, and thus the numerator is the square of a number between -1 and 2, and thus when divided by 4 it's at most 1. See Figure 3 for a plot indicating the location of its fixed points.

13.5. Newton's Method versus Divide and Conquer

In Theorem 13.4.2 we not only proved the existence of a fixed point (for a continuous $f : [0,1] \to [0,1]$) but also gave a procedure to find it. The idea of splitting our interval in half and then confining the analysis to one of the two parts is a powerful technique, and it is used in a variety of problems (see for example the dyadic decompositions in Exercises 1.7.21 and 1.7.22). We often use the phrase **Divide and Conquer** to refer to such approaches.

Let's revisit the ideas from §13.4 to find a root of a continuous function which is positive at 0 and negative at 1. By the Intermediate Value Theorem there must be a root somewhere between these two points. The method consists of splitting the interval in half and looking at the sign of the function at the point $1/2$. If it's positive, then we know a root occurs between $1/2$ and 1; if it's negative, we know there's a root between 0 and $1/2$. We repeat this method in the new interval as many times as we wish to keep improving our estimate by a factor of 2 for each iteration. This means that every 10 iterations, we gain approximately 3 decimal digits of accuracy (we have shrunk the original interval by a factor of 2^{10}, and $2^{10} = 1024 \approx 1000$). Thus if 10 iterations gives us three decimal digits, we would need to do another 10 to get our answer to six places.

We illustrate this method applied to finding $\sqrt{3}$. Let $g(x) = x^2 - 3$, and let's search for the root between 1 and 2. We denote the interval in the nth stage by $[a_n, b_n]$. We have

$$
\begin{aligned}
[a_1, b_1] &= [1.000, 2.000] \text{ as } g(1.000) = -2.00 < 0 \text{ and } g(2.000) = 1.0000 > 0, \\
[a_2, b_2] &= [1.500, 2.000] \text{ as } g(1.500) = -0.75 < 0 \text{ and } g(2.000) = 1.0000 > 0, \\
[a_3, b_3] &= [1.500, 1.750] \text{ as } g(1.500) = -0.75 < 0 \text{ and } g(1.750) = 0.0625 > 0, \\
[a_4, b_4] &= [1.625, 1.750] \text{ as } g(1.625) \approx -0.36 < 0 \text{ and } g(1.750) = 0.0625 > 0,
\end{aligned}
$$

and thus after four iterations we know $\sqrt{3} \in [1.625, 1.750]$. The true value is approximately 1.73205, and thus we have yet to capture the first decimal digit!

While Divide and Conquer is a decent method of getting accurate estimates for the roots of a function, we can do significantly better in many cases by using one of the most efficient methods in mathematics: **Newton's Method**. The reason Newton's Method trumps Divide and Conquer is that it uses more information about f; whereas Divide and Conquer only assumes continuity of f, Newton's Method will assume f to be continuously differentiable. The drawback, of course, is that Newton's Method is not applicable to as many functions as Divide and Conquer; however, it shouldn't be too surprising that by assuming more we are able to do more.

Say we are trying to find an estimate for the value of $\sqrt{3}$. We've seen that this is equivalent to finding the positive root of $f(x) = x^2 - 3$. To illustrate how powerful the method is we start with a decent but not outstanding initial guess for the solution to $f(x) = 0$, namely $x_0 = 2$. We now have the point $(2, f(2))$, as

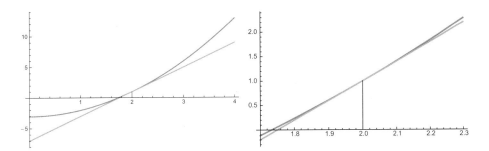

Figure 4. The first step in applying Newton's Method to find $\sqrt{3}$; the tangent line from our initial guess already closely approximates the root.

well as the slope of the tangent line $f'(2)$. Taking this tangent line to be a good approximation for f near the root, we can find where the tangent line

$$y - f(2) \;=\; f'(2)(x - 2)$$

intersects the x-axis (see Figure 4). We now have a new starting estimate, $x_1 = 7/4$, and we can repeat this same process as many times as we wish.

We can write these equations out more formally. At the nth stage we have the point $(x_n, f(x_n))$ and the slope $f'(x_n)$. The tangent line to f at x_n is

$$y - f(x_n) \;=\; f'(x_n)(x - x_n).$$

We want to find the x-intercept of this line, which happens when $y = 0$. This gives us our x_{n+1}, which is obtained by solving

$$-f(x_n) \;=\; f'(x_n)(x_{n+1} - x_n)$$

or

$$x_{n+1} \;=\; x_n - \frac{f(x_n)}{f'(x_n)}.$$

Notice we argued very generally so our result holds for all nice f; for $f(x) = x^2 - \alpha$ we would find

$$x_{n+1} \;=\; x_n - \frac{x_n^2 - \alpha}{2x_n} \;=\; \frac{1}{2}\left(x_n + \frac{\alpha}{x_n}\right).$$

Newton's Method converges *significantly* faster than Divide and Conquer in finding square roots. While it requires ten additional iterations of Divide and Conquer to gain just three more digits, every iteration of Newton's Method essentially doubles the number of digits (see Exercise 13.7.57). We show the amazing accuracy in Table 1.

Thus, not only do we have a rapidly converging sequence, but at each point we have a rational approximation to $\sqrt{3}$.

We end with two remarks connecting the above to other topics of the course. First, this is a chapter on fixed points. While we showed that we can recast solving equations as fixed point problems, there is also a nice explicit connection here. If we set

$$g(x) \;=\; \frac{1}{2}\left(x + \frac{\alpha}{x}\right),$$

n	x_n	x_n (approx)	$\lvert\sqrt{3}-x_n\rvert$
0	2	2	0.267949
1	$\frac{7}{4}$	1.75	0.017949
2	$\frac{97}{56}$	1.73214285714285714285714285714	0.000092
3	$\frac{18817}{10864}$	1.73205081001472754050073637702	$2.4 \cdot 10^{-9}$
4	$\frac{708158977}{408855776}$	1.73205080756887729525435394607	$1.7 \cdot 10^{-18}$
5	$\frac{1002978273411373057}{579069776145402304}$	1.73205080756887729352744634150	$8.6 \cdot 10^{-37}$

Table 1. First five iterations of Newton's Method to estimate $\sqrt{3}$, which is approximately 1.73205080756887729352744634150586723669428.

then the positive fixed point of $g(x)$ is $x = \sqrt{\alpha}$, precisely the point we are trying to find.

The second item connects Newton's Methods to cousins of the Mandelbrot set. By the Fundamental Theorem of Algebra, a polynomial of degree n has exactly n roots. Unfortunately the real numbers are not complete, and a polynomial with real coefficients can have complex roots; however, all the roots of a polynomial with complex coefficients are complex (equivalently, once we add $\sqrt{-1}$, the solution to $x^2 + 1 = 0$, there is no need to add anything further).

One of the most important questions we should ask, when applying Newton's Method, is whether or not it converges for a given starting seed. Once we know it converges, it's then natural to ask whether or not it converges to one of the roots, and if so which root. If we have a polynomial of degree n we can assign a different color to each root and then color each point in the complex plane based on what root it converges to (reserving one color for starting values which do not iterate to a root); the resulting object is called the **Newton fractal** associated to the polynomial. In Figure 5 we perform this analysis for the function $f(z) = z^3 - 1$ (we have switched to using z for the variable to emphasize the necessity of looking at this problem in the complex plane).

13.6. Equivalent Regions and Fixed Points

Frequently in mathematics we prove a result for one object and then show that it holds for all equivalent objects. For example, often proofs begin with the phrase "without loss of generality", where we make some simplifications and argue that it suffices to study one special case. Our goal here is to discuss how to transfer fixed point results in one setting to another.

To do so, we first review some notation. A function $f : S \to T$ is **one-to-one** (or **injective**) if $f(s_1) = f(s_2)$ if and only if $s_1 = s_2$, and it is **onto** (or **surjective**)

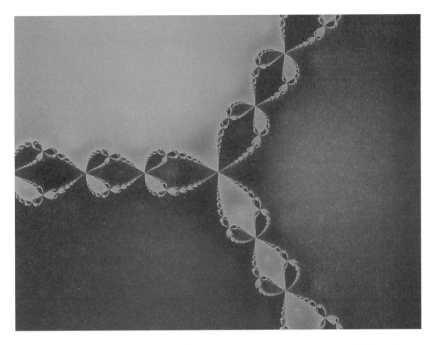

Figure 5. Newton fractal for z^3-1. Image from user Vibhijain via Wikimedia Commons.

if for every $t \in T$ there is an $s \in S$ such that $f(s) = T$. Thus, for injective functions, distinct inputs go to distinct outputs, while for onto functions every possible output is realized. A function that is both injective and surjective is said to be **bijective**.

If $g : S \to T$ is a bijection we will say S and T are **equivalent** under g. This leads to a huge subject: when are two spaces equivalent? For this to be a meaningful study, however, we should put some restrictions on g. For example, if g is not required to be continuous, then so long as S and T have the same cardinality they would be equivalent. Therefore, we assume our g is at the very least continuous, and we have entered the realm of **topology**, the study of properties of spaces that are preserved under continuous transformations. Note, though, that we are assuming more than just g is continuous; we're also assuming g is bijective. If we took g to be a constant function it would be continuous, but many properties of S would not be preserved under g.

We could of course require g to have additional properties. Maybe we want g to be differentiable, maybe we want g to be infinitely differentiable, maybe we want g to be complex differentiable. Not surprisingly, the more properties we assert for g, the fewer regions are equivalent. In later chapters we'll frequently study maps from a triangle (or higher-dimensional analogue, a simplex), and thus we might want to study regions equivalent to a triangle. Differentiability will be hard with a triangle at the corners. Thus maybe what we should do is lower our expectations and insist on differentiability in the interior of the triangle and the set. One of the major results of complex analysis, the **Riemann Mapping Theorem**, is that so long as an open set is a proper subset of the plane and is simply connected (i.e., it

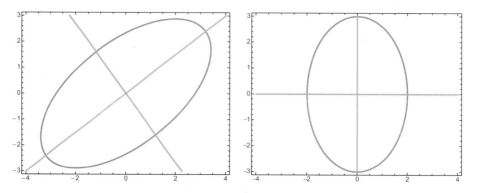

Figure 6. Left: The ellipse $\dfrac{13}{100}x^2 - \dfrac{9}{50}xy + \dfrac{73}{400}y^2 = 1$. Right: The ellipse $(u/2)^2 + (v/4)^2 = 1$.

does not have any holes in it), then there is an infinitely complex differentiable map from it to the unit disk; furthermore, in many situations the map can be extended continuously to the boundary. See [**La, Rud2, SS**].

Let's do an example to show the utility of equivalent regions; let's study ellipses. In §3.1 and Exercise 3.7.1 we saw we could write the equation of the ellipse (whose major and minor axes are parallel to the coordinate axes and whose center is the origin) $(x/a)^2 + (y/b)^2 = 1$ in matrix form:

$$(x \ \ y) \begin{pmatrix} 1/a^2 & 0 \\ 0 & 1/b^2 \end{pmatrix} \begin{pmatrix} x \\ y \end{pmatrix} = 1.$$

What if the ellipse axes are not parallel to the coordinate axes? For example, consider the ellipse

$$(x \ \ y) \begin{pmatrix} 13/100 & -9/100 \\ -9/100 & 73/400 \end{pmatrix} \begin{pmatrix} x \\ y \end{pmatrix} = \frac{13}{100}x^2 - \frac{9}{50}xy + \frac{73}{400}y^2 = 1,$$

which we show in Figure 6.

The major and minor axes can be computed (see Exercise 13.7.69), but their equations are significantly less transparent than when the ellipse is aligned with the coordinate axes. We can easily pass back and forth between this ellipse and the rotated one whose axes align with the coordinate axes by rotation. Explicitly, consider

$$\begin{pmatrix} x \\ y \end{pmatrix} = \begin{pmatrix} 3/5 & 4/5 \\ -4/5 & 3/5 \end{pmatrix} \begin{pmatrix} u \\ v \end{pmatrix};$$

the equation of the ellipses in xy-space,

$$\frac{13}{100}x^2 - \frac{9}{50}xy + \frac{73}{400}y^2 = 1,$$

becomes in uv-space

$$(u/2)^2 + (v/4)^2 = 1.$$

We end with an important result which substantiates all our time just spent on equivalence.

Theorem 13.6.1. *Let T and S be equivalent under a continuous bijection $g : T \to S$. If every continuous function from T to T has a fixed point, then every continuous function from S to S has a fixed point.*

Proof. This proof illustrates a great idea in mathematics, the idea of **transfer**. If you've never seen transfer in mathematics, you've almost surely seen it if you've ever listened to a politician answer a question. In a presidential debate, when a candidate is asked a question, first they kind of meander for 30 seconds and then they shift into their prepared remarks on something that's vaguely related to the question that was just asked. They transfer from what they were given to something they already know. That's exactly what we want to do here.[1] You want to know something about S, and you know everything there is to know about T. We want to transfer from S to something about T.

Thus, given some function $f : S \to S$, we must show it has a fixed point. We want to find a function related to f that lives on T, and thus it would have to take inputs in T and give outputs in T. There's a natural way to get from T to S; we use our bijection $g : T \to S$ and its inverse $g^{-1} : S \to T$ (see Exercise 13.7.64). We thus consider

$$h = g^{-1} \circ f \circ g : T \to T,$$

which means g^{-1} composed with f composed with g (see Exercise 13.7.68 on associativity, which proves the order of the parentheses do not matter; note how similar this is to the ellipse example); composing by the bijection g on the right and its inverse on the left is called **conjugation** (by g), and it is a terrific way to transfer from questions on S to questions on T.

As we assumed all continuous functions from T to T have a fixed point, then $h = g^{-1} \circ f \circ g$ has a fixed point which we denote by t^*. (Stars are often used to denote fixed points; another good choice is to write t_f to highlight the connection with f, but that becomes a bit harder to read.) Thus

$$h(t^*) = t^* \quad \text{or} \quad (g^{-1} \circ f \circ g)(t^*) = t^*.$$

Our goal is to show that f has a fixed point, which we now do by applying g to both sides of $g^{-1}(f(g(t^*))) = t^*$. By associativity of convolution, and noting $(g \circ g^{-1})$ is the identity, we obtain

$$f(g(t^*)) = g(t^*);$$

it's now trivial to notice that $g(t^*)$ is a fixed point of f, completing the proof. $\square$

13.7. Exercises

Definitions and Uses

Hint: if you are having trouble finding fixed points, are there any natural candidates, which are suggested by the form of the function, worthy of trying?

Exercise 13.7.1. *Find all fixed points for the function $f : [0,1] \to [0,1]$ given by $f(x) = x(1-x)$.*

Exercise 13.7.2. *Find all fixed points for the function $f : [0,1]^2 \to [0,1]^2$ given by $f(x,y) = (xy, (x^2 + y^2)/2)$.*

---

[1]Except we'll also answer the original question.

Exercise 13.7.3. *Find all fixed points for the function* $f : [0,1]^2 \to [0,1]^2$ *given by* $f(x,y) = (\sqrt{xy}, (x^2 + y^2)/2)$; *is there an interpretation for such points?*

Exercise 13.7.4. *Find all fixed points for the function* $f : [0,1]^3 \to [0,1]^3$ *given by* $f(x,y) = ((xyz)^{1/3}, \sqrt{(xy + yz + xz)/3}, (x + y+)/3)$; *is there an interpretation for such points?*

The next few problems involve the **Riemann zeta function**, given for $\mathrm{Re}(s) > 1$ by

$$\zeta(s) = \sum_{n=1}^{\infty} \frac{1}{n^s}.$$

It will be useful to recall the Fundamental Theorem of Arithmetic (see Exercise 2.6.59), which states every integer can be written uniquely as a product of prime powers (where the primes are listed in increasing order).

Exercise 13.7.5. *Prove the **geometric series formula**: if* $|r| < 1$, *then*

$$1 + r + r^2 + r^3 + \cdots = \frac{1}{1 - r}.$$

Exercise 13.7.6. *Prove the **Euler Product** representation of the zeta function:*

$$\zeta(s) = \prod_{p \text{ prime}} \left(1 - \frac{1}{p^s}\right)^{-1}.$$

Hint: use the Fundamental Theorem of Arithmetic and the geometric series formula. *This problem highlights the value of the zeta function; it connects the well-understood integers with the desired primes.*

The next two problems are **special value proofs** of the infinitude of primes; note these proofs show there must be infinitely many primes *but* they give no information as to what integers are prime.

Exercise 13.7.7. *Prove there must be infinitely many primes by noting that as* s *decreases to 1 from above,* $\zeta(s)$ *diverges to infinity.*

Exercise 13.7.8. *Euler proved that*

$$\sum_{n=1}^{\infty} \frac{1}{n^2} = \frac{\pi^2}{6};$$

*this is known as the **Basel problem**. Use this result to prove there are infinitely many primes.*

Examples

The set of problems below involve the dynamics of processes related to the $3x + 1$ map.

Exercise 13.7.9. *Consider the* $3x + 1$ *map given; prove the only positive integer fixed point below a million is 1.*

Exercise 13.7.10. *A more natural map to study is the map T_2, defined on the odd numbers, given by*
$$T_2(x) \; = \; \frac{3x+1}{2^k},$$
where k is the largest power of 2 dividing $3x+1$. Prove that the only positive integer fixed point below a million is 1.

Exercise 13.7.11. *Consider the $3x - 1$ map, defined on odd integers, by*
$$T_{-2}(x) \; = \; \frac{3x-1}{2^k},$$
with k the largest power of 2 dividing $3x - 1$. Find all integer fixed points whose absolute value is at most a million. What do you conjecture happens as we iterate this map?

One of the challenges in the computations above is quickly checking each integer; while the calculation isn't too bad for a given number, there are a lot to check. Imagine if instead of determining whether or not an integer is a fixed point of T_2 we instead want to know if it eventually iterates to 1; in other words, is the $3x + 1$ conjecture true for these inputs? The brute force approach is to check each number from 1 to N and see if it eventually iterates to 1. As this is quite time-consuming, and as a point of this book is to highlight efficient approaches, we sketch a better method below.

Let $I_N := \{1, 3, 5, \ldots, N\}$ (for an odd number N) and set $a(n) = \infty$ initially for $n \in I_N$. We update our assignments as follows. We set $a(1) = 1$ as $T_2(1) = 1$, and if $a(2k+1)$ has been updated for all $k \le n-1$ we adjust the value of $a(2n+1)$ to be the first odd integer in iterates of $2n + 1$ by T_2 which is smaller than $2n + 1$. Thus $a(3) = 1$ as $3 \to 5 \to 1$, $a(5) = 1$ as $5 \to 1$, $a(7) = 5$ as $7 \to 11 \to 17 \to 13 \to 5$, and so on.

Exercise 13.7.12. *Prove the assignment above is well defined if the $3x + 1$ conjecture is true.*

Exercise 13.7.13. *Prove the above assignment is faster than the brute force verification as it eliminates our need to perform the same computations multiple times.*

Exercise 13.7.14. *Our verification procedure can be made more efficient. Notice that when we computed $a(7)$ we dealt with iterations with input 7, 11, 17, and 13. Design a better method where we use this chain and not only assign a new value to $a(7)$ but also assign values to $a(11)$, $a(17)$, and $a(13)$.*

Exercise 13.7.15. *Modify the above method to efficiently calculate how long it takes to iterate to 1 for all odd starting seeds from 1 to a million for the $3x + 1$ problem. Can you do all odd seeds up to a billion? Up to a trillion?*

Exercise 13.7.16. *While most integer systems cannot handle numbers with millions, billions, or trillions of digits, it's easy to do so through lists. The $3x+1$ map only involves addition, multiplication, and division; show how to code this problem using a large list to represent numbers and iterates.*

Exercise 13.7.17. *In the problem above we need to add 1, multiply by 3, and divide by 2. If we are working with lists it would be much easier to divide by 10, as* *that can be coded by removing the rightmost digit and shifting. Thus if $\{1, 7, 0, 1, 0\}$ represents 17,010, to divide by 10 we just drop the rightmost digit to get $\{1, 7, 0, 1\}$. Show that we can replace division by 2 with multiplication by c followed by division by 10 for some c, and determine c.*

We end with a series of related problems investigating certain features of the $3x + 1$ problem. For a_n odd let $a_{n+1} = g(a_n) = (3a_n + 1)/2^\ell$, with $2^\ell \| 3a_n + 1$, be the $3x + 1$ map. If we apply g to a starting seed a_0 three times and pull out ℓ_1 powers of 2 the first time, ℓ_2 the second, and ℓ_3 the third time, we say the 3-path of g is (ℓ_1, ℓ_2, ℓ_3); if we were to apply g a total of m times the m-path would be $(\ell_1, \ell_2, \ldots, \ell_m)$. Below all starting seeds are *odd*. See [**KonSi, Si**] for solutions to the first three questions.

Exercise 13.7.18. *Find all starting seeds a_0 such that the 1-path is (1), such that the 1-path is (2), and such that the 1-path is (ℓ). What fraction of all starting seeds less than X (as $X \to \infty$) has 1-path (ℓ)?*

Exercise 13.7.19. *Find all starting seeds a_0 such that the 2-path is $(1, 1)$, such that the 2-path is $(1, 2)$, such that the 2-path is (ℓ_1, ℓ_2). What fraction of all starting seeds less than X (as $X \to \infty$) has 2-path (ℓ_1, ℓ_2)?*

Exercise 13.7.20. *Find all starting seeds a_0 such that the m-path is $(\ell_1, \ell_2, \ldots, \ell_m)$. What fraction of starting seeds less than X (as $X \to \infty$) has m-path $(\ell_1, \ell_2, \ldots, \ell_m)$?*

Exercise 13.7.21. *We say the first scale factor is a_1/a_0. Conjecture an average value for a_1/a_0. Run **extensive** computer simulation to estimate this. You must have at least 1,000,000 explorations for a run, and you must investigate numbers at least as large as 10^{100}.*

Exercise 13.7.22. *We say the maximal scale factor is a_n/a_0, where a_n is the largest term in the sequence starting at a_0. (Note that we do not know this quantity is finite!) Conjecture an average value for the maximal scale factor. Run **extensive** computer simulation to estimate this. You must have at least 10,000 explorations for a run, and you must investigate numbers at least as large as 10^{100}.*

We end with some problems on the Mandelbrot set.

Exercise 13.7.23. *To plot the Mandelbrot set we need to evaluate polynomials rapidly; discuss how to use Horner's algorithm from earlier in the book to do this. How much is the savings in evaluating the nth iterate?*

Exercise 13.7.24. *The Mandelbrot set is all c such that the iterates of 0 by $f_c(z) = z^2 + c$ are bounded. Prove there is an r such that if $|c| < r$, then c is in the Mandelbrot set. What is the largest r you can prove?*

Exercise 13.7.25. *The Mandelbrot set is all c such that the iterates of 0 by $f_c(z) = z^2 + c$ are bounded. Prove there is an r such that if $|c| > r$, then c is not in the Mandelbrot set. What is the smallest r you can prove?*

Exercise 13.7.26. *It is very easy to plot the Mandelbrot set in Mathematica; the zoom-in was done by inputting*

```
MandelbrotSetPlot[{-1.5765-.0015I,-1.574+.0015I}, MaxIterations->400]
```

(to get help on any command in Mathematica, type a question mark before it at the prompt; thus to get help on this function type ?MandelbrotSetPlot). Use this command, or search the web, to explore the Mandelbrot set's structure.

Real Analysis Preliminaries

Exercise 13.7.27. *Prove that if $x_n \to x$ and $x_n \to y$, then $x = y$.*

Exercise 13.7.28. *Prove the Cauchy Criterion.*

Exercise 13.7.29. *Let $a_n = r^n$ for some $|r| < 1$, and set $s_n = a_0 + \cdots + a_n$. Use the Cauchy Criterion to show $\{s_n\}$ converges.*

The next few problems involve the **exponential function** $\exp(x)$, often written e^x, by

$$e^x := \sum_{n=0}^{\infty} \frac{x^n}{n!} = 1 + x + \frac{x^2}{2!} + \frac{x^3}{3!} + \cdots .$$

Exercise 13.7.30. *Prove $\exp(x)$ converges for all x.*

Exercise 13.7.31. *Prove $e^{x+y} = e^x e^y$. Warning: this must be proved. If this were false, then our notation would be horrible, but this problem is asking you to show an infinite sum equals the product of two other infinite sums.*

Exercise 13.7.32. *If $x'(t) = ax(t)$, prove the general solution is $x(t) = e^{at}x(0)$, where $x(0)$ is the initial value.*

While you are always strongly encouraged to look at simpler cases to build intuition for the behavior in the more general case, you must always be careful not to assume what happens in one case carries over in another. Sometimes it will, and sometimes it will not. The next few exercises deal with the **matrix exponential**. As you're doing these exercises compare the results to the standard exponential.

In some of the problems below we need the **matrix commutator**, given by

$$[\mathbf{A}, \mathbf{B}] := \mathbf{AB} - \mathbf{BA}.$$

Exercise 13.7.33. *If $\mathbf{A}$ is an $n \times n$ matrix generalize the exponential function to*

$$e^{\mathbf{A}} := \sum_{n=0}^{\infty} \frac{\mathbf{A}^n}{n!} = \mathbf{I} + \mathbf{A} + \frac{\mathbf{A}^2}{2!} + \frac{\mathbf{A}^3}{3!} + \cdots ,$$

where $\mathbf{A}^0 = \mathbf{I}$ (the $n \times n$ identity matrix). Find a good notion of absolute value here, and show $e^{\mathbf{A}}$ converges for any $\mathbf{A}$.

Exercise 13.7.34. *Prove the following properties of the matrix commutator:*

(1) Square matrices $\mathbf{A}, \mathbf{B}$ commute if and only if $[\mathbf{A}, \mathbf{B}]$ is the zero matrix.

(2) If p is any polynomial, then $[p(\mathbf{A}), \mathbf{A}]$ is the zero matrix.

(3) $[\mathbf{A}, \mathbf{B}] = -[\mathbf{B}, \mathbf{A}]$.

(4) $[\mathbf{A}, \mathbf{BC}] = [\mathbf{A}, \mathbf{B}]\mathbf{C} + \mathbf{B}[\mathbf{A}, \mathbf{C}]$.

Exercise 13.7.35. *Prove that if* $[A, B]$ *is the zero matrix (i.e.,* A *and* B *commute), then* $e^{A+B} = e^A e^B$. *One useful special case is when* $B = p(A)$ *for some polynomial* p.

Exercise 13.7.36 (Baker–Campbell–Hausdorff formula). *If* $[A, B]$ *is not the zero matrix (i.e.,* A *and* B *do not commute), then* $e^{A+B} \neq e^A e^B$. *Choose two non-commuting matrices* A, B *and show for your choice that* $e^{A+B} \neq e^A e^B$.

The Baker-Campbell-Hausdorff formula is an explicit formula for the matrix $\log(e^A e^B)$; *note if the matrices commute this equals* $A + B$. *A useful analogue is the* **Zassenhaus** *formula: for* t *a complex number,*

$$e^{t(A+B)} \;=\; e^{tA} e^{tB} e^{-t^2[A,B]/2} e^{t^3(2[B,[A,B]]+[A,[A,B]])/6} \cdots .$$

Exercise 13.7.37. *In the previous problem the parameter* t *was introduced to help see how the different matrix commutators enter the expansion. Note the first terms are linear in* t *and involve just* A *or* B, *the next terms are quadratic in* t *and involve products of two matrices, and the term after that is cubic in* t *and involves products of three matrices. What do you conjecture for the quartic* t *term? What do you think the coefficient of* t^4 *will be? Note it was 1 for the linear case,* $-1/2$ *for the quadratic, and* $1/6$ *for the cubic....*

Exercise 13.7.38. *Let* $\vec{x}(t) = (x_1(t), \ldots, x_n(t))^T$ *(so it's a column vector with* n *components), and let* A *be an* $n \times n$ *matrix. Solve* $\vec{x}'(t) = A\vec{x}(t)$.

Exercise 13.7.39. *Let* $f^{(k)}(t)$ *denote the* k^{th} *derivative of* f. *Assume there are constants* $a_0, \ldots, a_{n-1}$ *such that*

$$f^{(n)}(t) \;=\; a_{n-1} f^{(n-1)}(t) + a_{n-2} f^{(n-2)}(t) + \cdots + a_1 f'(t) + f(t).$$

Show we can find a vector of functions $\vec{x}(t)$ *and a matrix* A *with* $\vec{x}'(t) = A\vec{x}(t)$ *such that we can pass from the solution to this matrix formulation to our original problem.*

Exercise 13.7.40. *Let* A *map the surface of the* n-*sphere to itself; thus if* $\vec{v} = (v_1, \ldots, v_n)$ *with* $v_1^2 + \cdots + v_n^2 = 1$, *then the sum of the squares of the components of* $A\vec{v}$ *is also 1. Must A have a fixed point?*

Below is a sketch of an alternate proof of the geometric series formula. Imagine two players are alternating shooting baskets, where the first one to get a basket wins. The first always makes a basket with probability p, the second with probability q, and all shots are independent. Let $r = (1 - p)(1 - q)$ denote the probability they both miss.

Exercise 13.7.41. *The geometric series formula follows by showing two items:*

- *Show the probability the first person wins on her* n^{th} *shot is* $r^{n-1} p$, *and thus the probability she wins,* x, *is*

$$x \;=\; p(1 + r + r^2 + \cdots) \;=\; p \sum_{n=1}^{\infty} r^{n-1} \;=\; p \sum_{n=0}^{\infty} r^n.$$

- *Show that the probability the first person wins is also*

$$x \;=\; x + rx.$$

Hint: we have a **memoryless process**; once both miss it's as if the game has just begun.

Deduce the geometric series formula.

Exercise 13.7.42. *The previous exercise proves the geometric series formula when $0 \le r < 1$. Can you use this result to extend to $-1 < r < 0$? What about complex r?*

One-Dimensional Fixed Point Theorem

Exercise 13.7.43. *Finish the proof of the Intermediate Value Theorem by rigorously showing that $\lim a_n = \lim b_n$, and if we denote this common value by c, then $f(c) = C$.*

Exercise 13.7.44. *Give an example of a continuous function $f : [0, 1] \to [0, 1]$ such that $f(0) = 0$, $f(1) = 1$ and for any $C \in [0, 1]$ there are exactly 1776 $c \in [0, 1]$ such that $f(c) = C$.*

Exercise 13.7.45. *Fix a positive integer M. Can you redo the previous problem so that there are always exactly M solutions? If not, for what M can you do this?*

Exercise 13.7.46. *For the last two problems: are your functions differentiable? Can you give differentiable examples (and if so, for what M)?*

Exercise 13.7.47. *If we remove the assumption that $f : [0, 1] \to [0, 1]$ must be continuous, are we still assured of the existence of a fixed point? If yes prove it, if no give a counter-example.*

Hint: if you are having trouble finding fixed points, are there any natural candidates, which are suggested by the form of the function, worthy of trying?

Exercise 13.7.48. *Find all fixed points of the **logistic map** $f : [0, 1] \to [0, 1]$ where $f(x) = rx(1-x)$ and r is a positive real number (obviously the existence and location of the fixed points could depend on the value of r). See [**Ma**] for one of the first papers to show the importance of this map.*

Exercise 13.7.49. *Find all fixed points of $f(x) = (1+4x\sin x)/2$; does your answer contradict Theorem 13.4.2?*

Exercise 13.7.50. *Consider the following proposed proof that if we have a continuous $f : [0, 1]^2 \to [0, 1]^2$, then f has a fixed point; either prove the steps are all valid and the result is true or find the mistake.*

Exercise 13.7.51. *We proved that if $f : [0, 1] \to [0, 1]$ is continuous, then f has a fixed point. Let $S_{a,b}$ be the set of all continuous functions from $[0, 1]$ to $[a, b]$. Find all a, b such that every $f \in S_{a,b}$ has a fixed point.*

Exercise 13.7.52. *A function $f : S \to T$ is **onto** (or **surjective**) if for any $t \in T$ there is an $s \in S$ such that $f(s) = t$; in other words, every element of the output space is hit. Redo the previous problem for $\mathcal{S}_{a,b}$, the set of surjective continuous functions from $[0, 1]$ to $[a, b]$. Thus find all a, b such that every $f \in \mathcal{S}_{a,b}$ has a fixed point.*

- *We can write $f(x, y) = (f_1(x, y), f_2(x, y))$ where each $f_i : [0, 1] \to [0, 1]$.*
- *For any fixed y there is an x_y such that $f_1(x_y, y) = x_y$.*
- *For any fixed x there is a y_x such that $_2(x, y_x) = y_x$.*
- *Taking $y = y_x$ in the second step and $x = x_y$ in the third step yields a fixed point, as now $f_1(x_y, y_x) = x_y$ and $f_2(x_y, y_x) = y_x$.*

Newton's Method vs. Divide and Conquer

Exercise 13.7.53. *Use Divide and Conquer to approximate π to 3, 6, and 9 decimal places by solving $\cos(x/4) = \sin(x/4)$ with $0 \le x \le 4$. Note: remember we don't need to evaluate these trig functions exactly; we just need to see which is larger than the other.*

Exercise 13.7.54. *Use Divide and Conquer to approximate $\sqrt{3}$ to 3, 6, and 9 decimal places by solving $x^2 - 3 = 0$.*

Exercise 13.7.55. *Use Newton's Method to find a formula for the cube root of $\alpha > 0$.*

Exercise 13.7.56. *Use Newton's Method to find a formula for the kth root of $\alpha > 0$.*

Exercise 13.7.57. *Let $f(x) = x^2 - \alpha$ for some $\alpha > 0$. We showed Newton's Method leads to the sequence*

$$x_{n+1} = \frac{1}{2}\left(x_n + \frac{\alpha}{x_n}\right).$$

Investigate the convergence of $|\sqrt{\alpha} - x_n|$ to zero.

Hint from [**Rud1**], Exercise 13, Chapter 3. *Let $\epsilon_n = x_n - \sqrt{\alpha}$ and show*

$$\epsilon_{n+1} = \frac{\epsilon_n^2}{2x_n^2} < \frac{\epsilon_n^2}{2\sqrt{\alpha}}$$

(where we assume the initial guess $x_1 > \sqrt{\alpha}$), and thus

$$\epsilon_{n+1} < 2\sqrt{\alpha}\left(\frac{\epsilon_1}{2\sqrt{\alpha}}\right)^{2^n}.$$

Exercise 13.7.58. *Consider $f(x) = x^2 - \alpha$, which has roots $\sqrt{\alpha}$ and $-\sqrt{\alpha}$. For each real number $x_0 \in \mathbb{R}$, define $g(x_0)$ to be 1 if when we apply Newton's Method with starting value x_0 we converge to $\sqrt{\alpha}$, -1 if we converge to $-\sqrt{\alpha}$, and 0 if we do not converge. Determine $g(x_0)$ for all x_0.*

Exercise 13.7.59. *Generalize the previous exercise to $f(x) = x^4 - 5x^2 + 4$.*

Exercise 13.7.60. *If $z = a + ib$ is a complex number, its **complex conjugate** $\bar{z}$ is defined to be $a - ib$ (where, as always, $i = \sqrt{-1}$). Prove that if $f(x)$ is a polynomial with real coefficients, then the roots occur in complex conjugate pairs. This means that either a root r is real or, if r is not real, then $\bar{r}$ is also a root.*

Exercise 13.7.61. *Prove if $f(x)$ is a polynomial of odd degree n with real coefficients, then f has a real root.*

Equivalent Regions and Fixed Points

Exercise 13.7.62. *Define* $f : [-1, 1] \to [-1, 1]$ *by* $f(x) = x^2$. *Is* f *injective? Is it surjective?*

Exercise 13.7.63. *Assume* $f : S \to T$ *is not surjective; prove that there is always a subset* T' *of* T *such that* $f : S \to T'$ *is surjective. What is* T'?

Exercise 13.7.64. *Let* $f : S \to T$ *be a bijection. Prove* f *is invertible. If* f *is continuous must its inverse be continuous?*

Exercise 13.7.65. *Let* $f : S \to T$ *and* $g : T \to U$ *be bijections. Prove the* **composition** $g \circ f : T \to U$ *is a bijection* ($g \circ f$ *is given by* $(g \circ f)(s) = g(f(s))$).

Exercise 13.7.66. *Let* $f : S \to T$ *and* $g : T \to U$. *If* f *is injective is the composition* $g \circ f$ *injective? If* g *is injective is the composition injective? If not give a counter-example.*

Exercise 13.7.67. *Let* $f : S \to T$ *and* $g : T \to U$. *If* f *is surjective is the composition* $g \circ f$ *surjective? If* g *is surjective is the composition surjective? If not give a counter-example.*

Exercise 13.7.68. *Prove function composition is associative:*

$$(f \circ g) \circ h = f \circ (g \circ h);$$

we thus write $f \circ g \circ h$ *as there is no confusion as to what this means.*

Exercise 13.7.69. *Consider the ellipse* $\frac{13}{100}x^2 - \frac{9}{50}xy + \frac{73}{400}y^2 = 1$. *Prove the major axis is the line* $(y - 2.4) = \frac{2.4}{3.2}(x - 3.2)$, *while the minor axis is the line* $(y - 1.6) = -\frac{1.6}{1.2}(x + 1.2)$.

Exercise 13.7.70. *Generalize the previous problem to find the line of the major and minor axis for a general ellipse given by*

$$(x \quad y) \begin{pmatrix} \alpha & \beta \\ \beta & \gamma \end{pmatrix} \begin{pmatrix} x \\ y \end{pmatrix} = 1,$$

where the two eigenvalues of the matrix are positive.

Exercise 13.7.71. *The matrix*

$$\boldsymbol{Q} = \begin{pmatrix} 3/5 & 4/5 \\ -4/5 & 3/5 \end{pmatrix}$$

is a **rotation matrix**; *these are real matrices that satisfy* $\boldsymbol{Q}^T \boldsymbol{Q} = \boldsymbol{Q}\boldsymbol{Q}^T = \boldsymbol{I}$ *and* $\det(\boldsymbol{Q}) = 1$. *Prove this matrix is a rotation matrix, and write down the general form of a* 2×2 *rotation matrix.*

Chapter 14

Contraction Maps

We introduced Fixed Point Theorems in Chapter 13 and saw examples where we could exploit additional information about our function to deduce properties of the fixed point (such as the rate of convergence). A great example of this was the difference between Divide and Conquer and Newton's Method; we used the differentiability of f to obtain an enormously faster rate of convergence in the latter.

In this chapter we concentrate on a special class of functions, contraction maps. Not surprisingly, this additional structure translates to stronger results than hold in general. Fortunately, in many problems of interest these conditions are met.

14.1. Definitions

Our goal is to find fixed points of functions on some space S. We briefly describe the spaces we'll study below, and thus take a quick detour through analysis and topology. We always assume S is a **metric space**. This means S is a collection of points and there is a distance function $d : S \times S \to \mathbb{R}$ satisfying the following properties:

- $d(x, y) \geq 0$ and $d(x, y) = 0$ if and only if $x = y$,
- $d(x, y) = d(y, x)$ (symmetry),
- $d(x, z) \leq d(x, y) + d(y, z)$ (**triangle inequality**).

The most common example is the set of real (or complex) numbers with the absolute value function as the distance function, and in fact we often denote the distance on a general space with the absolute value. Thus, when we write $|x - y|$ we really mean $d(x, y)$.

A metric space S is **complete** if every Cauchy sequence in S converges to an element in S. For us the most important examples are $\mathbb{R}^n$ and $\mathbb{C}^n$.

Before giving the definition of a contraction map we comment on one subtle aspect of it. It's very important to note below that we assume there's a fixed $C < 1$ in the statement below; it's not enough that there is some $C_{xy} < 1$ for each x, y (with C_{xy} allowed to depend on x and y). The reason this is so important is that it gives us a uniformity in the analysis. After reading this section look at Exercise 14.7.16, which illustrates what happens if there is not one C which works for all pairs.

Definition 14.1.1. *We say $f : S \to S$ is a* **contraction map** *if there exists a positive $C < 1$ such that, for all $x, y \in S$,*

$$|f(x) - f(y)| < C|x - y|.$$

Thus when we apply f to two points x and y, the two iterates are closer to each other than the two original points (hence the name 'contraction map'). Furthermore, the amount they approach each other is at least a fixed positive amount of their initial separation.

Given a contraction map, it's natural to ask if it has a fixed point. If yes, how many does it have, and can we find them? Before we prove a fixed point exists, we take a quick detour and give a simple proof that there is at most one fixed point.

Theorem 14.1.2. *If a contraction map f has a fixed point, then it is unique.*

Proof. Assume x_1, x_2 are fixed points of a contraction map f with contraction constant $C < 1$. Then

$$|f(x_1) - f(x_2)| < C|x_1 - x_2|;$$

however, as these are fixed points, $f(x_i) = x_i$. Thus, since $C < 1$,

$$|x_1 - x_2| < |x_1 - x_2|,$$

a contradiction.   □

We thus don't have to worry about there being more than one fixed point. What's left is for us to prove that there *is* one at all, and to show how we can find it.

14.2. Fixed Points of Contraction Maps

Theorem 14.1.2 shows that a contraction map has at most one fixed point. We now show one always exists, and in the course of proving this give a constructive way to find it.

Theorem 14.2.1. *Let $f : S \to S$ be a contraction map with $C < 1$ on a complete metric space S. Then there exists a unique fixed point for f.*

As remarked above, the proof not only gives existence but also provides a construction to find the fixed point. The idea is remarkably simple: choose *any* point $x_0 \in S$ as your initial seed. Keep iterating under f. The resultant sequence converges to the fixed point. Thus we can start anywhere and by iterating we'll approach the unique fixed point. Notice that we've assumed our space S is a complete metric space; we need this to ensure that a sequence below converges to an element in S.

It's worth noting that the condition above is a *sufficient* condition for the existence of a fixed point; we will see later it is not a *necessary* condition. In other words, if we have a contraction map, then we have a fixed point, but a map that is not a contraction may also have a fixed point.

To see the proof idea in practice, namely iterating from a chosen point, let's consider an example. Let $f : [0,1]^2 \to [0,1]^2$ be a map from the unit square to the unit square given by

$$f(x,y) = \left(\frac{7x^2y + 3xy^2 + 3/4}{15}, \frac{x+y+1/5}{3} \right);$$

while it is clear that this maps the unit square to itself (both components are non-negative and at most 1 for $x, y \in [0,1]$), whether or not it's a contraction map is more involved. However, we can chose a point in the square and start iterating. Taking $(x_0, y_0) = (1,1)$, we obtain a sequence (see Figure 1) that appears to converge to something near $(1/20, 1/8)$, though we know it cannot be that number exactly as

$$f(1/20, 1/8) = \left(\frac{4829}{96000}, \frac{1}{8} \right) \approx (.0503, 0.125).$$

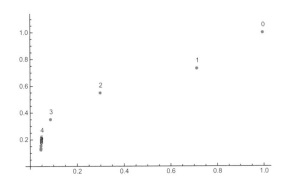

Figure 1. Iterates of $(1,1)$ under the map f.

Proof. Choose any $x_0 \in S$ and define the sequence $\{x_n\}$ by

$$x_{n+1} = f(x_n).$$

We first show that x_n converges. We use a common technique from real analysis and **write it as a telescoping sum**:

$$x_n = x_0 + (x_1 - x_0) + (x_2 - x_1) + \cdots + (x_n - x_{n-1})$$

$$= x_0 + \sum_{k=1}^{n} (x_k - x_{k-1});$$

the reason we do this is that the subsequent terms get smaller and smaller (as f is a contraction map with constant C), and we'll be able to show convergence by using the Cauchy Criterion, Theorem 13.3.2. Explicitly,

$$|x_n - x_m| = \left| \sum_{k=m+1}^{n} (x_k - x_{k-1}) \right|.$$

From the triangle inequality

$$|x_n - x_m| \leq \sum_{k=m+1}^{n} |x_k - x_{k-1}|.$$

By definition $x_j = f(x_{j-1})$, and thus

$$|x_k - x_{k-1}| = |f(x_{k-1}) - f(x_{k-2})| \leq C |x_{k-1} - x_{k-2}|.$$

Continuing we find

$$|x_k - x_{k-1}| \leq C^{k-1} |x_1 - x_0|,$$

and substituting this for each term yields

$$|x_n - x_m| \leq \sum_{k=m+1}^{n} C^{k-1} |x_1 - x_0| \leq \sum_{k=m+1}^{\infty} C^{k-1} |x_1 - x_0|.$$

We have thus bounded $|x_n - x_m|$ by an infinite geometric series with ratio $r = C < 1$. Using the geometric series formula gives

$$|x_n - x_m| \leq \frac{C^m |x_1 - x_0|}{1 - C}.$$

This means we can make $|x_n - x_m|$ arbitrarily small by simply picking large enough values of m; therefore the Cauchy Criterion is satisfied. Thus the sequence x_n converges to some limit point, which we shall call x, as we assumed our space S is a complete metric space.

All that remains is to show that the limit x is the fixed point of f (we can say 'the' fixed point, and not 'a' fixed point, as we know there can be at most one). We know that $x_k \to x$, and thus $x_{k+1} = f(x_k) \to x$. In other words,

$$\lim_{k \to \infty} f(x_k) = x.$$

Because f is continuous, we can pass the limit inside the function:

$$f(\lim_{k \to \infty} x_k) = x.$$

A final substitution now yields

$$f(x) = x.$$

Therefore f has a fixed point, which must be unique. Moreover, this fixed point can be found by taking the limit of the sequence $x_{n+1} = f(x_n)$ where we may choose *any* point as our initial seed.   □

Figure 2 illustrates a function with a fixed point; is it a contraction on $[-\pi, \pi]$?

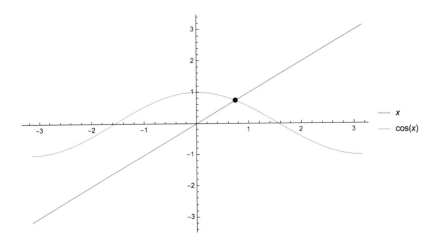

Figure 2. If we plot the functions $y = \cos(x)$ and $y = x$, we see that they intersect at a unique point at $x \approx 0.7$, a fixed point of the function $\cos(x)$.

14.3. Introduction to Differential Equations

One powerful application of contraction maps is proving existence and uniqueness theorems for differential equations. We'll briefly introduce the topic in this section. In addition to providing an excellent use of our contraction results, this topic provides motivation to review some of the results you should have seen in a real analysis class, and we'll see how they can be useful. We'll then end the chapter with a detailed analysis of differential equations; for now, our goal is to give a general flavor so that if you choose not to read the detailed exposition, you will have a sense of the connections.

We'll confine ourselves to differential equations involving functions of one variable. We often use y to represent the function, and t its argument; if you want, think of t as time and $y(t)$ as the location of a particle at time t, or how much of a quantity we have at time t, or the value of our portfolio at time t, ... In addition to proving a solution exists, we often generate an efficient way of approximating it. See [**BoDi**] for additional information.

14.3.1. Differential equations and difference equations. Differential equations are in general mathematic equations relating some function with its derivatives. According to Isaac Newton, there are three general forms of differential equations:

$$\frac{dy}{dx} = f(x), \quad \frac{dy}{dx} = f(x,y), \quad x_1\frac{\partial y}{\partial x_1} + x_2\frac{\partial y}{\partial x_2} = y.$$

Essentially, differential equations describe continuous systems, where rates of change (derivatives) are expressed in terms of other values in the system. One great example of a very widely used differential equation is the heat equation:

$$\frac{\partial u}{\partial t} = k\frac{\partial^2 u}{\partial x^2},$$

with boundary conditions given by

$$
\begin{aligned}
u(x,\ 0) &= f(x), \\
u(0,\ t) &= T_1(t), \\
u(L,\ t) &= T_2(t).
\end{aligned}
$$

Here $u(x,\ t)$ is the temperature at some point x at time t, and T_1, T_2 are initial temperatures for $u(0,\ t)$ (start point of some object) and $u(L,\ t)$ (end point of some object of length L), respectively. The heat equation is surprisingly applicable to many areas, including explaining behavior in the stock exchange. There are many more differential equations that are just as applicable, if not more, so they are certainly worth further study.

Differential equations are not to be confused with difference equations, which are a discrete analogue. The latter is ordered based on time, so each variable ends up having time subscripts. For example, a very simple difference equation is the **logistic equation**:

$$ x_t = a \cdot x_{t-1}(1 - x_{t-1}). $$

Fibonacci numbers are also a type of difference equation often used to model population growth.

The simplest problem to solve involving y' and y is $y'(t) = ay(t)$ for some fixed a. By inspection we see

$$ y(t) = Ce^{at} $$

works for any fixed C. We thus have one **free parameter**, which we can determine by incorporating an initial condition. As we only have one unknown, we only need one initial condition. A common choice is to look at $t = 0$, and we find $C = y(0)$ or

$$ y(t) = e^{at}y(0). $$

Remark 14.3.1. *Building on this success, we could look at*

$$ y''(t) = ay'(t) + by(t). $$

*This is a second order differential equation, as it involves two derivatives of y. We would now need two pieces of information to uniquely specify a solution; typically one uses $y(0)$ and $y'(0)$. As the coefficients of $y'(t)$ and $y(t)$ are constant, it turns out to be relatively easy to solve. We can use the method of **Divine Inspiration**. The above is similar to the constant coefficient fixed depth recurrence relations (such as the one for the Fibonacci numbers, $F_{n+1} = F_n + F_{n-1}$), which we studied earlier. For those we guessed $a_n = r^n$; here the guess becomes $y(t) = \rho^t$. As we want to take derivatives of this, we immediately rewrite our guess as $y(t) = e^{rt}$ (so $\rho = e^r$) to facilitate differentiation. We again get a simple characteristic polynomial:*

$$ r^2 e^{rt} = are^{rt} + be^{rt} \implies r^2 - ar - b = 0; $$

we can now solve this as before.

We now consider first order differential equations of the form

$$ y'(t) = f(t, y) $$

for some fixed function f; the key questions are what conditions on f are needed to ensure that a solution exists (at least for t in some interval), and if a solution exists is it unique? The main result in the subject is the following.

Theorem 14.3.2. *Let f and $\partial f / \partial y$ be continuous functions on the rectangle $R = [-a, a] \times [-b, b]$. Then there is an $h \leq a$ such that there is a unique solution to the differential equation $y'(t) = f(t, y)$ with initial condition $y(0) = 0$ for all $t \in (-h, h)$.*

One way to try and prove the above is to integrate, as integration is the inverse operation to differentiation. We obtain

$$\int_0^\tau y'(t) dt = \int_0^\tau f(t, y) dt,$$

which simplifies to solving

$$y(\tau) - y(0) = \int_0^\tau f(t, y) dt.$$

To simplify the algebra, we always assume $y(0) = 0$; this is harmless as it's a simple translation (think of it as adjusting our scale). We're left with

$$y(\tau) = \int_0^\tau f(t, y) dt.$$

As this is a chapter on contraction maps, not surprisingly the goal is to reformulate this problem so that we can appeal to those results. Let's iteratively construct a sequence of functions which, we hope, will converge to the solution to our differential equation. Let $y_0(t)$ be any nice function such that $y_0(0) = 0$ (a natural choice is to take $y_0(t)$ to be $y(0)$ everywhere), and define

$$y_{n+1}(\tau) = \int_0^\tau f(t, y_n(t)) dt.$$

The above is beginning to look a lot like our work earlier in the chapter. If we can show the above is a contraction map there will be a unique fixed point, and the hope is that function will solve the original equation.

This is a bit more complicated than the previous cases we've studied. We're no longer dealing with a function with real number inputs and outputs. Now, our inputs and outputs are functions as well, and we need to see whether these functions are closer or further apart after we apply our transformation.

Take functions y_n, w_n mapping to y_{n+1}, w_{n+1}. Then we have

$$y_{n+1}(\tau) - w_{n+1}(\tau) = \int_0^\tau [f(t, y_n(t)) - f(t, w_n(t))] \, dt.$$

We need a sense of a distance between our functions. The natural choice involves looking at the largest value of the absolute value of their difference. In the exercises you're asked to show that this does give a metric (Exercises 14.7.5 and 14.7.6).

Next, we want to compare $y_{n+1}(\tau) - w_{n+1}(\tau)$ to $y_n(\tau) - w_n(\tau)$. Where can we find a $y_n(\tau) - w_n(\tau)$ term in our integral expression? One idea then is to use the Mean Value Theorem, which tells us that $g(b) - g(a) = g'(c)(b - a)$ for some $c \in [a, b]$. We can write this as

$$g(b) = g(a) + g'(c)(b - a).$$

This looks a bit like a first-order Taylor series, the difference being that we have a $g'(c)$ instead of an $g'(a)$. This seems promising, but more work is needed.

We're going to assume that the derivative of f with respect to its second component is bounded and at most M. This is not a huge assumption, as many functions satisfy this condition. Now we have

$$|f(t, y_n(t)) - f(t, w_n(t))| \ \le \ M|y_n(t) - w_n(t)|.$$

Using this, we get

$$|y_{n+1}(\tau) - w_{n+1}(\tau)| \ \le \ M \int_0^\tau |y_n(t) - w_n(t)|dt.$$

We're starting to see how the contraction map comes into play. We'll want $\tau M < C < 1$, which means that we're going to be getting solutions in small intervals (whose size will depend on M) about the origin.

We won't push the analysis any further than this here, as it would require a significant amount of additional work and involve a good amount of real analysis. The important takeaway from this is to see how contraction maps can help us; those wanting to see the nuts and bolts are strongly encouraged to read on....

Remark 14.3.3. *We have elected to simplify notation and not state Theorem 14.3.2 in the greatest generality. We have performed two translations so that we assume the time interval is centered at 0 and the y values are centered at 0. There is no harm in such choices. To see this, consider instead the equation $du/d\tau = g_1(\tau, u(\tau))$ with $u(\tau_0) = u_0$. Clearly this is the most general such first order equation; we now show that we may transform this into the form of Theorem 14.3.2. Let $v(\tau) = u(\tau) - u(\tau_0)$. Note $v(\tau_0) = u(\tau_0) - u(\tau_0) = 0$, and as $dv/d\tau = du/d\tau$ we see $dv/d\tau = g_1(\tau, v(\tau) + u(\tau_0)) = g_2(\tau, v(\tau))$. This shows that there is no loss in generality in assuming the initial value is zero. A similar argument shows we may change the time variable to assume the initial time is zero.*

14.4. Real Analysis Review

In order to prove Theorem 14.3.2 we need several results from real analysis; thus, while the result is of interest in its own right, it provides a twoderful opportunity to review our real analysis and see how those results can be used.[1] We quickly review and collect these below.

Lemma 14.4.1. *Let $g : \mathbb{R} \to \mathbb{R}$ be a continuous function on a finite interval $[\alpha, \beta]$. Then there is some M such that $|g(x)| \le M$. If instead $g : \mathbb{R}^2 \to \mathbb{R}^2$ is a continuous function on a finite rectangle $[\alpha, \beta] \times [\gamma, \delta]$, then there is an M such that $|g(x,y)| \le M$.*

A nice feature of many analysis proofs is that the exact value of M doesn't matter, instead what is important is that there is some finite value for M which works. As an example, consider the function $g(x) = e^{-x^2} - x^4 + x^2\cos(2x)$ on the interval $[0, 2]$. We have

$$|g(x)| \ \le \ |e^{-x^2}| + |x^4| + |x^2| \cdot |\cos(2x)|;$$

---

[1] My children, Cameron and Kayla, are big fan's of Victor Borge's *Inflationary Language* routine, where you replace any number sound in a word with one higher; thus 'before' becomes 'befive', 'create' morphs to 'crenine', and hence 'wonderful' is now 'twoderful'. In their honor I've used this word; also this is a good test to see how carefully you're reading the text!

this is a *very* wasteful way to find an upper bound for g, but it will yield one. The largest the first term can be is 1, the largest the second is $2^4 = 16$, and the largest the last is $2^2 = 4$; thus $|g(x)| \leq 1 + 16 + 4 = 21$.

Definition 14.4.2 (Absolutely and Conditionally Convergent Series). *We say a series is absolutely convergent if $\sum_{n=0}^{\infty} |a_n|$ converges to some finite number a; if $\sum_{n=0}^{\infty} |a_n|$ diverges but $\sum_{n=0}^{\infty} a_n$ converges to a finite number a, we say the series is conditionally convergent.*

For example, the series $\sum_{n=0}^{\infty} 1/2^n$ is absolutely convergent, while the series $\sum_{n=0}^{\infty} (-1)^n/n$ is only conditionally convergent.

Lemma 14.4.3 (Comparison Test). *Let $\{b_n\}_{n=0}^{\infty}$ be a sequence of non-negative numbers whose sum converges; this means $\sum_{n=0}^{\infty} b_n = b < \infty$ (and this immediately implies $\lim_{n \to \infty} b_n = 0$). If $\{a_n\}_{n=0}^{\infty}$ is another sequence of real numbers such that $|a_n| \leq b_n$, then $\sum_{n=0}^{\infty} a_n$ converges to some finite number a.*

Remark 14.4.4. *In Lemma 14.4.3, we don't need $|a_n| \leq b_n$ for all n; it suffices that there is some N such that for all $n \geq N$ we have $|a_n| \leq b_n$. This is because the convergence or divergence of a series only depends on the tail of the sequence; we can remove finitely many values without changing the limiting behavior (convergence or divergence).*

For example, we know $\sum_{n=0}^{\infty} r^n = \frac{1}{1-r}$ if $|r| < 1$. Thus the series $a_n = (-1)^n/n!$ converges as $|a_n| \leq (1/2)^n$ for $n \geq 1$.

Other useful series to know are the p-series. Let $C > 0$ be any real number and let $p > 0$. Then $\sum_{n=1}^{\infty} C/n^p$ converges if $p > 1$ and diverges if $p \leq 1$.

Theorem 14.4.5 (Lebesgue's Dominated Convergence Theorem). *Let f_n be a sequence of continuous functions such that (1) $\lim_{n \to \infty} f_n(x) = f(x)$ for some continuous function f, and (2) there is a non-negative continuous function g such that $|f_n(x)|$ and $|f(x)|$ are at most $g(x)$ for all x and $\int_{-\infty}^{\infty} g(x)dx$ is finite. Then*

$$(14.1) \qquad \lim_{n \to \infty} \int_{-\infty}^{\infty} f_n(x)dx = \int_{-\infty}^{\infty} \lim_{n \to \infty} f_n(x)dx = \int_{-\infty}^{\infty} f(x)dx.$$

We have stated this result with significantly stronger conditions than is necessary, as these are the conditions that hold in our problem of interest. It is essential, though, that we have *some* conditions on f; see Exercise 14.7.49 for an example where we cannot interchange the limit and the integration.

The last result we need is the Mean Value Theorem.

Theorem 14.4.6 (Mean Value Theorem (MVT)). *Let $h(x)$ be differentiable on $[a, b]$, with continuous derivative. Then*

$$(14.2) \qquad h(b) - h(a) = h'(c) \cdot (b - a), \qquad c \in [a, b].$$

Remark 14.4.7 (Application of the MVT). *For us, one of the most important applications of the Mean Value Theorem is to bound functions (or more exactly, the difference between a function evaluated at two nearby points). For example, assume f is a continuously differentiable function on $[0, 1]$. This means that the derivative f' is continuous, so by Lemma 14.4.1 there is an M so that $|f'(w)| \leq M$ for all* *w ∈ [0, 1]. Thus we can conclude that $|f(x) - f(y)| \leq M|x - y|$. This is because the Mean Value Theorem gives us the existence of a $c \in [0, 1]$ such that $f(x) - f(y) = f'(c)(x - y)$. Taking absolute values and noting $|f'(c)| \leq \max_{0 \leq w \leq 1} |f'(w)|$, which by Lemma 14.4.1 is at most M, yields the claim.*

14.5. First Order Differential Equations Theorem

We sketch the proof of Theorem 14.3.2. Our argument follows closely the one in [**BoDi**]. We break the proof into four major steps. Unfortunately the third step appeals to Lebesgue's Dominated Convergence Theorem (Theorem 14.4.5), which is typically not covered until a course in measure theory. That said, the rest of the proof is fairly elementary, and this argument can be taken as motivation to delve into measure theory and the Lebesgue integral.

Step 1: Picard's Iteration Method

We begin by constructing a sequence of functions $\phi_n(t)$ by setting $\phi_0(t) = 0$ and

$$\phi_{n+1}(t) = \int_0^t f(s, \phi_n(s)) \, ds.$$

Note that $\phi_n(0) = 0$ for all n, which is good (as we are trying to solve the differential equation $y'(t) = f(t, y)$ with initial condition $y(0) = 0$). This is known as **Picard's iteration method** for solving differential equations.

We want to prove two facts: first, that $\phi_n(t)$ exists for all n, and second that it is continuous. If $\phi_n(t)$ exists for some n, then $\phi_{n+1}(t)$ exists as well. This is because

$$\phi_{n+1}(t) = \int_0^t f(s, \phi_n(s)) ds,$$

and the integral of a continuous function is continuous (regard $f(s, \phi_n(s))$ as some new function, say $g(s)$, and now we can use our results from first year calculus). The only problem is that we want $\phi_{n+1}(t)$ to always lie in the interval $[-b, b]$.

Recall that we are trying to solve the differential equation $y'(t) = f(t, y)$ for $t \in [-a, a]$ and $y \in [-b, b]$. As f is continuous, by Lemma 14.4.1 there is an M such that $|f(t, y)| \leq M$ for all $t \in [-a, a]$ and all $y \in [-b, b]$. If we restrict to $t \in [-h, h]$ for $h \leq \min(b/M, a)$, then the integral

$$\int_0^t f(s, \phi_n(s)) ds$$

is at most $M|t| \leq Mh \leq b$. We see now why we restricting to $t \in [-h, h]$ is potentially needed; this ensures that $\phi_{n+1}(t)$ takes on values in $[-b, b]$.

Step 2: Limit Exists and is Continuous

We want to show that $\lim_{n \to \infty} \phi_n(t)$ exists for all t *and is continuous!* We write $\phi_n(t)$ as

$$\phi_n(t) = \sum_{k=1}^n (\phi_n(t) - \phi_{n-1}(t)),$$

remembering that $\phi_0(t) = 0$. Thus

$$\phi_3(t) \;=\; (\phi_1(t) - 0) + (\phi_2(t) - \phi_1(t)) + (\phi_3(t) - \phi_2(t));$$

this is a telescoping sum. We've already seen these sums are often easy to analyze.

We now show that $\lim_{n\to\infty} \phi_n(t)$ exists for all t. *We can do this for any $r < 1$ by making sure h is sufficiently small (remember we have restricted to studying only $t \in (-h, h)$).* Let's fix some $t \in (-h, h)$. Assume we could show that there is some $r < 1$ such that

$$|\phi_k(t) - \phi_{k-1}(t)| \;\leq\; r^k \quad \text{for all } k.$$

Then for this t the limit exists by the Comparison Test (Lemma 14.4.3). (To use the Comparison Test, we let $b_n = r^n$ and $a_n = \phi_n(t) - \phi_{n-1}(t)$, and note that $\lim_{n\to\infty} \phi_n(t) = \sum_{n=1}^{\infty} a_n$.)

Thus we are reduced to showing that there is an $r < 1$ with

$$|\phi_k(t) - \phi_{k-1}(t)| \;\leq\; r^k.$$

This will follow from using the Mean Value Theorem to estimate the integrals for ϕ_k and ϕ_{k-1} and mathematical induction. Recalling that $\phi_k(t) = \int_0^t f(s, \phi_{k-1}(s)) ds$ and similarly for $\phi_{k-1}(s)$, we find

$$\phi_k(t) - \phi_{k-1}(t) \;=\; \int_0^t \left[f(s, \phi_{k-1}(s)) - f(s, \phi_{k-2}(s)) \right] ds.$$

We now apply the Mean Value Theorem (Theorem 14.4.6) to the function $g(y) = f(s, y)$. We take our two points to be $\phi_{k-1}(s)$ and $\phi_{k-2}(s)$. Thus there is some point $c_k(s)$ between $\phi_{k-1}(s)$ and $\phi_{k-2}(s)$ such that

$$g(\phi_{k-1}(s)) - g(\phi_{k-2}(s)) \;=\; g'(c_k(s)) \cdot (\phi_{k-1}(s) - \phi_{k-2}(s));$$

we chose to write the point as $c_k(s)$ to remind ourselves that it depends on k and s (in particular, we are doing this for *every* s in the integral). Noting that $g'(y) = \partial f / \partial y$, we see we have shown

$$(14.3) \quad \phi_k(t) - \phi_{k-1}(t) \;=\; \int_0^t \frac{\partial f}{\partial y}(s, c_k(s)) \cdot (\phi_{k-1}(s) - \phi_{k-2}(s))\, ds.$$

We now proceed by induction. We assume that we have already shown

$$|\phi_{k-1}(s) - \phi_{k-2}(s)| \;\leq\; r^{k-1}$$

for $s \leq t$. We substitute this into (14.3), and use Lemma 14.4.1 to bound $\partial f / \partial y$ by M and find

$$|\phi_k(t) - \phi_{k-1}(t)| \;\leq\; \int_0^t M r^{k-1} ds \;=\; M t r^{k-1} \;\leq\; M h r^{k-1}.$$

As long as we choose h so that $Mh < r$ (i.e., $h < r/M$), then we obtain the desired result!

We now want to show that

$$\phi(t) \;=\; \lim_{n\to\infty} \phi_n(t)$$

is continuous. (It had better be continuous, as we want it to be the solution to the differential equation, and if it isn't continuous, then it can't be differentiable!) To show ϕ is continuous, we must show that given any $\epsilon > 0$ there is a $\delta > 0$ such that

$|t_2 - t_1| < \delta$ implies $|\phi(t_2) - \phi(t_1)| < \epsilon$. For notational convenience assume $t_1 < t_2$. We have

$$
\begin{aligned}
\phi(t_2) - \phi(t_1) &= \lim_{n\to\infty} \phi_n(t_2) - \lim_{n\to\infty} \phi_n(t_1) \\
&= \lim_{n\to\infty} (\phi_n(t_2) - \phi_n(t_1)) \\
&= \lim_{n\to\infty} \int_{t_1}^{t_2} f(s, \phi_n(s)) ds
\end{aligned}
$$

(the last line follows from the fact that $\phi(t_1)$ is an integral from 0 to t_1 while $\phi_n(t_2)$ is an integral from 0 to t_2. By Lemma 14.4.1, there is an M such that $|f(s,y)| \le M$. Thus

$$
|\phi(t_2) - \phi(t_1)| \;\le\; \int_{t_1}^{t_2} M\, ds \;=\; M|t_2 - t_1| \;\le\; M\delta;
$$

therefore as long as we choose $\delta < \epsilon/M$ we see that $|\phi(t_2) - \phi(t_1)| < \epsilon$.

Step 3: Limit Function Satisfies Differential Equation

In this step we show that the limit function

$$
\phi(t) \;=\; \lim_{n\to\infty} \phi_n(t)
$$

satisfies the differential equation $y'(t) = f(t,y)$ with initial condition $y(0) = 0$ (i.e., taking $y(t) = \phi(t)$ gives a solution). From construction, it's clear that $\phi(0) = 0$ as each $\phi_n(0) = 0$. The difficulty is showing that $d\phi/dt = f(s, \phi)$. To see this, we argue as follows:

$$
\begin{aligned}
\phi(t) &= \lim_{n\to\infty} \phi_n(t) \\
&= \lim_{n\to\infty} \int_0^t f(s, \phi_{n-1}(s)) ds.
\end{aligned}
$$

We want to move the limit inside the integral; this can be done because the conditions of Lebesgue's Dominated Convergence Theorem (Theorem 14.4.5) are met (all functions are continuous, and we may take $g(x) = Mh$ to be the bounding function required by the theorem). Thus

$$
\begin{aligned}
\phi(t) &= \int_0^t \lim_{n\to\infty} f(s, \phi_{n-1}(s)) ds \\
&= \int_0^t f(s, \lim_{n\to\infty} \phi_{n-1}(s)) ds,
\end{aligned}
$$

where the last step (moving the limit inside the function) follows from the fact that f is continuous in each variable. Thus we have shown

$$
\phi(t) \;=\; \int_0^t f(s, \phi(s)) ds,
$$

and all functions are continuous. Therefore the Fundamental Theorem of Calculus now yields $\phi'(t) = f(s, \phi(t))$.

Step 4: The Solution is Unique

The last result to be shown is that the solution is unique. The proof of this is similar to Step 2. We assume there is another solution $\psi(t)$ and we find

$$\phi(t) - \psi(t) \;=\; \int_0^t \left(f(s, \phi(t)) - f(s, \psi(t)) \right) ds.$$

If the two functions are not the same, then there is an $\epsilon > 0$ such that, for some t, $|\phi(t) - \psi(t)| > \epsilon$.

Let

$$m \;=\; \max_{0 \le t \le h} |\phi(x) - \psi(x)|$$

and let M be a bound for $\partial f / \partial y$. Using the Mean Value Theorem we find

$$|\phi(t) - \psi(t)| \;\le\; \int_0^t M\,|\phi(s) - \psi(s)|\,ds \;\le\; M|t|m \;\le\; Mhm.$$

If we choose $h < \epsilon/2mM$, this implies that for all $t < h$, $|\phi(t) - \psi(t)| < \epsilon/2$, which contradicts the fact that there was some t where the difference was at least ϵ.

14.6. Examples of Picard's Iteration Method

We give two examples of Picard's method. Consider the differential equation and initial condition

$$\phi'(t) \;=\; 2\phi(t), \;\; \phi(0) \;=\; 1.$$

Applying the method, we find

$$
\begin{aligned}
\phi_1(t) \;&=\; \int_{s=0}^t f(s,\ \phi_0(s))\ ds \\
&=\; \int_{s=0}^t 2\phi_0(s)\ ds \\
&=\; \int_{s=0}^t 2\ ds \\
&=\; 2t,
\end{aligned}
$$

$$
\begin{aligned}
\phi_2(t) \;&=\; \int_{s=0}^t f(s,\ \phi_1(s))\ ds \\
&=\; \int_{s=0}^t 2\phi_1(s)\ ds \\
&=\; \int_{s=0}^t 2(2s)\ ds \\
&=\; 2t^2,
\end{aligned}
$$

and

$$\phi_3(t) \;=\; \int_{s=0}^{t} f(s,\ \phi_2(s))\ ds$$

$$=\; \int_{s=0}^{t} 2\phi_2(s)\ ds$$

$$=\; \int_{s=0}^{t} 2(2s^2)^2\ ds$$

$$=\; \frac{4t^3}{3}.$$

A pattern appears to be emerging, with the kth term being

$$\phi_k(t) \;=\; \frac{(2t)^k}{k!};$$

this can be proved with a little work, and yields the Taylor series

$$\phi(t) \;=\; 1 + 2t + 2t^2 + \frac{4t^3}{3} + \cdots.$$

Thus the *exact* solution is given by

$$\phi(t) \;=\; \lim_{n\to\infty} \phi_n(t)$$

$$=\; \sum_{k=1}^{\infty} \frac{(2k)^k}{k!}$$

$$=\; \sum_{k=0}^{n} \frac{(2k)^k}{k!}$$

$$=\; e^{2t},$$

which you can directly check satisfies the conditions.

We now turn to a harder example, and discuss how to use Picard's iteration method to solve

$$\phi'(t) \;=\; 2(\phi(t) + 1), \ \phi(0) \;=\; 0.$$

Assume the initial guess is $\phi_0(t) = 0$. We then have our initial condition at the origin. Then, we just apply the iteration to the differential question as such:

$$\phi_1(t) \;=\; \int_{s=0}^{t} f(s,\ \phi_0(s))\ ds$$

$$=\; \int_{s=0}^{t} 2(\phi_0(s) + 1)\ ds$$

$$=\; \int_{s=0}^{t} 2\ ds$$

$$=\; 2t.$$

Now we have our guess of $\phi_1(t) = 2t$. Do another iteration to get $\phi_2(t)$, and so on and so forth. The rest of the iteration is left as Exercise 14.7.62.

14.7. Exercises

Definitions

Exercise 14.7.1 (Triangle inequality). *Prove the triangle inequality: if $x, y, z \in \mathbb{R}$, then $|z - x| \leq |y - x| + |z - y|$.*

Exercise 14.7.2. *Let $d : \mathbb{R}^n \to \mathbb{R}$ be defined as follows: given $\vec{x}, \vec{y} \in \mathbb{R}^n$ set*

$$d(\vec{x}, \vec{y}) := \sqrt{(x_1 - y_1)^2 + \cdots + (x_n - y_n)^2}.$$

Prove that d is a distance function and $\mathbb{R}^n$ is a metric space.

Exercise 14.7.3. *Find a distance function which makes $\mathbb{C}^n$ into a metric space.*

Exercise 14.7.4. *Find a distance function which makes the set of $n \times n$ matrices with real entries a metric space.*

Exercise 14.7.5. *Let S be the set of all continuous functions from $[0, 1]$ to $[0, 1]$. Let*

$$d(f, g) := \sup_{x \in [0,1]} |f(x) - g(x)|.$$

Prove d is a distance function.

Exercise 14.7.6. *Redo the previous problem, but now assume S is the set of all bounded functions from $[0, 1]$ to $[0, 1]$.*

Exercise 14.7.7. *Prove $\mathbb{R}^n$ and $\mathbb{C}^n$ are complete metric spaces.*

Exercise 14.7.8. *Prove or disprove: the metric space S from Exercise 14.7.5 is complete.*

Exercise 14.7.9. *Prove or disprove: the metric space S from Exercise 14.7.6 is complete.*

Exercise 14.7.10. *Give an example of a function $f : S \to S$ where for any pair of points $x, y \in S$ there is a positive constant $C_{xy} < 1$ such that $|f(x) - f(y)| < C|x - y|$; however, there is a sequence of pairs (x, y) such that $C_{xy} \to 1$.*

Exercise 14.7.11. *Let $f : S \to S$ be a twice continuously differentiable function on a closed interval S. Assume $|f'(x)| < C < 1$ for all $x \in S$. Is f a contraction map?*

Exercise 14.7.12. *Redo the previous problem, but now just assume that $|f'(x)| < 1$ for all $x \in S$. Is f a contraction map?*

Exercise 14.7.13. *In the previous two problems we assumed f was a twice continuously differentiable function. Do we need f to be twice continuously differentiable, or can this assumption be weakened to say f is continuously differentiable?*

Exercise 14.7.14. *Let $f(x) = rx(1 - x)$ on the interval $[0, 1]$ for some $r \in (0, 4)$. Is f a contraction map?*

Exercise 14.7.15. *Let S be the set of all continuous functions from $[0, 1] \to [0, 1]$. Define $\mathcal{K} : S \to S$ by*

$$(\mathcal{K}(f))(x) := \int_0^1 e^{-xt} f(t) dt.$$

Is $\mathcal{K}$ a contraction map? Note this is an example of an **integral transform**, with **kernel** e^{-xt}.

Exercise 14.7.16. *Consider* $f(x) = x + e^{-x}$ *with* $x \geq 0$. *Show that for any* $x, y \geq 0$ *we have* $|f(x) - f(y)| < |x - y|$, *but for any* $C < 1$ *we can always find a pair* $x, y \geq 0$ *such that* $|f(x) - f(y)| > C|x - y|$. *Prove that* f *does not have a fixed point on the interval* $[0, \infty)$, *and thus* f *cannot be a contraction map (as contraction maps have fixed points).*

Fixed Point of Contraction Maps

Exercise 14.7.17. *Consider again*

$$f(x, y) = \left(\frac{7x^2y + 3xy^2 + 3/4}{15}, \frac{x + y + 1/5}{3} \right).$$

Prove or disprove: f *is a contraction map from the unit square to the unit square. Prove* f *has a fixed point, and estimate it.*

Exercise 14.7.18. *Can* $f(x) = x^2 - \alpha$ *be a contraction map for some* $\alpha > 0$ *on some interval* S *symmetric about the origin?*

Exercise 14.7.19. *Generalize the previous exercise to* $x, y, z \in \mathbb{C}$.

Exercise 14.7.20. *Prove that if* f *is continuous, then*

$$\lim_{x \to x_0} f(x) = f\left(\lim_{x \to x_0} x \right).$$

Exercise 14.7.21. *Let* $f_n : [0, 1] \to [0, 1]$ *be a sequence of continuous functions. Must*

$$\lim_{n \to \infty} \int_0^1 f_n(x)dx = \int_0^1 \lim_{n \to \infty} f_n(x)dx?$$

Exercise 14.7.22. *Approximate the fixed points of* $f(x) = \sin(\pi x/2)$ *and* $g(x) = \cos(\pi x/2)$ *to ten decimal places.*

Exercise 14.7.23. *Let* A *be an* $n \times n$ *matrix. What must be true about its eigenvalues for it to be a contraction map? For it to have a fixed point?*

Introduction to Differential Equations

Exercise 14.7.24. *Write down the general solution to*

$$y''(t) = ay'(t) + by(t),$$

using initial conditions $y(0)$ *and* $y'(0)$. *Using the interpretation of* $y(t)$ *as the location of our particle at time* t, *we can interpret the above as saying that if we know our initial location and speed, the particle's location is uniquely determined for all time.* Note that the form of the solution will depend on whether or not the associated characteristic polynomial has repeated roots.

Exercise 14.7.25. *Generalize the previous problem to*

$$y^{(n)}(t) = a_{n-1}y^{(n-1)}(t) + \cdots + a_1y'(t);$$

how many initial conditions do we need now? Note the form of the solution will depend on whether or not the associated characteristic polynomial has repeated roots.

Exercise 14.7.26. *Instead of differential equations, which are continuous, we can study **difference equations**, which are discrete. We've already seen one example: $F_{n+1} = F_n + F_{n-1}$ with initial conditions $F_0 = 0$, $F_1 = 1$. More generally consider a second order difference equation with constant coefficients, say $G_{n+1} = aG_n + bG_{n-1}$, and initial conditions $G_0 = \alpha$ and $G_1 = \beta$; what is the general solution? Warning: the form of the answer depends on whether or not the characteristic polynomial has distinct roots.*

Exercise 14.7.27. *Consider the logistic equation $x_t = a \cdot x_{t-1}(1 - x_{t-1})$. For what a is there a fixed point? For such a, find the fixed point.*

Exercise 14.7.28. *Instead of having a second order recurrence, consider $a_{n+1} = a_n + p(n)$ with a_0 given and $p(n)$ a fixed polynomial of degree d. What is the form of the general solution?*

Exercise 14.7.29. *Many problems can be placed in the form of the previous problem. For example, if $a_n = 0 + 1 + \cdots + n$, find the associated $p(n)$ and solve by Divine inspiration. Similarly if $a_n = 0^2 + 1^2 + \cdots + n^2$, find the associated $p(n)$ and solve by Divine inspiration. Can you use your answer from the previous problem to say something about the solution of $a_n = 0^k + 1^k + \cdots + n^k$?*

Exercise 14.7.30. *Let $f(t,y) = t^2 y + e^{t^2 y^2}$. Find a neighborhood for t such that the resulting integral transform is a contraction map for y near 0.*

Exercise 14.7.31. *Find the time change of variables to prove that we may assume the time variable is centered at 0 in Theorem 14.3.2.*

The next few problems introduce some of the most important types of differential equations and the techniques to solve them; there are many other forms where solutions exist, and many other methods (for example, look up *variation of parameters*).

Exercise 14.7.32 (Integrating factors). *For a differential equation of the form $y'(t) + p(t)y(t) = g(t)$, the general solution is*

$$y(t) = \frac{1}{\mu(t)} \left[\int \mu(t)g(t)dt + C \right],$$

where

$$\mu(t) := \exp\left(\int p(t)dt \right)$$

*is the **integrating factor** and C is a free constant (if an initial condition is given, then C can be determined uniquely). Prove the claimed solution is valid.*

Exercise 14.7.33. *Solve $y'(t) - 2ty(t) = \exp\left(t^2 + t \right)$.*

Exercise 14.7.34 (Separable). *For the **separable** differential equation $M(x) + N(y)dy/dx = 0$ the general solution is*

$$\int_{x_0}^{x} M(s)ds + \int_{y_0}^{y} N(s)ds = 0,$$

where we are using the shorthand notation $y_0 = y(x_0)$ and $y = y(x)$. We could also write the solution as

$$\int M(x)dx + \int N(y)dy = C,$$

and then determine C from the initial conditions. NOTE: if we can write the differential equation as $y' = F(v)$ for $v = y/x$, then we can convert this to a separable equation: $y = vx$ so $v + xv' = F(v)$ or $-\frac{1}{x} + \frac{v'}{F(v)-v} = 0$. Prove the claimed solution is valid.

Exercise 14.7.35. *Solve the equation $3x^2 + \cos(y)y' = 0$.*

Exercise 14.7.36 (Exact). *An equation $M(x,y) + N(x,y)dy/dx = 0$ with $\partial M/\partial y = \partial N/\partial x$ is called **exact**. Then there is a function $\psi(x,y)$ such that the solution to the differential equation is given by $\psi(x,y) = C$, with C determined by the initial conditions. One way to find ψ is as follows. Our problem implies that $\partial\psi/\partial x = M$ and $\partial\psi/\partial y = N$. Thus*

$$\psi(x,y) = \int M(x,y)dx + g(y),$$

$$\psi(x,y) = \int N(x,y)dy + h(x),$$

and then determine $g(y)$ and $h(x)$ so that the two expressions are equal. Prove the claimed solution is valid.

Exercise 14.7.37. *Solve $(3x^2 - 2xy + 2) + (6y^2 - x^2 + 3)dy/dx = 0$.*

Exercise 14.7.38. *Solve $\frac{dy}{dx} = 5y^2x^3$.*

Exercise 14.7.39. *Solve $\frac{d^2y}{dx^2} + \left(\frac{dy}{dx}\right)^2 + 4x - 2y = 6$.*

Exercise 14.7.40. *Solve $\left(\frac{dy}{dx}\right)^3 - x + 2y = 5\sin(x) - \sin(y)$.*

Real Analysis Review

Exercise 14.7.41. *Prove Lemma 14.4.1.*

Exercise 14.7.42. *The trivial upper bound estimate of 21 for $g(x) = e^{-x^2} - x^4 + x^2\cos(2x)$ isn't off by much; the actual maximum value is about 19. Determine the optimal value. Unfortunately if you try to use calculus and find the critical points, you end up with an extremely difficult problem to solve. You'll have to use Newton's Method or Divide and Conquer. Alternatively, if you can show the first derivative is positive for $x > 1$ you know the maximum value is at the endpoint. One must be careful as we care about the maximum of the absolute value, and thus you have to break the analysis into cases where g is positive and negative. This is one reason why we often just estimate crudely.*

Exercise 14.7.43. *If $\sum_{n=0}^{\infty} a_n$ converges absolutely, show $\lim_{n\to\infty} a_n = 0$.*

Exercise 14.7.44. *Prove $\sum_{n=1}^{\infty} \frac{1}{n^2+2n+5}$ converges.*

Exercise 14.7.45. *Prove $\sum_{n=1}^{\infty} x^n/n!$ converges for all x (or at least for $|x| < 1$).*

Exercise 14.7.46 (**Rearrangement Theorem**). *Let $\{a_n\}$ be a conditionally convergent but not absolutely convergent series. Given any two numbers $\alpha < \beta$ (note $\alpha = -\infty$ or $\beta = \infty$ are allowed) show that we can rearrange the series so that the new series has liminf equal to α and limsup equal to β.*

Exercise 14.7.47. *Prove Lemma 14.4.3, the Comparison Test.*

Exercise 14.7.48. *Prove $\sum_{n=1}^{\infty} 1/n^p$ converges for $p > 1$ and diverges for $p \leq 1$.*

Exercise 14.7.49. *Let $f_n(x) = 0$ if $x \leq n$, $n(x-n)$ if $n \leq x \leq n+1$, $n(n+2-x)$ if $n+1 \leq x \leq n+2$, and 0 otherwise; thus $f_n(x)$ is a triangle of height n and width 2 centered at $n+1$. Show that for any x, $\lim_{n\to\infty} f_n(x) = 0$. Why can't we use Lebesgue's Dominated Convergence Theorem to conclude that $\lim_{n\to\infty} \int_{-\infty}^{\infty} f_n(x)dx = 0$?*

Exercise 14.7.50. *Prove Theorem 14.4.5.*

Exercise 14.7.51. *Let $f(x) = e^{x^2-4} - x^2 \sin(x^2+2x) + \frac{x+1}{x^2+5}$. Prove $|f(x)-f(y)| \leq 6|x-y|$ whenever $x, y \in [0,2]$.*

Exercise 14.7.52 (Rolle's Theorem). *Let $f : [a,b] \to \mathbb{R}$ be a continuously differentiable function such that $f(a) = f(b) = 0$. Prove that there is a $c \in [a,b]$ such that $f'(c) = 0$.*

Exercise 14.7.53 (Mean Value Theorem). *Let $g : [a,b] \to \mathbb{R}$ be a continuously differentiable function. Prove that there is a $c \in [a,b]$ such that $g'(c)(b - a) = g(b) - g(a)$. Hint: apply Rolle's Theorem from the previous exercise to a function related to g.*

First Order Differential Equations Theorem

Exercise 14.7.54. *Give the details for the proof by induction from Step 2. In particular, do the basis case and the inductive step carefully.*

Exercise 14.7.55. *Using the iterative method from the chapter, solve $y'(t) = t(1 + y(t))$ with $y(0) = 0$.*

Exercise 14.7.56. *Using the iterative method from the chapter, solve $y'(t) = 1 + y(t)^2$ with $y(0) = 0$.*

Exercise 14.7.57. *Using the iterative method from the chapter, solve $y'(t) = t^2 + y(t)$ with $y(0) = 0$.*

Exercise 14.7.58. *Find a discontinuous f so that if we applied Step 1 with this f, then we would have some ϕ_n discontinuous.*

Exercise 14.7.59 (Extended Mean Value Theorem). *Let $f, g : [a,b] \to \mathbb{R}$ be continuously differentiable functions. Prove there is some $c \in (a,b)$ such that*

$$(f(b) - f(a)) \cdot g'(c) = (g(b) - g(a)) \cdot f'(c).$$

Exercise 14.7.60. *Explore the Mean Value Theorem for functions of two variables. Does a generalization with equality hold? If not, is there a generalization with an inequality?*

Exercise 14.7.61. *Consider the sequence of bounded, continuous functions given by*

$$f_n(x) = \begin{cases} 1 - 2n\left(x - \frac{1}{n}\right) & \text{if } |x - 1/n| < 1/2n, \\ 0 & \text{otherwise.} \end{cases}$$

Prove

$$\lim_{n\to\infty} \int_0^1 f_n(x)dx \;\neq\; \int_0^1 \lim_{n\to\infty} f_n(x)dx;$$

does this contradict the Dominated Convergence Theorem?

Examples of Picard's Iteration Method

Exercise 14.7.62. *Solve $\phi'(t) = 2(\phi + 1)$, $\phi(0) = 0$ using Picard's iteration.*

Exercise 14.7.63. *Solve the heat equation $\frac{\partial u}{\partial t} = k\frac{\partial^2 u}{\partial x^2}$ given the following initial conditions:*

$$
\begin{aligned}
u(x,\ 0) &= f(x), \\
u(0,\ t) &= 0, \\
u(L,\ t) &= 0,
\end{aligned}
$$

and

(1) $f(x) = 4\sin(\pi x/L)$ or

(2) $f(x) = 11\sin(6\pi x/L) + 3\sin(5\pi x/L)$.

Chapter 15

Sperner's Lemma

Brouwer's fixed point theorem says that if you have a continuous function f from a "nice" region S to itself, then f has a fixed point. In order to better visualize what's happening we often concentrate on two dimensions. The generalization is straightforward, though notationally involved at times. We put the word nice in quotes as Exercise 15.5.1 shows that a natural candidate region fails, and thus we must make sure our region possesses certain properties.

Sperner's lemma, the focus of this chapter, provides an elementary path to a proof of Brouwer's fixed point theorem (note that elementary does not mean easy; it means non-calculus, it means counting arguments); see [**Fr**] for an accessible account of this and other approaches. Fixed point theorems have a wealth of applications, including fair division (such as rental harmony), solving differential equations, and Nash equilibria in game theory to name a few. A major difficulty in using these for applications is constructive versus non-constructive proofs. While it's nice to know a winning strategy exists in a game or conflict, it's even better to know what that strategy is! Sadly, many of the results in the subject are only existence ones. Fortunately, however, some are constructive (such as the iteration method from Chapter 13 for contraction maps).

We'll prove Sperner's lemma in this chapter, and then show how it implies Brouwer's fixed point theorem in the next. One advantage of this approach is that it provides a great opportunity to see how many results from real analysis are used, as well as see some powerful proof techniques.

15.1. Statement of Sperner's Lemma

We start with the two-dimensional case of Sperner's lemma, which we illustrate in Figure 1. We'll then quickly move to the one-dimensional case, which, after seeing what occurs in the two-dimensional setting, we'll be able to properly appreciate. We'll prove the one-dimensional case twice; the first proof is very straightforward but doesn't generalize well; the second approach is similar to the first but extends

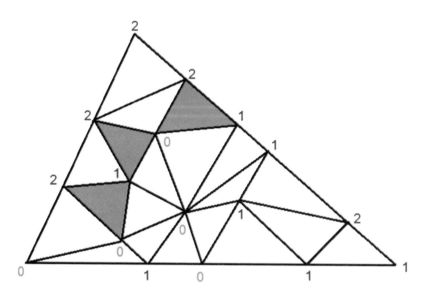

Figure 1. A two-dimensional example of Sperner's lemma.

to the higher dimensional cases. As a rule of thumb, it's often an excellent idea to start with a simple case, understand it well, and try to generalize.

Lemma 15.1.1 (Sperner's Lemma (2D)). *Consider a triangle whose three vertices are labeled 0, 1, and 2. Subdivide into non-intersecting triangles, and label the vertices from $\{0, 1, 2\}$, subject to each vertex along the three outer edges of the triangle having its label from among those of the edge's endpoints (thus the vertices along the $1 - 2$ edge all have label either 1 or 2). For the internal vertices, choose any labels among 0, 1, or 2. For every such labeling, there will always be a triangle whose vertices have distinct labels and which has no further division points inside it or along its edges.*

In other words, there is a subtriangle whose vertices are 0, 1 and 2. Interestingly, in the example in Figure 1 we find not one but *three* such triangles. If we were to create a large number of random triangulations and labelings we would soon notice an interesting fact: there are always an *odd* number of triangles with all distinct labels. This is not a result of luck, and forms the basis of our proof. Specifically, we show the number of such triangles is odd, and as the smallest odd number is positive we immediately get that there must be at least one such triangle!

The one-dimensional case is much easier.

Lemma 15.1.2 (Sperner's Lemma (1D)). *Consider a line segment with left point labeled 0 and right point labeled 1. Subdivide into non-intersecting segments, with the division points labeled 0 or 1. For every such labeling, there will always be an edge whose endpoints have distinct labels and which has no further division points inside it.*

Figure 2. A one-dimensional example of Sperner's lemma.

First proof of Sperner's lemma (1D). Take a line with endpoints 0 and 1 and add internal division points. We must show that one of the new line segments has 0 and 1 as its labels. To see this, we start at the left with a 0. We keep moving to the right until we first encounter a label of 1, at which point we have a segment whose vertices are labeled 0 and 1. Note that there *must* be a first occurrence of 1 as the rightmost vertex is labeled 1.   □

We illustrate this in Figure 2. While the proof we gave does not generalize well, fortunately a close analogue does. Notice, again, that in our example we have an odd number of segments with distinct labels. This is not a coincidence, and a similar behavior occurs in higher dimensions. This is the basis of our proof in general.

Second proof of Sperner's lemma (1D). We will prove that the number of segments in our subdivision whose endpoints have distinct labels is odd; as this number is non-negative we see it must be at least one.

There are three possibilities for a segment: the labels are both 0 (we denote the number of these by $S(0,0)$), the labels are both 1 (we denote the number here by $S(1,1)$), or one label is 0 and one label is 1 (we write $S(0,1)$ for the number of these).

Let's count how many times the label 0 occurs. Any internal 0 is counted twice, once as a left endpoint and once as a right endpoint; of the two endpoints of the initial segment, one is a 0 and one is a 1. Notice that each segment in $S(0,0)$ contributes two 0's, each in $S(0,1)$ contributes one 0, and those in $S(1,1)$ contribute none. Thus, if we count that left endpoint of 0 an extra time, we'll have double counted all the points labeled zero. In other words:

$$1 + 2 \cdot S(0,0) + 1 \cdot S(0,1) + 0 \cdot S(1,1) \;=\; \text{twice the number of vertices labeled } 0.$$

(I'm a huge fan of cumbersome notation. While we don't need to write $1 \cdot S(0,1)$, it helps by calling attention to the fact that we are weighing $S(0,0)$ by a factor of 2 and $S(0,1)$ by a factor of 1.) As the right hand side is even, the left hand side must be even. There are three terms on the left. Since the first is odd and the second is even, the only way the left can be even is if $S(0,1)$ is an odd number. In other words, there are an odd number of segments with two distinct labels.   □

For the example in Figure 2, we see $S(0,0) = 2$, $S(0,1) = 3$, and $S(1,1) = 1$, and we verify the proof's claim that $S(0,1)$ is odd.

Notice this suggests the *outline* of a proof strategy for the two-dimensional Sperner's lemma. We'll have $T(0,1,2)$ denote the number of triangles with three distinct labels, and somehow show that there are an odd number of these. Another possible ingredient is a proof by induction, somehow building on the one-dimensional case. There are other possible approaches; we discuss the ones above, and some others, in the next section.

15.2. Proof Strategies for Sperner's Lemma

How do we prove Sperner's lemma? Remember that Sperner's lemma (at least in two dimensions) says the following: If you subdivide a triangle into subtriangles such that each vertex is labeled 0, 1, or 2 and the vertices on the outside edges are labeled with one of the two labels from the endpoints of that side, then there is at least one triangle which is not further subdivided having all three labels.

Linear Programming Approach.

Here's one way to view the above. Think of it as a game you're playing with somebody where you take turns labeling (in the manner prescribed). Can you play defense in such a way as to prevent your opponent from ever getting a small triangle where all the vertices are labeled differently? Since Sperner's lemma is true, the answer to that question is no. Or you can flip the problem around and offer to go second and have your opponent try to prevent *you* from getting a triangle with all different labels; you can even offer to let her go twice for every move you do. Interestingly, since Sperner's lemma is true you have a very simple strategy: do whatever you want! You cannot help but win!

This gives us a way to view the problem; how could we prove it? We could set it up as a linear programming problem. This of course would be a difficult problem as it would be an integer programming problem, but we can have $x_v \in \{0, 1, 2\}$ denote the label of vertex v and set up constraints to ensure the labeling is legal. Each triangle which cannot be further subdivided and has all vertices differently labeled adds one to the objective function, while all other triangles add zero. Sperner's lemma is now equivalent to the minimum value of this sum being at least one. Unfortunately, while this works for a given triangulation, it only works after a possibly lengthy computation and must be run for each problem. Thus, while this gives us an approach for a specific case, it does not give us a way to attack the general problems.

Counting Approach.

We thus revisit our idea from counting. One way to show something exists is to show there are an odd number of occurrences. We give another example of the power of this method in Exercise 15.5.2, where we show a real polynomial of odd degree must have a real root. These counting arguments are very powerful and useful in many settings. Sometimes instead of showing that there are an odd number of items we show there are many items, which also implies that there must be at least one. While this second approach may seem to be both absurdly overkill and impractical, in many cases this is the only way we know how to show the existence of a desired object. One of the more interesting examples concerns primes in arithmetic progression. Dirichlet proved that if a and m are relatively prime,

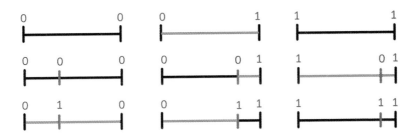

Figure 3. The various cases from the 1-dimensional Sperner proof. Notice in each of the six cases, the change in the number of 0–1 segments is even.

then there are infinitely many primes congruent to a modulo m. For some choices of a and m we can prove this directly (see Exercise 15.5.8, which is a modification of Euclid's argument from §6.5); interestingly, however, his argument cannot be generalized to arbitrary relatively prime a and m (see [**Mu**]).

Monovariant Approach.

We end with one last approach. Sometimes there are simple proofs that work in special cases that do not generalize; we've already seen an example of this phenomenon with one-dimensional fixed point theorems. We used the Intermediate Value Theorem to show a continuous $f : [0, 1] \to [0, 1]$ must have a fixed point; unfortunately that argument doesn't generalize to higher dimensions. In a similar spirit we can give a very simple proof of Sperner's lemma in one dimension. In fact, not only will we prove the claim but we'll also show the number of 0–1 segments is odd; unfortunately the approach does not extend to higher dimensions.

Consider a decomposition of $[0, 1]$ into a set of points $0 = p_0 < p_1 < \cdots < p_n = 1$ where each p_i is labeled 0 or 1 for $1 \le i \le n - 1$ and p_0 is labeled 0 and p_n is labeled 1. The idea is to add the points one at a time and keep track of the number of 0–1 intervals. We have three types of intervals: 0–0, 0–1, 1–1. Let's add the internal points one at a time, in any order, and keep track of what happens as each are inserted to the number of 0–1 segments. We depict this in Figure 3.

- 0–0: If we split a 0–0 segment by adding an interior point, if that point is labeled 0 we do not increase the number of 0–1 segments, while if it is labeled 1 we increase the number of 0–1 segments by 2.

- 0–1: If we split a 0–1 segment by adding an interior point, no matter what label we choose the number of 0–1 segments is unchanged.

- 1–1: If we split a 1–1 segment by adding an interior point, if that point is labeled 1 we do not increase the number of 0–1 segments, while if it is labeled 0 we increase the number of 0–1 segments by 2.

Thus, in each of the three cases, the addition of a new labeled point changes the number of 0–1 segments by an *even* number. As we begin with an odd number of 0–1 segments (when we just have p_0, labeled 0, and p_n, labeled 1), we must end with an odd number of such segments. The beautiful idea worth isolating here is that of an **invariant**. An invariant is a quantity which is unchanged throughout the process; sometimes it is useful to consider a **monovariant**, which is a quantity whose value either never increases or never decreases throughout a process. Here our invariant quantity is the number of 0–1 segments modulo 2 (i.e., the parity of the number of such segments). As the starting case is trivial to determine, we know the parity at all subsequent stages. We give additional examples of these concepts in the exercises.

Unfortunately, this approach does not generalize to higher dimensions. That said, it's worth seeing as it introduces a powerful perspective which can be used for a variety of other problems.

15.3. Proof of Sperner's Lemma

We now prove the two-dimensional Sperner's lemma; the challenge in proving the higher-dimensional case is not so much conceptual as it is notational. Our starting point is our second proof of the one-dimensional case, where we showed there existed a segment with labels 0 and 1 by showing the number of such segments is odd. We did this by double counting the number of times a label of 0 occurred, and getting a relation connecting this with $S(0,1)$, the number of segments with labels 0 and 1.

For the two-dimensional case, let $T(i,j,k)$ (with $0 \le i \le j \le k \le 2$) denote the number of triangles whose labels are i, j and k. We want a $(0,1,2)$ subtriangle. The idea of the proof is to somehow reduce back to the one-dimensional case.

Proof of Sperner's Lemma (2D). We count the number of segments whose endpoints are 0 and 1, which we call the 0–1 segments. There are none along the side of the initial triangle whose vertices' labels are drawn from $\{0,2\}$ or from $\{1,2\}$, while from the one-dimensional case we know there are an odd number along the outside edge with labels from $\{0,1\}$; the number of these is $S(0,1)$. This takes care of all 0–1 edges on the outside; all other edges are internal and are thus counted twice.

For example, consider the triangle from Figure 1, which for convenience we repeat here as Figure 4. There are three 0–1 segments along the bottom side, and 10 internal 0–1 segments (which we have highlighted). Each of the internal segments is part of exactly two internal triangles.

As we want 0–1 segments, the *only* triangles we need to study are those labeled $(0,0,1), (0,1,1)$, and $(0,1,2)$ (no other triangle can contribute a 0–1 segment). The first two each contribute two 0–1 segments, while the third contributes only one. Any internal 0–1 segment is counted twice, while those on the perimeter of the initial triangle are counted only once. Thus, if we add the number of 0–1 segments on the perimeter (which is $S(0,1)$) to the number of 0–1 segments from the triangles

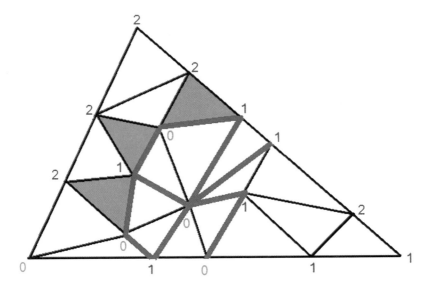

Figure 4. A two-dimensional example of Sperner's lemma, with the 0–1–2 triangles marked and the internal 0–1 segments highlighted.

we will have double counted all 0–1 segments:

$$2 \cdot T(0,0,1) + 2 \cdot T(0,1,1) + 1 \cdot T(0,1,2) + S(0,1) \;=\; 2(\# \text{ of } 0\text{–}1 \text{ segments}).$$

As the right-hand side is even the left-hand side must be even as well. The first two terms there are even, the fourth term is odd (from the one-dimensional case), and thus the third term, $T(0,1,2)$, must be odd. In other words, there are an odd number of triangles with distinct labels, and therefore there must be at least one.   □

While the proof of the general case is similar and proceeds by induction, it's worth remarking on notation. When the number of dimensions is small, it's fine to use notation such as $S(0,1)$ to denote the number of segments with labels 0 and 1, and $T(0,1,2)$ the number of triangles with labels 0, 1 and 2; as the dimensions increase and we have more components, this becomes unwieldy as we need to go through too many letters. It's thus useful to overload symbols a bit. We can have $S(0,1)$ denote the number of segments with a 0 and a 1, $S(0,0,1)$ the number of triangles with labels 0, 0, and 1, $S(0,1,2)$ the number of triangles with labels 0, 1, and 2, $S(0,1,2,3)$ the number of tetrahedra with labels 0, 1, 2, and 3, and so on. If you're uncomfortable about having S take tuples with a differing number of components, you can add a subscript to make it clear: thus write $S_2(0,1)$ for $S(0,1)$, $S_3(0,0,1)$ for $S(0,0,1)$, and so on. This is similar to overloading the absolute value, as we did in §14.1 and §13.3 where we used $|x - y|$ to denote the distance between two points x and y.

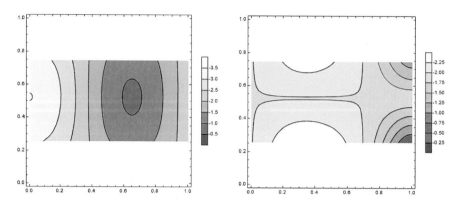

Figure 5. The valuations of two people on our bar; the left is how much the first person values each part, while the right is for the second.

15.4. Rental Harmony

This section is inspired by a terrific, and extremely readable, paper by Francis Su [**Su**]: *Rental Harmony: Sperner's Lemma in Fair Division* (available online at https://www.math.hmc.edu/~su/papers.dir/rent.pdf); see also [**FH-E**]. The paper begins with a simple and fascinating question.

> My friend's dilemma was a practical question that mathematics could answer, both elegantly and constructively. He and his housemates were moving to a house with rooms of various sizes and features, and were having trouble deciding who should get which room and for what part of the total rent. He asked, "Do you think there's always a way to partition the rent so that each person will prefer a different room?"

Su proceeds to show how this can be done by using Sperner's lemma, which he also proves. As he deals with that problem so nicely, we'll concentrate below on a simpler two person version which illustrates many of the issues in the subject, and encourage the interested reader to turn to [**Su**] next.

15.4.1. Preliminaries. Let's consider a block of length 1 and height $1/2$ which is divided between two people as follows: we make a vertical cut at x and everything from 0 to x goes to the first person, and everything from x to 1 to the second. Note this easily generalizes to two cuts to divide among 3 people, or in general $n-1$ cuts to divide among n; if we wanted, we could consider the harder case of allowing curves instead of straight line cuts (or, even more ambitiously, partitioning the block into any collection of subsets to distribute). Each person assigns different values to different parts of the bar, and we can then compute and see if they are happy with the outcome. We give an example in Figure 5; note the two players can disagree on the total value of the bar, as it could be worth different values to each.

The value person i assigns to the point $(x, y) \in [0, 1] \times [1/4, 3/4]$ is

$$f_1(x, y) \quad = \quad \frac{4 + \cos(x^2 + 4x)\exp(\cos x)\,|\sin(3y)|}{1.9177017188914786683},$$

$$f_2(x, y) \quad = \quad \frac{2 + \sin(x^3 + 4x)\exp\left(x\cos^2(3x)\right)\left(\frac{\cos^2(3y)}{1+\sqrt{y}}\right)^{1/2}}{1.0076865962803707166};$$

note that these densities are non-negative. They also integrate to 1 (or at least to a very, very close approximation); this means we have normalized each person's scale so each assigns the same value, namely 1, to having the entire bar, and no person considers having any part of the bar a liability.

Our goal is to have each person believing they received at least half the value of the bar, as if this is the case they will have no reason to switch sides. Of course, this begs the question as to whether or not this is a reasonable model of division. Specifically, we are assuming resources are continuously divisible, but this is not always the case; think of assets such as a house, a car, or a ticket to the Super Bowl. One way around this difficulty in the real world is to liquidate some of the assets, or just have a 'cash' asset, that is used in the division. We'll assume that's done here, and not comment on it any further, though it is possible to define an objective function and solve the division problem with binary integer programming.

15.4.2. Solution. While we use the functions from the previous subsection for the plots here, the argument applies in general. As we're only considering vertical cuts, we can introduce two functions: $u_1(x)$ for the first person and $u_2(x)$ for the second. These are utility functions, representing how much each player values obtaining the vertical strip at x of the bar. The valuation person 1 assigns to having the bar from 0 to x is just

$$\int_{t=0}^{x} u_1(t)\,dt \quad = \quad \int_{t=0}^{x}\int_{y=1/4}^{3/4} f_1(t, y)\,dy\,dt;$$

similarly the valuation person 2 assigns to having the bar from x to 1 is

$$\int_{t=x}^{1} u_2(t)\,dt \quad = \quad \int_{t=x}^{1}\int_{y=1/4}^{3/4} f_2(t, y)\,dy\,dt.$$

See Figure 6 for plots of the density and valuation. In our example, notice that the two people value different parts of the block differently. Maybe one likes frosting more than the other. Or perhaps one prefers the family car to the family boat. Or it could be one correctly sees the value of the complete works of Newton over that of Shakespeare....

We want to construct a division such that each person is satisfied. In other words, we have an **envy-free** allocation, where neither person wants what the other is receiving. As mentioned, we choose some point x from which to cut our bar. We claim the following conditions should be satisfied for x to be a good cut:

- $\int_0^x u_1(t)\,dt$ is at least as large as $\int_x^1 u_1(t)\,dt$; this means that the first person values their piece at least as much as the other. In particular, we immediately see they must value their cut as $1/2$ or more (since the two pieces have a combined value of 1).

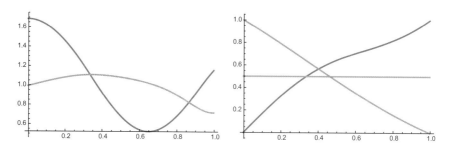

Figure 6. (Left) Plot of the value density for the two people. (Right) Plot of the valuation each player assigns for a cut at x.

- $\int_x^1 u_2(t)dt$ is at least as large as $\int_0^x u_2(t)dt$; this means that the second person values their piece at least as much as the other. As before, we find they must value their cut as $1/2$ or more.

Are there other conditions? For example, in all these discussions we're assuming the two people know each other's value functions and *the people are honest!* Note there is a strong temptation to lie; if you really want the part near your end but assign it a very low value, you could end up getting more than you "deserve". For the sake of this introductory discussion, we'll assume all players are honest and that there is perfect information; removing these assumptions lead to fascinating negotiation problems.

There are a lot of ways to try and solve this problem, and as Figure 6 shows there can be a continuum of x that work. The following is a simple condition that should seem very fair: if

$$R(x) \ := \ \int_0^x u_1(t)dt - \int_x^1 u_2(t)dt,$$

find x such that $R(x) = 0$. Note that if $R(x) = 0$, then the first person and the second person assign the same valuation to their respective pieces, and thus neither person feels the other is getting a better deal. This does *not* mean that the two are getting the best deal possible, as we'll see in a moment, but this issue will be easily fixed.

We first show that an x exists with $R(x) = 0$. Clearly $R(0) = -1$, because the second person has everything, and $R(1) = 1$, because now the first person has it all. By the Intermediate Value Theorem there must be some $x \in (0, 1)$ such that $R(x) = 0$. In fact, it's easy to find this x. Just start from 0 and slowly increase x until the two integrals are equal. Another efficient approach is to just use binary search, which involves choosing the midpoint each iteration, and if the midpoint is less than 0, ignore the left half and continue the iteration on the right half; otherwise continue on the left half if the midpoint is greater than 0. If the midpoint is 0, then we have found our x.

Unfortunately, there's a problem with this approach. Even if the two people get equal amounts relative to each other, they may not necessarily each get what they consider to be at least $1/2$ of the total. For example, suppose that person 1 values the left half as $1/4$ and the right half at $3/4$, whereas the values are swapped for the other. If we split the bar at the midpoint then each person gets a piece they value at 1/4, and hence each believes their deal is just as good as the other.

Fortunately there's a simple tweak which finishes the solution. All we need to do is switch the locations of the two players and then do the division again. Now each gets a piece worth 3/4.

15.5. Exercises

Statement of Sperner's Lemma

Exercise 15.5.1. *Let S be the region $\{(x, y) : 1 \leq x^2 + y^2 \leq 4\}$; thus S is the region from the circle of radius 1 centered at the origin to the circle of radius 2 centered at the origin (with the two circles included). Find a continuous map $f : S \to S$ such that f does not have a fixed point. Thus this S cannot be "nice". What property do you feel S is missing that allowed this example to be constructed?*

Exercise 15.5.2. *Let $f(x)$ be a polynomial with real coefficients and odd degree. Show that f has an odd number of real roots, and thus it must have a real root. Give an example of such a polynomial with degree 5 with exactly one real root, another with exactly three real roots, and a third with exactly five real roots.*

Exercise 15.5.3. *Let $f(x)$ be a polynomial with real coefficients and even degree. Show that f has an even number of real roots, but it need not have a real root. Give an example of such a polynomial with degree 4 with exactly zero real roots, another with exactly two real roots, and a third with exactly four real roots.*

Exercise 15.5.4. *Consider the proof of Sperner's lemma in one dimension. Must $S(0,0)$ be odd? Must it be even? What about $S(1,1)$? Must it be odd? Even? If the quantity is sometimes even and sometimes odd, give an example of each case.*

Exercise 15.5.5. *Consider the proof of Sperner's lemma in one dimension. Must $S(0,0) + S(1,1)$ always be odd? Even? If it is sometimes even and sometimes odd, give an example of each case.*

Exercise 15.5.6. *What do you think the statement of Sperner's lemma will be in three dimensions? How do you think the proof will proceed?*

Proof Strategies for Sperner's Lemma

Exercise 15.5.7. *Write down the integer Linear Programming problem to prove Sperner's lemma for a given subdivision of a triangle.*

The next few problems are beautiful examples of proofs where we prove that an object with a desired property exists by showing there are many such objects (and thus there must be at least one).

Exercise 15.5.8. *Euclid proved there are infinitely many primes as follows: if not say the set is finite: $\{p_1, p_2, \ldots, p_N\}$. Consider $p_1 \cdots p_N + 1$; either this is prime or it's divisible by a prime not in our list. Generalize this argument to show there are infinitely many primes congruent to 3 modulo 4, and discuss why this argument does not generalize to show there are infinitely many primes congruent to 1 modulo 4.*

Exercise 15.5.9. *Generalize the previous problem and determine other pairs of relatively prime integers a and m such that Euclid's argument can be extended to prove there are infinitely many primes congruent to a modulo m. See* [**Mu**] *for more on how far we can push this method.*

Exercise 15.5.10 (Dirichlet's Theorem on Primes in Arithmetic Progression). *Let $I(p^k, a; m)$ equal 1 if p^k is congruent to a modulo m, and 0 otherwise. Using a generalization of the Riemann zeta function (see §13.1 and the subsequent exercises), Dirichlet was able to prove that if a and m are relatively prime, then*

$$\sum_{p\ prime} \sum_{k=1}^{\infty} \frac{I(p^k, a; m)}{kp^{ks}}$$

tends to infinity as s decreases down to 1. Use this to prove there are infinitely many primes congruent to a modulo n. For a proof of this relation, see [**Da, MT-B**].

The purpose of the next problem is to highlight, yet again, that it's not enough to have an answer to a question; we need a way to compute it. The integral below gives the answer to one of the most important, open problems in number theory; while recent breakthroughs have finally allowed us to calculate it sufficiently well to resolve the problem of representing odd numbers as the sum of three primes, we cannot approximate well enough the corresponding integral for sums of two primes.

Exercise 15.5.11 (Ternary Goldbach Problem). *Euler conjectured that every sufficiently large even number is the sum of two primes, and every sufficiently large odd number is the sum of three. While the two prime case is still open (note: we believe 'sufficiently large' means at least 4), the ternary case with three primes is now a theorem: every odd integer which is at least seven is a sum of three primes. The main idea in the proof is the **Circle Method**; which we briefly describe. Let $a_3(m; N)$ denote the number of ways to write N as a sum of three primes, each at most N. Prove*

$$a_3(m; N) = \int_0^1 \left(\sum_{\substack{p\ prime \\ p \leq N}} e^{2\pi i p x} \right)^3 e^{-2\pi i m x} dx,$$

which expresses our desired quantity as an integral. As $a_3(m; N)$ is an integer, if we can show the integral is positive we win. This is where the Circle Method enters the story; we break the integral into two parts (the 'major arcs' and the 'minor arcs'), and show the contribution over the major arcs is positive and larger than the other term.

The final few problems concern invariants and monovariants. For these problems it often helps to find either a quantity that is conserved throughout a process or one that changes in only one direction.

Exercise 15.5.12. *Consider the integers $1, 2, \ldots, n$. Randomly choose two elements $x < y$ from the list; remove them and add $y - x$ to the list. Repeat this process, choosing two numbers and replacing the two with the larger minus the smaller, until our list contains just one number; is that number even or odd?*

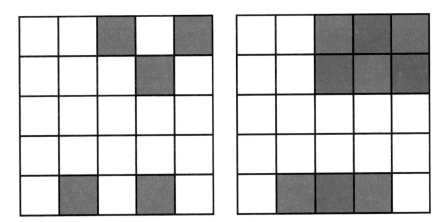

Figure 7. The spread of the infection. Left: initial configuration with five infected cells. Right: the spread of the infection one moment later (which is also the final state).

Exercise 15.5.13. *This is Problem A2 from the 2008 Putnam Math Competition. Start with a finite sequence $a_1, a_2, \ldots, a_n$ of positive integers. If possible, choose two indices $j < k$ such that a_j does not divide a_k, and replace a_j and a_k by $\gcd(a_j, a_k)$ and $\operatorname{lcm}(a_j, a_k)$, respectively. Prove that if this process is repeated, it must eventually stop and the final sequence does not depend on the choices made. (Note: gcd means greatest common divisor and lcm means least common multiple.)*

Exercise 15.5.14. *The following problem has been attributed to Tom Rike. Start with the set $\{3, 4, 12\}$. A move consists of choosing two elements of your set, a and b, and replacing them with $0.6a - 0.8b$ and $0.8a + 0.6b$. Can you transform the initial set into $\{4, 6, 12\}$ in finitely many moves?*

Exercise 15.5.15. *Assume there are N students in m classrooms. If there are students in multiple rooms, then one student is chosen at random to move from their room to a room with at least as many people. Prove that eventually everyone is in the same room.*

Exercise 15.5.16. *Consider an $n \times n$ checkerboard. The zombie virus has arrived, and once you are infected you are infected forever. Zombies can also create new zombies as follows: if a cell on the board shares at least two sides with infected cells, then it becomes infected (see Figure 7). Prove there is a configuration of n infected cells such that the infection will spread to every cell, but that there is no configuration of $n - 1$ infected cells such that the infection will spread everywhere.*

Exercise 15.5.17. *You have $2N$ coins of varying denominations (each is a nonnegative real number) in a line. Players A and B take turns choosing one coin from either end. Prove A always has a strategy that ensures she end up with at least as much as B.*

The monovariant in the next problems is extremely important and leads to a powerful technique known as the Fermat's **method of descent** (sometimes called the method of infinite descent). The idea is to associate a quantity which is a non-negative integer, and then show that if we have one solution we have another where the quantity is strictly smaller. As the smallest non-negative integer is zero, the process cannot go on infinitely often and we have a contradiction to the existence of a solution.

Exercise 15.5.18. *Prove $\sqrt{2}$ is irrational.* Hint: assume $\sqrt{2} = a/b$ with a, b positive integers. First prove a is even, then prove b is even, so we can write $a = 2a_1$ and $b = 2b_1$ with a_1, b_1 positive integers, and obtain $\sqrt{2} = a_1/b_1$. Repeat the process and obtain $a_1 = 2a_2, b_1 = 2b_2$ with a_2, b_2 integers and $\sqrt{2} = a_2/b_2$. Deduce that a contradiction as this process cannot be performed infinitely often, and thus our original assumption that a solution exists is false.

Fermat's Last Theorem asserts that if $x^n + y^n = z^n$ with $x, y, z \in \mathbb{Z}$ and $n \geq 3$ a positive integer, then either x, y or z is zero. This is now a theorem due to Wiles [**Wi**] and Taylor-Wiles [**TaWi**]; however, special cases have been known for a long time.

Exercise 15.5.19. *Prove there are no solutions to $x^4 + y^4 = r^2$ with $xyr \neq 0$; note the lack of solutions here implies there are no non-trivial solutions to $x^4 + y^4 = z^4$ (just take $r = z^2$).* Hint: assume a solution exists and prove a solution exists which is "smaller" relative to some property.

Proof of Sperner's Lemma

Exercise 15.5.20. *Recall $T(i, j, k)$ denotes the number of triangles with labels $0 \leq i \leq j \leq k \leq 2$. How many such triangles are there?*

Exercise 15.5.21. *Generalize the previous problem to four dimensions. How many tetrahedra are there with labels $0 \leq i \leq j \leq k \leq \ell \leq 3$?*

Exercise 15.5.22. *Consider a line segment. If we add k internal points is it always possible to use those, and only those, points to create a subdivision into line segments? If so is there only one possible way, or are there many ways? If there are many ways, how many ways are there to do so? If you cannot fully solve the problems look at some special cases to gather data and make conjectures.*

Sometimes looking at simpler problems, or lower dimension analogues, helps build intuition; sometimes, however, it does not help at all as it's too simple. Does the previous problem help with the next problem?

Exercise 15.5.23. *Consider an equilateral triangle. If we add k internal points is it always possible to use those, and only those, points to create a subdivision into triangles? If so is there only one possible way, or are there many ways? If there are many ways, how many ways are there to do so? If you cannot fully solve the problems look at some special cases to gather data and make conjectures.*

Exercise 15.5.24. *State and prove Sperner's lemma in four dimensions. In proceeding by induction is it enough to know Sperner's lemma in three dimensions, or do we also need it for one and two dimensions? If we need it for all lower dimensions, this is an example of **strong induction** (normal induction assumes the*

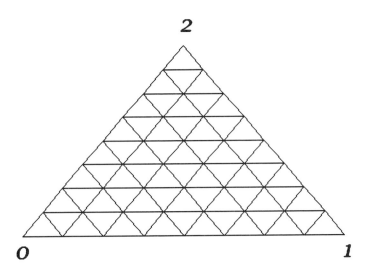

Figure 8. The 7th equal subdivision of a triangle.

claim for n and then shows it holds for $n + 1$, while strong induction assumes it holds for all $k \leq n$ and then proves it holds for $n + 1$).

Exercise 15.5.25. *State and prove Sperner's lemma in n dimensions.*

The next few problems assume some familiarity with probability. We say X is a **discrete random variable** if there is an at most countable set $\{x_i\}_{i \in I}$ such that $\text{Prob}(X_i = x_i) = p_i \in [0, 1]$, with $\sum_{i \in I} p_i = 1$. The **mean** (or **expected value**) of X, denoted μ_X or $\mathbb{E}[X]$, is

$$\mu_X = \sum_{i \in I} x_i p_i.$$

The **variance** of X, denoted σ_X^2, satisfies

$$\sigma_X^2 = \mathbb{E}[(X - \mu_X)^2] = \sum_{i \in I}(x_i - \mu_X)^2 p_i.$$

Consider an equilateral triangle. Call the nth equal subdivision the result of drawing n equi-spaced lines parallel to each of the three sides; Figure 8 shows the 7th equal subdivision. We consider below the probability space where each internal vertex is equally likely to be chosen from $\{0, 1, 2\}$, and the vertices on the i–j edge are drawn from $\{i, j\}$ with equal likelihood.

Exercise 15.5.26. *Prove σ_X^2 also equals $\mathbb{E}[X^2] - \mathbb{E}[X]^2$.*

Exercise 15.5.27. *Consider a line segment with n internal division points (and thus $n + 2$ points overall as there are two initial endpoints). Label the left endpoint 0, the right endpoint 1, and have each internal point be equally likely to be labeled 0 or 1. We proved $S(0, 1)$, the number of segments labeled 0–1, is always odd no matter what labeling is chosen; let's write $S_{n,\mathcal{L}}(0, 1)$ to emphasize the dependence on n and the labeling. Assume all labelings are equally likely. What is the expected value of $S_{n,\mathcal{L}}(0, 1)$?*

Exercise 15.5.28. *With notation as in the previous problem, what is the variance of $S_{n,\mathcal{L}}(0,1)$?*

Exercise 15.5.29. *If the mean and the variance of $S_{n,\mathcal{L}}(0,1)$ (from the last two problems) exist, denote them by μ_n and σ_n^2, and consider $X_{n,\mathcal{L}} = (S_{n,\mathcal{L}}(0,1) - \mu_n)/\sigma_n$. Do you think this converges to a limiting distribution as $n \to \infty$? If yes what distribution? Gather some evidence and try to prove your conjecture.*

Exercise 15.5.30. *Generalize Exercise 15.5.27 to two dimensions and an nth equal subdivision; specifically, given a valid labeling of a subdivision of a triangle (the three initial vertices are labeled 0, 1, and 2, the vertices on the external side with vertices 0 and 1 must be labeled 0 or 1, and so on), let $T_{n,\mathcal{L}}(0,1,2)$ denote the number of triangles labeled 0–1–2. Assume all legal labelings are equally likely. What is the expected value of $T_{n,\mathcal{L}}(0,1,2)$?*

Exercise 15.5.31. *With notation as in the previous problem, what is the variance of $T_{n,\mathcal{L}}(0,1)$?*

Exercise 15.5.32. *If the mean and the variance of $T_{n,\mathcal{L}}(0,1,2)$ (from the last two problems) exist, denote them by μ_n and σ_n^2, and consider $X_{n,\mathcal{L}} = (T_{n,\mathcal{L}}(0,1,2) - \mu_n)/\sigma_n$. Do you think this converges to a limiting distribution as $n \to \infty$? If yes what distribution? Gather some evidence and try to prove your conjecture.*

Rental Harmony

Exercise 15.5.33. *Consider the functions from §15.4.1. What is the value of the piece each gets using our method?*

Exercise 15.5.34. *Consider the functions from §15.4.1. If instead of making a vertical cut we instead can make any cut so long as it's a line, what is the valuation each will get according to our method?*

Exercise 15.5.35. *Generalize the previous problem where now we can make any curve of the form $y = g(x)$ for some continuously differentiable function $g(x)$. Note: this is quite a challenge, as our functions are not integrable in closed form.*

Exercise 15.5.36. *We assumed when dividing the bar that the two people were honest about their valuations. Show that if one person is honest and one lies that the liar can increase the value of their piece. There is a simple defense the other person can take against this.* Hint: make two cuts.

Exercise 15.5.37. *Assume we have a continuous rod $[0,1]$ and three people, Arroyo, Boyd, and Clemens. The value Arroyo assigns to a segment $[\alpha, \beta] \subset [0,1]$ is $\int_\alpha^\beta \mathfrak{f}_A(t)dt$, and similarly for the other two. Assume for $t \in [0,1]$ that*

$$
\begin{aligned}
\mathfrak{f}_A(t) &= 1, \\
\mathfrak{f}_B(t) &= 2t, \\
\mathfrak{f}_C(t) &= \frac{1/2 - (1/2 - t)^2}{5/12}.
\end{aligned}
$$

What divisions, if any, of the stick will all accept as fair?

Exercise 15.5.38. *Modify the previous problem so that now for $t \in [0, 1]$ we have*

$$\mathfrak{f}_A(t) \;=\; 1,$$

$$\mathfrak{f}_B(t) \;=\; \frac{\pi}{2}\sin\left(\frac{\pi t}{2}\right),$$

$$\mathfrak{f}_C(t) \;=\; 2\sin^2(t).$$

What divisions, if any, of the stick will all accept as fair?

Exercise 15.5.39. *More generally, modify the previous two problems so that there are N people, and person $n \in \{1, \ldots, N\}$ has the value function*

$$\mathfrak{f}_n(t) \;=\; nx^{n-1}.$$

What divisions, if any, of the stick will all accept as fair for $N = 4$? If you can, solve the problem for general N (if you cannot do it theoretically can you simulate an answer and conjecture how much the first or last player should take)?

Exercise 15.5.40. *Generalize these methods to dividing a block among three people. How do you determine where the cuts should be? Is there a similar issue as before in that each person could get a piece valued at less than $1/2$, and if so can this be resolved?*

Exercise 15.5.41. *Use Sperner's lemma to divide the rent for three people and three rooms. What reasonable assumptions should you make? See [**Su**] for the solution.*

Exercise 15.5.42. *Imagine there are P people and N items, and for each person p there is a non-negative function u_p such that they assign utility $u_p(n)$ to item n, with $\sum_{n=1}^{N} u_p(n) = 1$. Assume each item cannot be divided and must be assigned to one of the people. Choose variables, parameters, and constraints to set this up as almost a linear programming problem: the constraints must be linear but the objective function need not.*

Exercise 15.5.43. *If possible, redo the previous problem with a linear objective function.*

Chapter 16

Brouwer's Fixed Point Theorem

We now turn to the proof of Brouwer's fixed point theorem. We established Sperner's lemma in the previous chapter. Now we first review a needed theorem from real analysis, the Bolzano-Weierstrass Theorem and then introduce barycentric coordinates; those two tools will suffice for our proof.

16.1. Bolzano-Weierstrass Theorem

In this section we prove the Bolzano-Weierstrass Theorem. In addition to playing a key role in our passing from Sperner's lemma to the Brouwer fixed point theorem, it's of great use in a variety of analysis problems. One reason for its utility is that we can use it to show a continuous function attains its maximum and minimum value on a compact set, which is a useful ingredient in many optimization problems. See Exercises 16.6.7 to 16.6.9.

Theorem 16.1.1 (Bolzano-Weierstrass Theorem). *Let S be a compact (i.e., closed and bounded) subset of $\mathbb{R}^n$, and $\{x_n\}_{n=1}^{\infty}$ a sequence of points in S. Then there is a subsequence $\{x_{n_k}\}_{k=1}^{\infty}$ which converges to a point in S.*

Proof. We give the proof when $S \subset \mathbb{R}$ and leave the general case to the exercises. We first assume that $S = [0, 1]$.

- Split $[0, 1]$ into $[0, 1/2]$ and $[1/2, 1]$; without loss of generality at least one of these two subsets has infinitely many elements of our sequence (if both choose the left segment), and denote that segment S_1. Let n_1 be the smallest index such that $x_{n_1} \in S_1$.

- Split S_1 in half (if $S_1 = [0, 1/2]$ they are $[0, 1/4]$ and $[1/4, 1/2]$; if $S_1 = [1/2, 1]$ they are $[1/2, 3/4]$ and $[3/4, 1]$). By the same logic as before at least one of the two subintervals must have infinitely many elements of our sequence (if both choose the left segment), and denote that segment S_2. Let n_2 be the smallest index greater than n_1 such that $x_{n_2} \in S_2$.

- We continue by induction. Assume S_k has been constructed by induction (with corresponding index n_k); split S_k in half and at least one half contains infinitely many elements of our sequence. Denote that segment by S_{k+1}, and let n_{k+1} be the smallest index greater than n_k such that $x_{n_{k+1}} \in S_{k+1}$.

Consider now the sequence $\{x_{n_k}\}$. Note $x_{n_k} \in S_k$ and $|S_k| = 1/2^k$. As the sequence $\{x_{n_k}\}$ is Cauchy (see Theorem 13.3.2) it converges to a unique value x; as S is closed we must have $x \in S$, completing the proof when $S = [0,1]$.

For general $S \subset \mathbb{R}$, as S is compact it's bounded. Thus we may translate and dilate S so that $S \subset [0,1]$. Notice in the proof above we never needed $S = [0,1]$; the only argument used was that $S \subset [0,1]$, and thus a similar analysis handles general compact $S \subset \mathbb{R}$. See Exercise 16.6.6 for the general case.   □

16.2. Barycentric Coordinates

In linear algebra we've seen the power of different coordinate systems (a particular nice example concerns the ellipse not aligned with the coordinate axes from §13.6). This is a major theme of linear algebra: choose a coordinate system advantageous to the problem at hand. This leads to changing bases, where all bases have the same dimension.

In this section we introduce a new coordinate system. Interestingly, it uses $n+1$ variables to describe an n-dimensional object. At first this should seem to contradict results from linear algebra, as two bases of a finite-dimensional vector space must have the same dimension. It's not a contradiction, however, as there is a constraint among the variables and thus only n of them are free.

We now define the **barycentric coordinates**. Consider $n+1$ vectors $\vec{v}_0, \ldots, \vec{v}_n$ such that $\vec{v}_1 - \vec{v}_0, \ldots, \vec{v}_n - \vec{v}_0$ are **linearly independent** (a set of vectors $\{\vec{w}_1, \ldots, \vec{w}_k\}$ is linearly independent if $a_1 \vec{w}_1 + \cdots + a_k \vec{w}_k = \vec{0}$ implies $a_1 = \cdots = a_k = 0$). We consider the set

$$\{x_0 \vec{v}_0 + \cdots + x_n \vec{v}_n : x_0 + \cdots + x_n = 1, \ x_i \geq 0\}.$$

If we place equal masses at each $\vec{v}_i$, the center of mass corresponds to the point where each $x_i = \frac{1}{n+1}$; that point is called the barycenter of the region, hence the name of these coordinates. In two dimensions we have a triangle (three dimensions would be a tetrahedron, and so on). For a triangle, $(1,0,0)$ corresponds to being at the vertex $\vec{v}_0$, $(0,1,0)$ at $\vec{v}_1$, and $(0,0,1)$ at $\vec{v}_2$. We illustrate these coordinates in Figure 1. Note the bottom line corresponds to $x_2 = 0$; in other words, it's all points $(x_0, x_1, 0)$ with $x_0 + x_1 = 1, x_i \geq 0$.

The constraint $x_0 + \cdots + x_n = 1$ means $x_n = 1 - (x_0 + \cdots + x_{n-1})$, and thus there are only n free variables. We therefore have an n-dimensional space. Geometrically, when $n = 1$ we have a line segment from $\vec{v}_0$ to $\vec{v}_1$; when $(x_0, x_1) = (1,0)$ we are at $\vec{v}_0$, when $(x_0, x_1) = (0,1)$ we are at $\vec{v}_1$, and $(1/2, 1/2)$ corresponds to the midpoint of the line from $\vec{v}_0$ to $\vec{v}_1$.

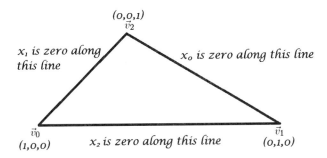

Figure 1. Barycentric coordinates for a 2-simplex.

Barycentric coordinates provide a useful way to parametrize a region. As the x_i are non-negative and sum to 1, we can view a point $\sum_i x_i \vec{v}_i$ as a weighted combination of the $\vec{v}_i$. We have seen the power of weights earlier, in §10.1 when we discussed multi-objective linear programming. There we had multiple items which we simultaneously cared about, and we assigned different non-negative weights according to how much they matter. When deciding how to weigh, we might as well have the weights sum to 1; if they summed to 50 we could just re-scale everything by dividing by 50 without changing the *relative* importance of the items.

If we have freedom, we often translate our region so that $\vec{v}_0$ is the zero vector. This is a natural choice and is possible if we're only studying one region. If, however, we are looking at multiple regions simultaneously, then in general this is not possible. This is why we allow ourselves the extra flexibility that comes with a general $\vec{v}_0$, and do not assume $\vec{v}_0 = \vec{0}$. Also, one of the reasons linear algebra is so useful is that it provides a clean way to handle computations without being bogged down in coordinate calculations. We have a similar gain here with barycentric coordinates.

16.3. Preliminaries for Brouwer's Fixed Point Theorem

We state the Brouwer fixed point theorem in the case of the standard n-**simplex**:

$$\Delta_n := \{\vec{x} \in \mathbb{R}^{n+1} : x_0 + \cdots + x_n = 1, x_i \geq 0\}.$$

Thus, what geometrically the 1-simplex corresponds to an interval and the 2-simplex to a triangle; see Figure 2.

We'll give the proof in two dimensions, as that's enough to see what's going on, but still small enough to keep the notation manageable and allow us to have good visual clues. This is similar to our fixed point analysis from before; one dimension was too simple to really illustrate what is going on (we obtained the fixed point by invoking the Intermediate Value Theorem). It's also similar to our analysis of Sperner's lemma, where in one dimension we could give some very simple proofs that don't generalize (and we could also give one that does).

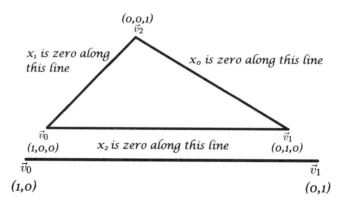

Figure 2. Barycentric coordinates for a 1-simplex and a 2-simplex.

Theorem 16.3.1 (Brouwer fixed point theorem). *Any continuous $f : \Delta_n \to \Delta_n$ has a fixed point.*

Before giving the proof, we describe the key idea. We'll assume f has no fixed points and deduce a contradiction (if f had a fixed point the claim would be easy to show!). Thus $f(x_0, \ldots, x_n) \neq (x_0, \ldots, x_n)$ for all points in Δ_n. In particular, we can define an **index function** Ind_f associated to f at $(x_0, \ldots, x_n) \in \Delta_n$ as the first index i such that the ith coordinate of $f(x_0, \ldots, x_n)$ is less than x_i, *remembering that our coordinates are labeled as the zeroth, then the first, then the second, and so on.* Thus if

$$f(1/4, 1/2, 1/4) \ = \ (1/4, 1/6, 7/12)$$

we have $\mathrm{Ind}_f(1/4, 1/2, 1/4) = 1$, as the zeroth coordinate of $f(1/4, 1/2, 1/4)$ equals the zeroth coordinate of our point, while the first coordinate differs from x_1; if instead

$$f(1/4, 1/2, 1/4) \ = \ (1/6, 4/7, 11/42),$$

then $\mathrm{Ind}_f(1/4, 1/2, 1/4) = 0$. The index function is well defined (see Exercise 16.6.14).

When f has no fixed points, the index function is related to our labeling in Sperner's lemma. Let's consider the 2-simplex from Figure 2. Since we're assuming f has no fixed points, we must have

$$\mathrm{Ind}_f(1,0,0) \ = \ 0, \quad \mathrm{Ind}_f(0,1,0) \ = \ 1, \quad \mathrm{Ind}_f(0,0,1) \ = \ 2.$$

For example, consider $\mathrm{Ind}_f(0,1,0)$. As f has no fixed points, $f(0,1,0) \neq (0,1,0)$. Thus the middle coordinate of $f(0,1,0)$ (corresponding to index 1) cannot be 1. Since the sum of the barycentric coordinates is 1, if it is not 1 it *must* be less than 1, and this will be the first index where the coordinate of $f(0,1,0)$ is less than the corresponding coordinate of $(0,1,0)$. The proof for the other two claims is similar. Thus if we use Ind_f to assign a label of 0, 1, or 2 to a point, we would label the vertex $\vec{v}_0$ with 0, the vertex $\vec{v}_1$ with a 1, and $\vec{v}_2$ with a 2; see Figure 3.

We continue and consider how we would label any point on the boundary of the triangle. For example, consider the bottom edge where $x_2 = 0$; these points are

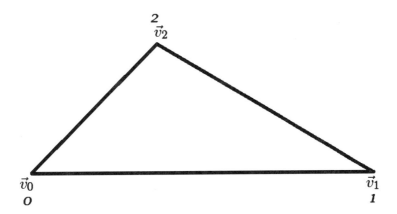

Figure 3. Labeling the vertices with the index function when f has no fixed points.

of the form $(x_0, x_1, 0)$ with $x_0 + x_1 = 1$ and $x_i \geq 0$. Let

$$f(x_0, x_1, 0) = (y_0, y_1, y_2) \neq (x_0, x_1, 0);$$

we need to determine $\text{Ind}_f(x_0, x_1, 0)$, the smallest i such that $y_i < x_i$. Could $i = 2$? If so, then

$$x_0 \leq y_0 \quad \text{and} \quad x_1 \leq y_1.$$

If we add these two inequalities and recall that $x_0 + x_1 = 1$ we find

$$1 = x_0 + x_1 \leq y_0 + y_1.$$

As $(y_0, y_1, y_2) \in \Delta_2$ we must have $y_0 + y_1 + y_2 = 1$. Thus the inequality above forces $y_0 + y_1$ to equal 1, and thus $y_2 = 0$; as $x_2 = 0$, we see that this cannot be the first index where $y_i < x_i$ (as they are equal here!). Thus on the line $x_2 = 0$ we must have $\text{Ind}_f(x_0, x_1, 0) \in \{0, 1\}$. We similarly find $\text{Ind}_f(x_0, 0, x_2) \in \{0, 2\}$ and $\text{Ind}_f(0, x_1, x_2) \in \{1, 2\}$. We collect these results in Figure 4. Notice this is precisely the setting for Sperner's lemma; the vertices are labeled 0, 1, and 2, and between vertices 0 and 1 we can only label with a 0 or a 1, and so on.

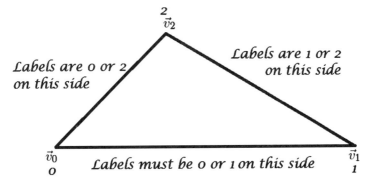

Figure 4. Labeling the vertices and sides with the index function when f has no fixed points.

We give the proof of Brouwer's theorem in the next section. The idea is to subdivide Δ_n into smaller and smaller unions of triangles, use Ind_f to label the subdivision points, and then show that a sub-sequence of triangles with distinct labels converges to a fixed point of f.

16.4. Proof of Brouwer's Fixed Point Theorem

We now prove Brouwer's fixed point theorem for a continuous $f : \Delta_2 \to \Delta_2$; the proof for general n follows similarly. Once we have the result for maps from Δ_n to itself we can extend to regions that are equivalent to Δ_n by the methods of §13.6.

Given any triangle we can subdivide it into four smaller triangles by connecting the midpoints of the three sides. We can then subdivide these new triangles in the same way, and continue the process as many times as we wish. We illustrate this in Figure 5. Let $\mathcal{T}_m$ be the collection of triangles arising from m subdivisions; note each triangle in $\mathcal{T}_m$ has area equal to $1/4^m$ of the area of the initial triangle and perimeter $1/2^m$ of the original triangle.

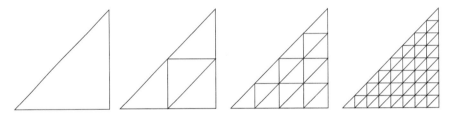

Figure 5. Subdividing a triangle through the midpoints of the sides (first three iterations).

We assume f does not have a fixed point, and use the index function Ind_f from §16.3 to label the vertices of the original triangle and the subdivision points. The labeling must be as described in Figure 4; in particular, vertex $\vec{v}_i$ is labeled i, and along the outside edge of the initial triangle running from $\vec{v}_i$ to $\vec{v}_j$ the labels are either i or j. For all internal vertices of triangles in $\mathcal{T}_n$, the label may be either 0, 1, or 2.

By Sperner's Lemma, for each m there must be a triangle T_m in $\mathcal{T}_m$ whose vertices have distinct labels. Let p_m be the midpoint of triangle T_m. By the Bolzano-Weierstrass Theorem there is a subsequence of $\{p_m\}$, say $\{p_{m_k}\}$, which converges to a point $p \in \Delta_2$. We also have three other useful sequences, coming from the vertices of the triangles $\{T_m\}$. For $i \in \{0, 1, 2\}$ let $\nu_m^{(i)}$ be the vertex of T_m which is labeled i; we also have each $\nu_m^{(i)} \to p$ (see Exercise 16.6.20).

We claim that p is our fixed point. The key idea is extracted in Exercise 16.6.18: as we're using barycentric coordinates if we have two points (x_0, x_1, x_2) and (y_0, y_1, y_2) with $x_i \geq y_i$ for all i, then $x_i = y_i$.

Let
$$p = (p_0, p_1, p_2) \quad \text{and} \quad f(p_0, p_1, p_2) = (q_0, q_1, q_2) \quad \text{with } q_0 + q_1 + q_2 = 1, q_i \geq 0.$$
If p is not a fixed point of f, then there is a first i with $q_i < p_i$. We now deduce a contradiction.

As the triangles converge to p (remember each triangle is half the size of the triangles at the previous level), their vertices converge to p:

$$\nu^{(0)}_{m_k}, \nu^{(1)}_{m_k}, \nu^{(2)}_{m_k} \; \to \; p.$$

By the definition of $\nu^{(i)}_{m_k}$, we see

$$\mathrm{Ind}_f(\nu^{(i)}_{m_k}) \; = \; i;$$

in other words, the smallest index where f at the vertex $\nu^{(i)}_{m_k}$ is smaller than the ith coordinate of $\nu^{(i)}_k$ is i. Therefore the ith coordinate of $\lim_{k \to \infty} f(\nu^{(i)}_{m_k})$ cannot be larger than the ith coordinate of the limit $\lim_{k \to \infty} \nu^{(i)}_{m_k} = p$. Thus the zeroth coordinate of $f(p)$ cannot be larger than the zeroth coordinate of p, and similarly the first and second coordinates of $f(p)$ cannot be larger than the first and second coordinates of p. We therefore have

$$q_0 \; \le \; p_0, \quad q_1 \; \le \; p_1, \quad q_2 \; \le \; p_2.$$

As these are barycentric coordinates their sum is 1, and thus

$$1 \; = \; q_0 + q_1 + q_2 \; \le \; p_0 + p_1 + p_2 \; = \; 1;$$

the only way this can hold is if all the inequalities are equalities. In other words, $q_i = p_i$ for all i, and thus p is a fixed point, completing the proof. $\qquad\square$

16.5. Nash Equilibrium

16.5.1. Introduction and Terminology. One of the most famous and useful applications of Brouwer's fixed point theorem is in the proof of the existence of Nash Equilibria, which is an important result in game theory. John Nash was a prolific mathematician and professor at Princeton in the mid-twentieth century. He was eventually awarded the Nobel Prize in Economics for his discovery of non-cooperative equilibria, which was his graduate thesis [**Na**]; at under 30 pages, it's worth the time to read the original, though as his results are used so widely there is vast literature on the subject. Much of this chapter is drawn from some online references aimed at students, such as [**JL-B, Kh, Ku, Sp**].

Before Nash's work, there was an extensive literature on two person zero-sum games (so if one person wins X, the other loses X); the classic work on the subject was [**vNM**]. Nash's monumental contribution was to show that in many multiplayer, non-cooperative, non-zero sum games there is an equilibrium where no player has an incentive to change their strategy. The purpose of this section is to briefly introduce the reader to this field.

We begin by introducing game theory. **A finite, n-person game** is a tuple (N, A, u), where

- N is a finite set of n players;
- $A = (A_1, \dots, A_n)$ with A_i the set of actions available to player i;

- vectors $a = (a_1, \ldots, a_n) \in A$ called profiles, where each a_i is the action taken by player i in the profile a;

- $u = (u_1, \ldots, u_n)$ where $u_i : A \to \mathbb{R}$ is a utility function for player i.

Essentially, a game presents each player with a set of choices, known as pure strategies. Each of these leads to an outcome when in combination with other players' pure strategies. Each outcome gives the player an amount of utility, and every player is trying to maximize their utility.

Nash's Theorem concerns a subset of games known as non-cooperative games. A game that allows each player to confer with the others, enabling them to conspire to maximize utility, is a cooperative game. Any game that is not cooperative is **non-cooperative**.

Imagine all major airlines got together and agreed to raise their prices collectively. Then each company makes more money, and therefore greater utility, than they would have otherwise. This is a real-world example of a cooperative game and is known in economics as a cartel. However, if the companies had not conspired before the price adjustment, one airline could have kept prices low and taken customers from the others. This would maximize their own profit, and thus utility, but the global utility would not be maximized. Such a situation is an example of a non-cooperative game. Many classic games fall in the non-cooperative category, such as backgammon, bridge, chess, checkers if we stay with just the first few letters of the alphabet (though this means we miss great examples such as tic-tac-toe and all its variants); fortunately not everything falls in this category (if it did, international diplomacy would be very different).

For any player i, the set of mixed strategies is $S_i = \pi(A_i)$, where π is the set of probability distributions over any set. Denote a single mixed strategy as s_i, and $s_i(a_i)$ as the probability player i chooses strategy a_i in mixed strategy s_i.

In other words, a mixed strategy is just a vector of probabilities that a player chooses each pure strategy, such that the sum of the probabilities is one. For example, in the game of rock, paper, scissors, if I choose rock half of the time, paper one-third of the time, and scissors one-sixth of the time, my mixed strategy is $(1/2, 1/3, 1/6)$. This game easily illustrates many parts of the theory, and we will revisit it later. For the extension of this game to rock, paper, scissors, lizard, Spock, see `https://youtu.be/x5Q6-wMx-K8` and the additional weapons (Section 7.3) subpage of `https://en.wikipedia.org/wiki/Rock-paper-scissors`.

We end with an important remark that sets the stage for the entrance of the Brouwer fixed point theorem. *Note that because a mixed strategy is a vector of non-negative numbers that sum to 1, it can be interpreted as a point in a simplex.* A cartesian product of simplexes, such as the one described in the next subsection, is known as a simplotrope. The proof we gave of Brouwer's theorem can be easily extended to apply to simplotropes.

16.5.2. Statement of Nash's Theorem. We begin by defining the type of equilibria we will find. Any cartesian product of mixed strategies (or mixed strategy profile) $s_1 \times s_2 \times \cdots \times s_n$ is a Nash Equilibrium if the payoff u_i for each player i

when he uses s_i is as great or greater than any other mixed strategy, given that all other players are using their respective mixed strategies.

Equivalently, if a game is in Nash Equilibrium, no player would want to change mixed strategies given the strategies of the other players.

Next we define the **expected utility of a mixed strategy** u_i for a player i in a mixed strategy s as

$$u_i(s) \; = \; \sum_{a \in A} u_i(a) \prod_{j=1}^{n} s_j(a_j).$$

We also need to define a player i's **best response** to a mixed strategy profile not including his own strategy (denoted s_{-i}) as some $s_i^* \in S_i$ such that $u_i(s_i^*, s_{-i}) \geq u_i(s_i, s_{-i})$ for all $s_i \neq s_i^*$

Before we state and prove Nash's Theorem, let's examine a simpler example in order to cement our understanding of the concept. This game is called the prisoners' dilemma, and is a classic example of Nash Equilibrium, devised by Merrill Flood and Melvin Dresher in 1950, and named by Albert Tucker (see for example [**Ku**]). The result is surprising in that the equilibrium point is not the global optimal point.

Imagine we have a great video recording of two criminals robbing a convenience store; our exposition follows that in [**Kh**]. Let's call them Flood and Dresher. When interrogating them, the district attorney (DA) suspects that earlier that month they committed a murder in an unsolved case. Knowing that the robbery case is open-and-shut, the DA approaches the two criminals individually with a plea deal.

- If neither confess to the murder, they'll each get two years of jail time for the robbery.
- If both confess, they'll each get three years of jail time for both crimes, with some leniency shown for confessing.
- If one confesses and the other does not, the confessor will get one year of time in his plea deal, and the non-confessor will get ten years for staying quiet.

Clearly, the global optimal (for the two criminals!) is that both offenders stay quiet, resulting in a relatively short two years each in prison, given that they've committed several large crimes. However, as we will see, the Nash Equilibrium is actually that both Flood and Dresher will confess (and it does not matter whether or not they killed the person!).

If you're Flood, you have two possible strategies: Either confess, or don't. You know that Dresher has the same options. Imagine Dresher confesses. Which option is better for you? Clearly, it's better to confess so that you end up with three years instead of ten. What if Dresher doesn't confess? Well, then you're ecstatic, since you can confess and end up with only one year! So, regardless of what strategy Dresher chooses, it's better for Flood to confess. However, as the situation is symmetric the exact same analysis for Dresher has him arrive at the same result: confess. Thus, two confessions is the Nash Equilibrium in this game (it turns out that all other strategies are not equilibria, which can be shown relatively simply).

We conclude by stating Nash's result, which we prove in the next subsection.

Theorem 16.5.1 (Nash [**Na**]). *Every finite game has an equilibrium point (in mixed strategies).*

16.5.3. Proof of Nash's Theorem. We begin the proof by setting

$$\Phi_{i,a_i}(s) := \max\{0, u_i(a_i, s_i) - u_i(s)\},$$

where s is a strategy profile in S, $i \in N$, and $a_i \in A_i$. Then, we let $f : S \to S$ be such that $f(s) = s'$, where

$$s'_i(a_i) := \frac{s_i(a_i) + \Phi_{i,a_i}(s)}{\sum_{b_i \in A_i} s_i(b_i) + \Phi_{i,b_i}(s)} = \frac{s_i(a_i) + \Phi_{i,a_i}(s)}{1 + \sum_{b_i \in A_i} \Phi_{i,b_i}(s)}.$$

We can interpret f as mapping s to a new profile that gives better responses higher probabilities than in s.

Note that f is continuous because each Φ is continuous and that S is convex and compact. Applying Brouer's theorem to f, we see it has at least one fixed point. We hypothesize that these fixed points *are* Nash Equilibria, and now show that this is in fact the case.

Let s be an arbitrary fixed point. Let a'_i be an action with non-zero probability in s such that $u_{i,a'_i}(s) \leq u_i(s)$. Note $\Phi_{i,a'_i}(s) = 0$. Because s is a fixed point, $s'_i(a'_i) = s_i(a'_i)$, and hence

$$s_i(a'_i) = \frac{s_i(a_i) + \Phi_{i,a_i}(s)}{1 + \sum_{b_i \in A_i} \Phi_{i,b_i}(s)} = \frac{s_i(a'_i) + 0}{1 + \sum_{b_i \in A_i} \Phi_{i,b_i}(s)}.$$

We also know $s_i(a'_i)$ is positive because we defined it as such, so we can conclude that the denominator of the above fraction equals 1. This allows us to say that $\sum_{b_i \in A_i} \Phi_{i,b_i}(s) = 0$, and since all values of $\Phi_{i,b_i}(s)$ are positive, they must also be 0. So, under s, no player can *improve* their expected utility payoff by switching to a pure strategy! Therefore, s is a Nash Equilibrium. $\square$

16.5.4. Example: Rock, Paper and Scissors. We conclude by re-examining the rock, paper, scissors game from earlier (see for example [**Sp**]). I'll assume that the rules of the game are familiar. If not, search engines are your friend. We'll also assume in this game that the utility payoff for winning is $+1$ and for losing is -1, while a draw receives 0 utility. It's reasonable to posit that the Nash Equilibrium in this game is the mixed strategy profile in which each player plays each action with probability $1/3$. This happens to be the *only* Nash Equilibrium, but proving that is time-consuming and not interesting, so we'll just show that this is *a* Nash Equilibrium. We can do so by simply finding the expected utility for each player, and confirming there is no pure strategy that one player can switch to that will raise his expected utility.

First, we calculate expected utility. Since each player's mixed strategy is the same, we need only do it once. It should be clear that for each move, player A has a $1/3$ chance of winning, losing, or drawing, and that the expected payoff will sum to 0 over all the moves. To explain it another way, consider the following: Both player A and player B have a $1/3$ chance to play rock, thus there is a $1/9$ chance of a tie as the result of double-rock. Similarly, there is a $1/9$ chance of any other combination. Multiply each probability by the resulting utility for player A, then add to get 0. A similar analysis will see that player B's expected payoff is also 0.

Now we examine the possibility that switching to a pure strategy will raise player A's utility, given player B remains in the mixed strategy above. Remember, in order to prove this point is *not* a Nash Equilibrium, we must *raise* the expected utility of one of the players. There are three pure strategies available to player A, so let's examine them. First, imagine player A switches to playing only paper. His expected payoff is now

$$1(\frac{1}{3})(\text{paper}_A)(\text{rock}_B) + 1(\frac{1}{3})(\text{paper}_A)(\text{paper}_B) + 1(\frac{1}{3})(\text{paper}_A)(\text{scissors}_B)$$

$$= 1(\frac{1}{3})(1) + 1(\frac{1}{3})(0) + 1(\frac{1}{3})(-1) = \frac{1}{3} + 0 - \frac{1}{3} = 0.$$

Therefore, switching to the pure strategy paper does not increase player A's expected utility. We can run a similar analysis on rock and scissors, and indeed on all of player B's strategies, to see that our original mixed strategy is indeed a Nash Equilibrium.

16.6. Exercises

Bolzano-Weierstrass Theorem

Exercise 16.6.1. *In the Bolzano-Weierstrass Theorem we assumed our set S was compact (closed and bounded). Show the theorem's claim can fail if our set is not closed.*

Exercise 16.6.2. *In the Bolzano-Weierstrass Theorem we assumed our set S was compact (closed and bounded). Show the theorem's claim can fail if our set is not bounded.*

Exercise 16.6.3. *Consider $x_n = \cos(n)$ (measured in radians). The Bolzano-Weierstrass Theorem asserts it has a subsequence which converges to a point in $[-1, 1]$. Explicitly find such a subsequence. Is it easier, harder, or the same to prove the analogous statement for $y_n = \sin(n)$? Do so.*

Exercise 16.6.4. *The previous exercise concerned convergent subsequences of $x_n = \cos(n)$ and $y_n = \sin(n)$. More is true. Prove any $x \in [-1, 1]$ such that there is a subsequence of $\{x_n\}$ converging to x, and similarly for any $y \in [-1, 1]$ and $\{y_n\}$.*

Exercise 16.6.5. *Consider a sequence of finite closed intervals $\{I_k\}$ such that $I_{k+1} \subset I_k$ and $|I_{k+1}| = \frac{1}{2}|I_k|$ for all k. Prove $\bigcap_{k=1}^{\infty} I_k$ is a unique point. Further, if $\{y_k\}_{k=1}^{\infty}$ is a sequence such that $y_k \in I_k$, then $\lim_{k \to \infty} y_k$ exists and is the unique point of intersection of all the sets. Hint: we can write I_k as $[a_k, b_k]$ and the sequences satisfy $a_k \leq a_{k+1} \leq b_{k+1} \leq b_k$.*

Exercise 16.6.6. *We sketch the proof of the general case of the Bolzano-Weierstrass Theorem. We first do $n = 2$. Show without loss of generality we may assume $S \subset [0, 1] \times [0, 1]$. Divide into four squares by adding the midpoint $(1/2, 1/2)$, and show that at least one of these four squares contains infinitely many terms in our sequence. How does the general case proceed?*

Exercise 16.6.7. *Prove that if f is continuous on a compact set S, then f attains its maximum and minimum on S; in other words, there are $x_{\max}$ and $x_{\min}$ such that for all $x \in S$ we have $f(x_{\min}) \leq f(x) \leq f(x_{\max})$.*

Exercise 16.6.8. *Give an example to show the result of the previous exercise, namely that f attains its maximum and minimum values on S, and fails if S is not compact.*

Exercise 16.6.9. *Give an example to show the result of Exercise 16.6.7, namely that f attains its maximum and minimum values on S, and fails if f is not continuous.*

Barycentric Coordinates

Exercise 16.6.10. *Let $\{\vec{v}_1, \ldots, \vec{v}_n\}$ and $\{\vec{w}_1, \ldots, \vec{w}_m\}$ be two different bases of the same subspace of $\mathbb{R}^r$. Prove $n = m$.*

Exercise 16.6.11. *Assume $\vec{v}_1 - \vec{v}_0, \ldots, \vec{v}_n - \vec{v}_0$ are linearly independent; notice that we have made $\vec{v}_0$ play a special role. What if instead of subtracting $\vec{v}_0$ we subtract a different vector; would the resulting n vectors also be linearly independent?*

Exercise 16.6.12. *Given*

$$\vec{v}_0 = (0,0,0), \quad \vec{v}_1 = (1,2,3), \quad \vec{v}_2 = (6,5,9), \quad \vec{v}_3 = (4,8,2),$$

write down the Cartesian coordinates corresponding to the point whose barycentric coordinates are (x_0, x_1, x_2, x_3). What is the center of the region in barycentric coordinates? In Cartesian coordinates?

Exercise 16.6.13. *Consider the barycentric coordinates for the triangle with*

$$\vec{v}_0 = (0,0), \quad \vec{v}_1 = (1,0), \quad \vec{v}_2 = (1/2, \sqrt{3}/2).$$

What is the center of the triangle in barycentric coordinates? In Cartesian coordinates? Draw the region, and draw the lines corresponding to $x_0 = 0$, $x_1 = 0$, $x_2 = 0$, $x_0 = x_1$, $x_0 = x_2$, and $x_1 = x_2$.

Preliminaries for Brouwer's Fixed Point Theorem

Exercise 16.6.14. *Prove the index function Ind_f is well defined.*

Exercise 16.6.15. *Let $f : \Delta_2 \to \Delta_2$ be given by $f(x_0, x_1, x_2) = (x_1, x_2, x_0)$. Determine the values of Ind_f; in other words, where in Δ_2 we have $\mathrm{Ind}_f(x_0, \ldots, x_2)$ equalling 0, 1, and 2.*

Exercise 16.6.16. *Assume $f : \Delta_n \to \Delta_n$ has no fixed points. What is $\mathrm{Ind}_f(\vec{e}_i)$, where $\vec{e}_i$ is the $n+1$ tuple which has a 1 in the ith index and 0 elsewhere; thus $\vec{e}_0 = (1, 0, \ldots, 0)$ and $\vec{e}_n = (0, \ldots, 0, 1)$.*

Exercise 16.6.17. *Assume $f : \Delta_n \to \Delta_n$ has no fixed points. What are the possible values for Ind_f on the face $x_k = 0$?*

Proof of Brouwer's Fixed Point Theorem

Exercise 16.6.18. *Consider two points in barycentric coordinates, $(x_0, \ldots, x_n)$ and $(y_0, \ldots, y_n)$; note $x_i, y_i \geq 0$ and*

$$x_0 + \cdots + x_n = y_0 + \cdots + y_n = 1.$$

Prove if $x_i \geq y_i$ for all i, then $x_i = y_i$.

Exercise 16.6.19. *Generalize the proof of Brouwer's fixed point theorem to arbitrary n.*

Exercise 16.6.20. *Prove that $\nu_{m_k}^{(i)} \to p$ for $i \in \{0, 1, 2\}$.*

Exercise 16.6.21. *How many vertices are there in $\mathcal{T}_m$? How many edges? How many triangles?*

Exercise 16.6.22. *Find all fixed points of $f(x_0, x_1, x_2) = (x_1 x_2, x_0 x_2, x_0 x_1)$.*

Exercise 16.6.23. *Find all fixed points of*

$$f(x_0, x_1, x_2) = \left(\frac{x_0 + x_1}{2}, \sqrt{x_1 x_2}, 1 - \frac{x_0 + x_1}{2} - \sqrt{x_1 x_2} \right).$$

Exercise 16.6.24. *Find all fixed points of*

$$f(x_0, x_1, x_2) = \left(\frac{x_0^2 + 1}{4}, \frac{x_0 x_2}{2}, 1 - \frac{x_0^2 + 2}{4} - \frac{x_0 x_2}{2} \right);$$

if you cannot solve this in closed form approximate the solution.

Exercise 16.6.25. *Consider the map from the unit disk (all points $(x, y) : x^2 + y^2 \leq 1$) to itself given by*

$$f(x, y) = \left(\frac{(y - 1/2)^2}{8}, \frac{(x - 1/2)^2}{8} \right).$$

Does this map have any fixed points? Why or why not? If yes find or approximate it.

Exercise 16.6.26. *Can our proof of Brouwer's fixed point theorem be turned into a constructive argument to find the fixed point? Why or why not?*

Nash Equilibrium

Exercise 16.6.27. *Imagine your opponent in rock, paper, scissors plays rock 40% of the time, paper 35% of the time, and scissors 25% of the time. What should your strategy be?*

Exercise 16.6.28. *Consider a weighted version of rock, paper, scissors where if you win with rock you get $3, if you win with paper you get $2, and if you win with scissors you get $1. What should your strategy be now? Is there a Nash equilibrium?*

The Prisoner's Dilemna problem below is one of the most famous games. It was originally proposed by Merrill Flood and Melvin Dresher in 1950; the version below is Albert Tucker's phrasing (taken from the Wikipedia article https://en.wikipedia.org/wiki/Prisoner%27s_dilemma).

Exercise 16.6.29 (The Prisoner's Dilemna). *Two members of a criminal gang are arrested and imprisoned. Each prisoner is in solitary confinement with no means of communicating with the other. The prosecutors lack sufficient evidence to convict the pair on the principal charge. They hope to get both sentenced to a year in prison on a lesser charge. Simultaneously, the prosecutors offer each prisoner a bargain. Each prisoner is given the opportunity either to betray the other by testifying that the other committed the crime or to cooperate with the other by remaining silent. The offer is:*

- *If A and B each betray the other, each of them serves 2 years in prison.*

- *If A betrays B but B remains silent, A will be set free and B will serve 3 years in prison (and vice versa).*

- *If A and B both remain silent, both of them will only serve 1 year in prison (on the lesser charge).*

Analyze the strategies of the two players. Are there any equilibria? Any Nash equilibria?

Exercise 16.6.30. *Create a generalized Prisoner's Dilemna with 3 people, or with n.*

The data for the following problem is taken from Joel Wiles' honors thesis, *Mixed Strategy Equilibrium in Tennis Serves*, available at `https://sites.duke.edu/djepapers/files/2016/10/Wiles.pdf`.

Exercise 16.6.31. *To simplify a game of tennis, we shall concentrate only on the serve (though a similar analysis would work for penalty kicks in soccer). The server may choose to serve to the left or right of the returner, who may choose to prepare moving left or right. Assume if both choose left the server wins 61% of the time, if they both choose right the server wins 56% of the time, if they choose opposite sides the server wins 73% of the time if he chooses left and 78.9% of the time if he chooses right. Analyze the strategies of the two players. Are there any equilibria? Any Nash Equilibria?*

Part 5

Advanced Topics

Chapter 17

Gale-Shapley Algorithm

There are many optimization questions we can study which have important applications to real world problems and also introduce us to interesting mathematics. We discussed fixed points and Nash Equilibrium in §16.5. In this chapter we study another problem, which interestingly, and probably not coincidentally given the theme in our book, led to a mathematician sharing a Nobel Prize in Economics. Briefly, the problem is what is the 'best' way to match objects from two different sets. Of course, the answer depends greatly on what metric we use to measure the quality of the matching. While integer linear programming would solve the problem, it is computationally very expensive, and the Gale-Shapley algorithm below will find matchings significantly faster. In the first formulation we lose the ability to efficiently search our candidate solutions for one that's "best", however.

17.1. Introduction

There are many situations where we want to match elements from one set with another. It could be new hires and jobs in a company, soldiers and posts across the world, customers and tellers at a bank, marrying men and women, or, the example we'll concentrate on right now, students and schools. We'll follow a familiar pattern in economics and mathematics. We'll first greatly simplify the problem by removing many constraints that we would want if it is to accurately model the real world, and then add them back in.

This case was beautifully explored by David Gale and Lloyd Shapley in their 1962 paper, "College Admissions and the Stability of Marriage" [**GS**]. This paper appeared in the *American Mathematical Monthly*, a journal known for its quality expositions; this work is no exception, and you are *strongly* urged to read it. In this section we describe the problem. We then analyze some of the issues, describe their algorithm to efficiently find good matchings, and end the chapter with a discussion of the applications of their method which led to Shapley sharing a Nobel prize in Economics.

Suppose n students apply to Wossamotta University, which wishes to enroll q. The college wants to get the strongest class it can; however, it's unlikely that the proper strategy is admitting the top q people as the students are almost surely applying to multiple colleges, and some might prefer another institution to Wossamotta. Therefore, to end with a class of size q, the college will likely have to admit more than q students.

One strategy for the college is to create a waiting list, where strong applicants who are initially not offered admission can be placed. Now if someone declines an offer, the college just moves to the first person on the list. Notice, however, that right now we have adopted the perspective of what is best for the college. As most of the readers of this book are students, many of you might prefer a different approach. What is best for the students? For example, if an applicant is waitlisted at Wossamotta but accepted to Grand Lakes University (a great institution, but not as good as Wossamotta), what should they do? Should they play it safe and accept Grand Lakes or take a chance that Wossamotta will admit them? Or perhaps they should accept Grand Lakes *but* if Wossamotta offers them a spot later break their word and not go to Grand Lakes? While going back on your word may not be ethical, it is often legal and can leave a school in a lurch, as it could be too late to secure another quality student.

The real issue here is stability: as we match students with colleges, does either member of a pair have an incentive to switch? Of course, for this to happen there must be at least one other pair (if there isn't, the matching problem becomes trivial!). In the notation below we'll use lowercase letters for students and uppercase for college. It's worth taking the time to set good notation; you want to be able to glance down and quickly see the relationships (this is similar to using f to denote a function in calculus and f' its derivative and F its anti-derivative).

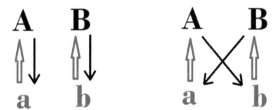

Figure 1. Two possible preference sets with two students and two schools. Arrows point towards which of the two alternatives is preferred. Thus in the left with have a and A preferring each other and b and B preferring each other; in the right a still prefers A but now A prefers b (and similarly for b and B).

Let's consider students a and b and colleges A and B. Assume a prefers A to B and b prefers B to A; furthermore, assume A prefers a to b and B prefers b to a (see Figure 1, left). In this case, it's very easy to assign students to colleges: pair a with A and b with B, and no party has any incentive to switch. If we had instead paired a with B and b with A, then we would have an **unstable** matching, as all parties would prefer the party they were not matched to.

We now turn to a more interesting set of preferences (see Figure 1, right). Assume a would rather go to A than B and b would rather go to B than A. If we only cared about student preferences the solution is clear: have a go to A and b go to B; let's denote this by a—A and b—B. However, this doesn't take into account what the colleges want. To make things interesting, let's assume A prefers student b to a, while B prefers student a to b. If we match to make the colleges happy, we would have a—B and b—A. Notice the answer is different than before.

Thus the two pairings (a—A, b—B and a—B, b—A) have very different behavior. In each pairing, half of the participants are less happy than they would be in the alternative. Which pairing should we choose? Who should be given preference, students or colleges? Notice that both matches have a good property: they're **stable**. This means that no two players will unilaterally switch from their assigned partner to each other. For example, in the matching a—A, b—B, while college B would much rather have student a, student a prefers A to B and won't change.

17.2. Three Parties

Things get more interesting with three players. We analyze this case in great detail, as it gives a good sense of the issues. For simplicity, we assume for now that each college enrolls one and only one student. Consider students a, b, c and colleges A, B, C. A **matching** is an assignment so that each student is paired with one and only one college.

When we looked at the case of two parties, we saw a preference set that created an unstable matching; however, that matching immediately corrected into a stable one. Does this always happen? Given any set of preferences can we always find a **stable matching**? Explicitly, a matching $\mathcal{M}$ is unstable if there are students a and b assigned respectively to colleges A and B such that A prefers b to a and b prefers A to B (if a matching is not unstable, then it is **stable**). Notice in a stable matching no two players can unilaterally improve their situation by leaving their current partner for another.

Returning to the case of three students and three schools, let's look at different preference lists. We assume each student is willing to go to each school and each school is willing to take each student; however, not all schools or students are equally preferred. For simplicity we assume there are no ties in the valuations, and for each party we list how they view their options (from most preferred to least preferred).

Let's consider the following set of students, a, b, c, and colleges, A, B, C, with preferences:

$$a : \{A, B, C\} \qquad A : \{b, c, a\}$$
$$b : \{B, C, A\} \qquad B : \{c, a, b\}$$
$$c : \{C, A, B\} \qquad C : \{a, b, c\}.$$

We have $3! = 6$ possibly matchings; let's explore a—A, b—B, c—C. Note that each student is paired with their first choice, while each college is paired with their last choice. This is a stable matching; the easiest way to see the conditions are satisfied is to note that clearly no student has any incentive to change (even though each college does!). Another stable matching is the one the colleges prefer: A—b, B—c, C—a.

Our simple example here is already rich enough to prove an important point: *a stable matching is not necessarily unique.* Far from being a disappointing result, this indicates that this subject belongs in a course on optimization. Since we can have multiple feasible solutions, we can now search for an optimal solution. Of course, what solution is optimal will depend on how we evaluate the different matchings. If we place a high value on student happiness, the first candidate looks good, while the second option is superior if our first concern is the colleges.

We end by considering an unstable matching. Assume now the preference set is

$$a : \{A, B, C\} \qquad A : \{a, b, c\}$$
$$b : \{B, A, C\} \qquad B : \{a, b, c\}$$
$$c : \{B, A, C\} \qquad C : \{c, b, a\}$$

and the matching is a—A, b—C, c—B. Notice that b prefers B to C and B prefers b to c. Thus this matching is not stable for the given preference list; it's natural to ask if there is a stable matching for this choice.

In the next section we describe the Gale-Shapley algorithm, which shows that not only will a stable matching always exist (under very weak assumptions), but that we have a constructive way to find it quickly. This last part is quite important, as if we have n colleges and n students, then there are $n!$ possible matchings to consider!

17.3. Gale-Shapley Algorithm

As the title of the Gale-Shapley paper is "College Admissions and the Stability of Marriage", in this section we consider the task of matching n men with exactly n women, though we could have unequal numbers and be left with some unmatched people at the end. The mathematics is no different than matching students with colleges, except now we use Western norms and disallow multiple people to be matched to the same person; for college assignment this was a possibility, though we had not yet extended our discussion to include that case.

We want each person to be paired with exactly one member of the opposite gender. We assume everyone rates all n candidates, and there are no ties in the rankings. Let M be the set of n men and W the set of n women. To help clarify the roles, we use lowercase for men and uppercase for women.

We first describe the idea behind their method, then state it explicitly, and conclude by showing why it works and estimating the run-time. As there are $n! \approx n^n e^{-n} \sqrt{2\pi}$ matchings, when n is large it is infeasible to explore them all to find good candidates. A key reason behind the utility of the Gale-Shapley algorithm is that its run-time is bounded by n^2 and does not grow super-exponentially like $n!$.

Briefly, the idea is as follows. We have n men and n women. We will have the men propose and the women accept or reject; one can have the women propose and the men decide and explore how that changes the answer. The men have ranked all the women, and the women rank all the men.

We first describe in words what happens, and then in pseudocode.

- Each man proposes to his most desired mate; each woman who receives a proposal *tentatively* accepts the one she prefers most, and *permanently* rejects all the others.

- All men who are unmatched propose to their highest-ranked woman whom they have not already asked (if they had already asked her we know she cannot prefer him to one with whom she is currently matched). Women again look at all the offers they have, accepting the man they prefer most and *permanently* rejecting all the others.

- Each round has at least one woman accepting an offer (slowest case is if all men have the same preferences, and always ask the same person), and thus the process terminates after at most n rounds. Once everyone is matched, the women now permanently accept whomever they are currently with.

We write the process more formally in pseudocode. For convenience below we have the men proposing one at a time, but one could instead have the free men all propose together. The phrasing below is a modification of that on the Wikipedia entry.

Pseudocode for the Gale-Shapley algorithm.

Initially, all $m \in$ M and $W \in$ W are free.
WHILE there are free men who have not yet proposed to every woman:
    Choose one such man, m.
    Choose W, his highest-ranked woman to whom he hasn't proposed.
    IF W is free then m proposes to her and she temporarily accepts
    ELSE W is currently matched to m' say
        IF W prefers m' to m then
            W rejects m and m remains free
        ELSE W rejects m' and accepts m
            (m, W) are now matched
            m' is now free
            (i.e., m proposes to W, W accepts m, frees m')
      END of IF statement
  END of IF statement
END of WHILE statement
All women now accept their current partner.
Return the set S of matched pairs.

We now show why the algorithm works and analyze how quickly it runs. If a woman is unmatched, we say the the rank of the associated man to her is 0; her lowest-ranked man is given a rank of 1, while her most desired match is given a rank of n.

Our first lemma shows that, as time passes, things can only get better for the women.

Lemma 17.3.1. *The rank of the man that a woman is paired never decreases (i.e., it is a **monovariant** and is non-decreasing).*

Proof. Suppose in a round that a man m proposes to the woman W. There are two cases: woman W is unmatched, or woman W is matched to another man, say m'.

- Case 1: If woman W is unmatched, then she tentatively accepts the proposal. As any matching has a higher rank than zero, her man-rank increases.

- Case 2: If woman W is matched, then either m' is lower-ranked or higher-ranked than m (recall we assume there are no ties). If m' is lower-ranked, then according to our algorithm W rejects m' in favor of m, increasing her rank. If m' is higher-ranked, however, then m is rejected in favor of the current matching, maintaining her rank.

$\square$

Lemma 17.3.2. *Each man proposes at most n times.*

Proof. It suffices to show if a man proposes $n-1$ times, then he proposes at most once more. If after $n-1$ proposals the man is never free, he never proposes again and the claim follows. We are thus reduced to the case when after $n-1$ proposals a man eventually becomes free again, and must propose again.

Thus, we have a man who's already proposed $n-1$ times and is now free. Each of these proposals is unique by construction (i.e., he never proposes twice to the same person, as once someone rejects him he knows he has no chance with her). Thus, it must be that those other $n-1$ women are all paired. Since there are n men, then this man is the only free man. Furthermore, there is exactly one free woman. He proposes to her, she is forced to accept, and the algorithm terminates with him proposing exactly n times. $\square$

We are now ready to prove the Gale-Shapley algorithm has the desired properties.

Theorem 17.3.3. *The Gale-Shapley algorithm returns a stable matching.*

Proof. Suppose after the algorithm terminates that there exists two pairs (m, W) and (m', W') in our matching and m prefers W' to W, and W' prefers m to m'; we illustrate this in Figure 2. If this happened, then m and W' could unilaterally improve their situations by leaving their partners and accepting each other. We must show this cannot happen.

If m prefers W' to W, then our algorithm requires m to propose to W' before he can propose to W. As we are assuming m prefers W' to W, we know that at an earlier time he proposed to W'; however, since he is not matched with W' at some

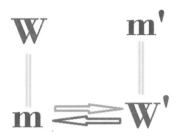

Figure 2. Arrows point towards which of the two alternatives is preferred. Bonds indicate marriage in the hypothetical assignment.

point she rejected him for someone she prefers. We showed the rank of the man matched with a woman is a non-decreasing monovariant; thus, once W' rejects m she will never accept someone she values less than m. Therefore whomever W' has at the end must be preferable to m, and thus she would not dump her match for m.

We've found a contradiction, and thus the matching is stable. $\square$

Lemma 17.3.4. *The run-time of our algorithm is n^2.*

Proof. As each man proposes at most n times, and there are n men, there are at most n^2 proposals. $\square$

17.4. Generalization

We can extended this approach to college admissions. First, all students apply to their first-ranked college. A college with a quota of q places the top q students on their waiting list (or all, if less than q students), and rejects the rest. Rejected applicants then apply to their second-ranked schools, and each college selects the top q from the combined pool of new applicants and those already on its waiting list for a new waiting list. This algorithm terminates when every applicant is on a waiting list or has been rejected by every college that they are willing attend. Then each college admits everyone on their waiting list, they all enroll, and we achieve stable assignment. This is called the **deferred-acceptance procedure**.

Let $\mathcal{M}_{\mathrm{GS}}$ be the stable matching generated by running the Gale-Shapley algorithm. Interestingly, not only is $\mathcal{M}_{\mathrm{GS}}$ stable, but it also gives an **optimal assignment** of applicants; this means that there is no other stable matching where any applicant ends up with a school they prefer over the school they have in $\mathcal{M}_{\mathrm{GS}}$. We of course have to be careful with words such as "optimal". Here, it's optimal with respect to the applicants, not necessarily with respect to the colleges. Notice that it's the applicants who initiate the process by applying to colleges....

Theorem 17.4.1. *Every applicant is at least as well off under $\mathcal{M}_{\mathrm{GS}}$ (i.e., the assignment given by the deferred-acceptance procedure in the Gale-Shapley algorithm) as he would be under any other stable assignment $\mathcal{S}$.*

Note this does *not* mean each applicant gets their top choice; it means no applicant prefers a school they would have in $\mathcal{S}$ to the school they have in $\mathcal{M}_{\text{GS}}$.

Proof. We must show each applicant is at least as well off with their partner in $\mathcal{M}_{\text{GS}}$ as they are with their partner in some stable matching $\mathcal{S}$. We label a college *possible* for an applicant if there exists *a* stable matching such that this applicant is accepted to this college; notice we do not require the applicant to be accepted by this college in *every* stable matching, only that there is at least *one* such stable matching.

We show that the deferred-acceptance procedure only rejects students from colleges that are impossible for them (i.e., schools which they are never paired with in a stable matching). If this is not the case, then there must be a first round where an applicant is declined by a possible school which they prefer to the one they got in $\mathcal{M}_{\text{GS}}$. Let's assume it's student a and they want to go to college A, but college A rejects them as they have a full slate of preferable candidates $a_1, a_2, \ldots, a_n$. We deduce a contradiction; specifically, we'll prove A is impossible for a as this would lead to an unstable matching (which cannot exist as our matching is stable).

We know that each a_i prefers college A to all other colleges *except* those that have already rejected them, but these are impossible for them from this point onward as a school will never revisit a rejected applicant as they have better people on the wait list or proposing. Therefore, consider an assignment in which a goes to A, and everyone else attends a possible school. Thus *at least* one of the a_i's attends a school other than A. This is an unstable match, because this a_i prefers A to its current pairing, and A prefers a_i to a. Therefore, this hypothetical assignment is unstable and cannot exist.

Therefore, we conclude that our algorithm only rejects applicants from colleges in which there is not a possible stable assignment at termination. Thus, we do return an optimal assignment. □

Note that our algorithm is *student-oriented*, i.e., every *applicant* is at least as well off under this assignment. We could invert the process to get a *college-oriented* set of assignments; however, in each of these approaches we favor one side over the other. It is a very active area of research to look at all stable matchings and choose a "best" solution. Of course, the challenges are in determining what we mean by best, and then efficiently finding it.

17.5. Applications

Alvin Roth and Lloyd Shapley worked together to extend this algorithm into game theory, for which they won the 2012 Nobel Prize in Economics.

One terrific application is the **National Resident Matching Program** (NRMP), which uses a modified version of the Gale-Shapley algorithm to match its medical students to residency programs. This algorithm is *applicant-based*, meaning it uses applicant proposals rather than hospital proposals. Hospitals provide a simple rank-ordered list rating applicants from most to least preferable. For students not matching in couples, it similarly requires a list of hospitals from most to least preferable. However, for students in couples (i.e., students in relationships who want to be placed together), it asks for a rank-ordered list of pairs of schools, or a school for one applicant and "no match" for the other, such that the couple can be treated as one unit. To build these preference lists, hospitals not only review the academic qualifications of the applicants, but interview them in person.

It is important to note here that adding couples to the problem complicates it significantly, to the point that not only might there exist no stable solution, but that determining if one exists (and finding it) is NP-complete. However, in simulation, the algorithm always terminates quickly, and there is an enormous increase in utility in helping couples stay together.

The NRMP notes that generally couples who match together end up both having less desirable programs – their paired matching is essentially achieved by individual matching, and then turning down their higher-ranked programs until they both end up at a suitable pair on the preference list.

A similar application is used in the New York metropolitan schools – students list up to 12 schools. Since inception in 2004, about half of the applicants were matched to their first-choice schools, and another third were matched to their second or third choices.

Roth applied Shapley's theoretical work to existing markets, such as the kidney shortage in the US. He noted that most kidney donations come from a deceased person, yet the demand for kidney donations is growing rapidly. It is possible to donate one kidney and live with the other. While many family members are willing to donate a kidney to the loved one in need, in approximately a third of these pairs there is a biological barrier (such as mismatched blood types or incompatible tissues). Roth theorized about a kidney market, where these mismatched pairs would enter, and find a matching mismatched pair. For example, suppose there is a pair with mismatched blood types, donor having A and recipient having B, and suppose there is an unrelated, yet nearby pair with donor having B and recipient having A. Then, through Roth's market, this pair who know nothing about each other can swap donors, and both recipients can receive the live kidney donation they need.

The following is excerpted from "Nobel winner Roth helped spark kidney donor revolution" by Edward Krudy (available at http://www.reuters.com/article/ nobel-prize-roth-kidney-idUSL1E8LFFW320121015), an article written in 2012 which describes the impact of this Nobel Prize-winning work.

> But arguably its greatest impact has been matching kidney donors to patients in a system that was first applied in New England hospitals under the New England Program for Kidney Exchange (NEPKE), a scheme Roth helped found in 2004-2005.
>
> The computerized pairing of groups of donors and patients that Roth's models inspired has revolutionized the way kidney transplants are handled in the United States and has actually increased the possible number of transplants.
>
> Throughout the United States nearly 2,000 patients have received kidneys under the system developed on Roth and Shapley's models that would otherwise not have received them, according to Ruthanne Hanto,

who has worked with Roth since 2005 after being co-opted to manage NEPKE.

In 2003, the year before the system was implemented, there were just 19 kidney transplants from live donors in the United States nationally, said Hanto. That number rose to 34 when the system was introduced in 2004. Last year it reached 443.

These numbers are staggering, and illustrate the incredible power mathematics has to attack problems in the real world. The trick is finding the right perspective, finding what math is appropriate.

17.6. Exercises

Introduction

Exercise 17.6.1. *If there are n colleges and n students, show there are $n!$ ways to match the students to the college so that each college gets one and only one student.*

Exercise 17.6.2. *Imagine there are n students and k colleges, where college i wants to admit n_i students and $n_1 + \cdots + n_k = k$. How many ways are there to match the students with the colleges?*

Exercise 17.6.3. *Show that $n!$ grows super-exponentially: $\lim_{n \to \infty} n!/c^n = \infty$ for any fixed c.*

Exercise 17.6.4. *Is it possible to have three students and three schools and the preferences so chosen so that we continuously cycle? Explicitly, if we have some initial assignment, then one student and one college would prefer each other to their current match and would pair together. In the new match that then forms there would be another student-college pair that would prefer each other to their current match, who would then break off and pair together....*

Three Parties

Exercise 17.6.5. *Consider three students and three colleges, where each college wants exactly one student. If we look at all possible preference lists of students and colleges, how many stable matches does each preference set have? On average how many stable matches does each preference set have? What is the variance?*

Exercise 17.6.6. *Repeat the previous problem with three replaced by four, and then with four replaced by five.*

Exercise 17.6.7. *If there are n students and n colleges, and a preference set is chosen uniformly at random from all possible preferences, how many stable matchings do you expect there to be? Note: this problem almost surely requires coding and numerical exploration!*

Gale-Shapley Algorithm

Exercise 17.6.8. *Find all stable matchings (if any exist) for the given preference set:*

$$A : [a, d, c, b, e] \qquad a : [D, E, A, B, C]$$
$$B : [e, d, b, c, a] \qquad b : [A, D, E, B, C]$$
$$C : [c, d, e, b, a] \qquad c : [B, A, C, D, E]$$
$$D : [a, b, c, e, d] \qquad d : [A, B, C, D, E]$$
$$E : [d, b, c, a, e] \qquad e : [E, C, A, D, B].$$

Exercise 17.6.9. *Find all stable matchings (if any exist) for the given preference set:*

$$A : [d, e, a, b, c] \qquad a : [E, D, A, C, B]$$
$$B : [c, a, e, b, d] \qquad b : [C, B, A, D, E]$$
$$C : [a, e, c, d, b] \qquad c : [B, C, E, D, A]$$
$$D : [a, c, e, d, b] \qquad d : [A, C, D, E, B]$$
$$E : [e, b, a, c, d] \qquad e : [B, C, A, E, D].$$

Exercise 17.6.10. *Come up with several possible objective functions that can be used to compare two stable matchings, and use them to compare some of your matchings from the previous two problems.*

Exercise 17.6.11. *Prove or disprove: In every instance of the Stable Matching Problem there is a stable matching containing a pair (m, W) such that m is ranked first on the preference list of W and W is ranked first on the preference list of m.*

Exercise 17.6.12. *Prove or disprove: Consider an instance of the Stable Matching Problem in which there exists a man m and a woman W such that m is ranked first on the preference list of W and W is ranked first on the preference list of m. Then the pair (m, W) belongs to every possible stable marriage for this instance.*

Generalization

Exercise 17.6.13. *Imagine n students are applying to n colleges and everything is as before except for one person, who is upset that we proved they cannot do better among stable matches than their outcome in the deferred-acceptance approach. They therefore decide to lie, and always propose to their second best choice (if available). Can this help them?*

Exercise 17.6.14. *Imagine everyone in a student-college matching is honest save for one person. If they lie about their preferences, can they improve the rank of their match?*

Exercise 17.6.15. *Generalize the college problem and allow students (or colleges) to declare that they will not accept certain matchings. Must a stable matching still exist? Must a matching now exist? What if only students can say there are colleges they won't attend, but each college must take their highest ranked applicant; does a matching now exist?*

Exercise 17.6.16. *In addition to ranking alternatives, assume people and colleges now assign a utility to each pairing. Discuss how you can use that information to choose among the stable matchings a "best" one.*

Applications

Exercise 17.6.17. *The application of the Gale-Shapley algorithm to kidney matching helps two pairs swap kidneys. Generalize this idea to more than just two pairs. What do you think should be maximized?*

Exercise 17.6.18. *Write down some complications that could exist in kidney matching. For example, perhaps it is not as simple as either two people are a match or they are not, but maybe one learns that there is an 85% chance that they could successfully donate a kidney to someone. How does this affect the analysis?*

Chapter 18

Interpolating Functions

The main theme of this book has been efficiency. We don't just want to know if solutions exist; we want to construct methods for arriving at solutions, or something sufficiently close to a solution, in as little time as possible. Here we extend that same desire for efficiency to some more basic computational methods. In particular, we explore reconstructing functions from a few data points.

There are a variety of methods one can use. The simplest, of course, is to sample our function so often that we obtain an excellent description; unfortunately, in practice we frequently cannot do this as it quickly becomes computationally too expensive. Thus, the idea is to transmit a small amount of information quickly through our network to reconstruct the function (or a close approximation) rapidly at the receiving end. This has the dual advantage of decreasing the amount of data we need to transmit and taking advantage of the computational power often available; probably the most familiar application is to streaming video, where we might be trying to send the red-green-blue color functions.

We are thus reduced to the problem where we want to send a sequence of values and use these to approximate the function f. One option is to Taylor expand f and send the first few Taylor coefficients. Or perhaps, if we know some Fourier analysis, we can send the Fourier or Fejer series; this represents f as a linear combination of sines and cosines instead of the polynomials x^n from the Taylor series. While these approaches have many good properties, in this chapter we explore another approach: sending pairs $(x_i, f(x_i))$ and using these to interpolate f. In doing so we have enormous freedom in choosing *where* to sample f; i.e., where we should locate the x_i. Not surprisingly, these places will be a function of the number of samples we take.

Before delving into the analysis, it's worth noting that if f is a polynomial of degree d, then $d+1$ points suffice to uniquely specify it. Thus, the problem is only interesting if f is not a polynomial, or for polynomial f if the number of data points is less than or equal to the degree.

18.1. Lagrange Interpolation

Since we have fast algorithms to multiply, polynomials seem like a good class of functions to use for interpolating. Hence, in setting up our approximation problem, we have an unknown target function $f(x)$, a sequence of observations $(x_i, f(x_i))$ which we call the **nodes**, and an approximating polynomial $p(x)$ which we construct and require to pass through the nodes. We frequently write y_i for $f(x_i)$.

Definition 18.1.1 (Interpolating Polynomial). *Given $n+1$ nodes (x_i, y_i) and $x_i \neq x_j$ for $i \in \{0, \ldots, n\}$ pre-computed from $f(x)$, $p(x)$ is an interpolating polynomial if $p(x_i) = y_i$ for all $i \in \{0, \ldots, n\}$ and $\deg(p(x)) = n$.*

We will soon see that interpolating polynomials are unique, due to the degree restriction; see Figure 1 for an example.

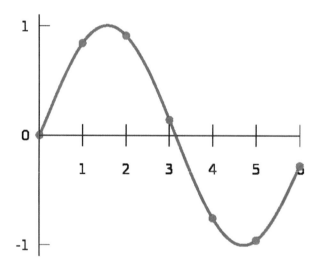

Figure 1. Example of a polynomial interpolation with 7 nodes.

Our requirement that the polynomial must be such that $p(x_i) = y_i$ allows us to convert this problem into a system of equations. If

$$p(x) = a_n x^n + a_{n-1} x^{n-1} + \cdots + a_0,$$

then we must determine $a_0, \ldots, a_n$ such that

$$
\begin{aligned}
p(x_0) &= a_n(x_0)^n + a_{n-1}(x_0)^{n-1} + \cdots + a_1(x_0) + a_0, \\
p(x_1) &= a_n(x_1)^n + a_{n-1}(x_1)^{n-1} + \cdots + a_1(x_1) + a_0 \\
&\vdots \\
p(x_{n-1}) &= a_n(x_{n-1})^n + a_{n-1}(x_{n-1})^{n-1} + \cdots + a_1(x_{n-1}) + a_0, \\
p(x_n) &= a_n(x_n)^n + a_{n-1}(x_n)^{n-1} + \cdots + a_1(x_n) + a_0,
\end{aligned}
$$

where $p(x_i) = y_i$.

We can re-write this as a linear algebra problem:

$$\begin{pmatrix} x_0^n & x_0^{n-1} & \cdots & x_0 & 1 \\ x_1^n & x_1^{n-1} & \cdots & x_1 & 1 \\ \vdots & \vdots & \vdots & \ddots & \vdots \\ x_n^n & x_n^{n-1} & \cdots & x_n & 1 \end{pmatrix} \begin{pmatrix} a_n \\ a_{n-1} \\ \vdots \\ a_1 \\ a_0 \end{pmatrix} = \begin{pmatrix} y_0 \\ y_1 \\ \vdots \\ y_{n-1} \\ y_n \end{pmatrix}.$$

We can write this as a matrix equation

$$\mathbf{V}\vec{a} = \vec{y}.$$

$\mathbf{V}$ is called the **Vandermonde matrix**.

Recall from linear algebra that the solutions to $\mathbf{V}\vec{a} = \vec{y}$ are related to the determinant of $\mathbf{V}$; if the determinant is non-zero, then there is a unique solution given by $\vec{a} = \mathbf{V}^{-1}\vec{y}$. The following lemma shows that $\mathbf{V}$ is invertible under our assumptions.

Lemma 18.1.2. *Let $\mathbf{V}$ be the $(n+1) \times (n+1)$ Vandermonde matrix arising from the $n+1$ numbers $x_0, \ldots, x_n$. Then*

$$\det(\mathbf{V}) = \prod_{0 \le j < i \le n} (x_j - x_i),$$

and if the x_i's are distinct, then the determinant is non-zero and there is a unique interpolating polynomial.

Proof. As the determinant of a matrix is a polynomial of its entries, we re-write $\det(\mathbf{V}) = V(x_0, \ldots, x_n)$ for some function V. Notice that if two of the x_i's are equal, say $x_i = x_j$, then the corresponding columns are equal, which implies the determinant is zero. Thus the determinant $V(x_0, \ldots, x_n)$ must be divisible by $x_j - x_i$ for all distinct pairs i, j, and there must be some function $\mathfrak{V}(x_0, \ldots, x_n)$ such that

$$V(x_0, \ldots, x_n) = \mathfrak{V}(x_0, \ldots, x_n) \prod_{0 \le j < i \le n} (x_j - x_i).$$

As the degree of V is $n(n+1)/2$ (see Exercise 18.5.1), which is the degree of the product on the right-hand side above, we see that $\mathfrak{V}(x_0, \ldots, x_n)$ must be constant. Let $x \lll y$ mean that x is *much less than* y. By choosing $x_0 \lll x_1 \lll x_2 \lll \cdots \lll x_n$, we see this constant is non-zero; by multiplying the expressions out and matching terms we see that it must in fact be 1 (see Exercise 18.5.2). The rest of the claims follow immediately. $\square$

While the existence of a degree n polynomial that passes through a chosen set of $n+1$ points isn't that surprising, a useful result of the above is that we're free to choose the computational method used to construct $p(x)$ from many possible options. Since the $p(x)$ passing through those nodes is unique, we can choose the method of construction that is most efficient in a given context.

The most direct construction is simply to invert the Vandermonde matrix, but this may be an expensive calculation for large n. Another option is to use **Lagrange polynomials**.

Definition 18.1.3 (Lagrange Interpolating Polynomials). *Given a set of $n + 1$ nodes (x_i, y_i) with the x_i distinct, the Lagrange Interpolating Polynomial is the linear combination*

$$L(x) = \sum_{i=0}^{n} y_i L_i(x),$$

where

$$L_i(x) = \prod_{\substack{0 \le m \le n \\ m \ne i}} \frac{x - x_m}{x_i - x_m}.$$

In Exercise 18.5.3 you're asked to prove these polynomials are interpolating; note that $L_i(x_j) = 0$ for $j \ne i$.

18.2. Interpolation Error

Now that we're guaranteed to have an interpolation polynomial, and we've developed some methods for finding such a polynomial, let's see how well it works. By definition it will perfectly match the value of our function at the nodes, but how close will it be elsewhere?

Figure 2 shows the graph for $\log(x)$ and the interpolating polynomials for 5, 10, and 50 evenly spaced nodes on the interval from 0 to 50. As we might expect, the more nodes, the more information we get on the shape of $\log(x)$ and the closer the fit for the interpolating polynomial. However, the trade off is that we have to interpolate with a polynomial of a much higher degree.

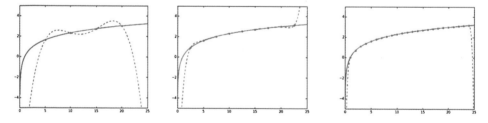

Figure 2. Lagrange Interpolation for $\log(x)$ for 10, 20 and 50 evenly spaced nodes.

Recall that our goal is to do these computations quickly, and so in terms of the economy of computational costs, if we have to interpolate with a more complicated polynomial we should at least get a more precise approximation of our target function. Unfortunately, this is not always the case. Consider the seemingly innocuous function $f(x) = \frac{1}{1+x^2}$. This is often called **Runge's function** (after Carl Runge); if we multiply by $1/\pi$ it integrates to 1 over the real line and is the **Cauchy probability distribution**. This is one of a class of functions whose interpolating polynomials exhibit extreme oscillation at the end points of the the interval of interest. Not surprisingly the oscillation at the endpoints is called **Runge's phenomenon**; we illustrate this in Figure 3.

In terms of the study of interpolation polynomials, this is an extremely important result. On the one hand, it shows that increasing the number of nodes, and hence the degree of the interpolating polynomial, needn't increase the accuracy of

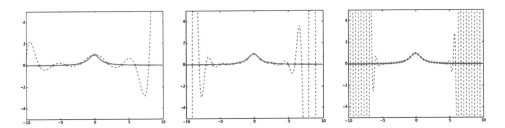

Figure 3. Lagrange Interpolation for the Runge functions for 5, 10, and 50 evenly spaced nodes.

the interpolation. On the other hand, this essentially means that the problem of interpolation is an interesting one. It isn't simply a matter of using enough computing power to calculate with high degree polynomials, rather there is an opening for a closer analysis of what exactly contributes to interpolation error.

The Runge phenomenon, and the divergence of some functions despite an increase in nodes, compels us to formalize this notion of divergence in the context of polynomial interpolation. The more times we can differentiate our function, the more we can say about the error. The following result is standard and is often covered in a calculus course when studying Taylor series; its utility arises when the derivatives are bounded, as then the factorial in the denominator leads to rapid decay.

Theorem 18.2.1 (Interpolation Error). *Let $f(x)$ be an $n+1$ times differentiable function on $[a, b]$, and $p_n(x)$ the interpolating polynomial for the $n+1$ distinct nodes (x_i, y_i) for $i \in \{0, 1, \ldots, n\}$. For every $x \in [a, b]$ there exists $c \in [a, b]$ (note in general c depends on x) such that the error $E(x)$ between the value of $f(x)$ and our approximation $p_n(x)$ satisfies*

$$E(x) := f(x) - p_n(x) = \frac{f^{n+1}(c)}{(n+1)!} \prod_{i=0}^{n} (x - x_i).$$

Proof. First note that if $x = x_i$, then the product of the $x - x_i$ vanish, and thus $f(x) = p_n(x)$ and there is no error.

Suppose now $x \neq x_i$. We proceed by constructing a helper function:

$$F(t) := f(t) - p_n(t) - g(x) \prod_{i=0}^{n} (t - x_i),$$

where

$$g(x) := (f(x) - p_n(x)) \prod_{i=0}^{n} (x - x_i)^{-1}.$$

Remember x is the point where we wish to compare our approximation to the actual value; it is fixed and thus $g(x)$ is fixed; t is the free variable. Thus the functions above are well defined, as the exponent of -1 is to a fixed, non-zero quantity.

The function $F(t)$ vanishes at $n+2$ distinct roots; a straightforward calculation shows it's zero at $x_0, x_1, \ldots, x_n$ (as $f(x_i) - p_n(x_i) = 0$), and then we have rigged things so it also vanishes at x. Since f and p_n are $n + 1$ times differentiable, so too is $F(t)$.

If we apply Rolle's Theorem $n + 1$ times (see Exercise 18.5.7) we find there is a $c \in (a, b)$ such that $F^{(n+1)}(c) = 0$. If we take $n + 1$ derivatives of $g(x) \prod_{i=0}^{n}(t - x_i)$ with respect to t we get $g(x) \cdot (n + 1)!$, and thus we find

$$F^{(n+1)}(c) = f^{n+1}(c) - g(x)(n + 1)!;$$

as $F^{(n+1)}(c) = 0$ we immediately obtain

$$g(x) = \frac{f^{n+1}(c)}{(n + 1)!}.$$

Substituting this value back into our original formulation for $g(x)$ yields that the error $E(x)$ satisfies

$$E(x) = f(x) - p_n(x) = \frac{f^{n+1}(c)}{(n + 1)!} \prod_{i=0}^{n}(x - x_i),$$

as claimed.   □

Armed with the above result, let's revisit the Runge phenomenon. There are two sources to the interpolation error: the value of the product, and the value of the derivative $f^{(n+1)}(c)$ at the point c which depends on x. One of the challenges in the analysis is that we don't have an explicit formula for c, and hence in practice we're typically stuck with bounding $f^{(n+1)}(c)$ by the largest value it attains. We do, however, have freedom in determining the location of the x_i. It is natural to choose them to minimize the product; unfortunately it's not clear where those locations are. One easy choice to work with is having the x_i evenly spaced in our interval of interest (see Exercise 18.5.9). Thus if $x_0 = a$ and $x_i = a + ik$ with steps $k = \frac{b-a}{n}$, then

(18.1)
$$E(x) \leq \frac{k^{n+1}}{4(n + 1)} \max_{x \in [a,b]} |f^{(n+1)}(c)|.$$

While the above analysis has many nice features, it's important to notice that we've lost the factorial in the denominator and thus convergence could be a tricky issue. These computations motivate our investigations for the rest of the chapter, specifically placing the nodes to minimize the product $\prod_{i=0}^{n}(x - x_i)$. These investigations will lead us to the Chebyshev polynomials.

18.3. Chebyshev Polynomials and Interpolation

The nth degree **Chebyshev polynomial** is defined recursively by the following:

- $T_0(x) = 1$,
- $T_1(x) = x$, and
- $T_{n+1}(x) = 2xT_n(x) - T_{n-1}(x)$.

These are sometimes referred to as the Chebyshev polynomials of the first kind (the second kind is defined similarly, with $U_0(x) = 1$, $U_1(x) = 2x$, and $U_{n+1}(x) = 2xU_n(x) - U_{n-1}(x)$). We denote these with a "T" from another accepted transliteration of Chebyshev's name, Tchebyshev. Figure 4 shows the first five Chebyshev polynomials. In this section we introduce some of their properties, ending with a result indicating their utility in constructing good interpolating polynomials.

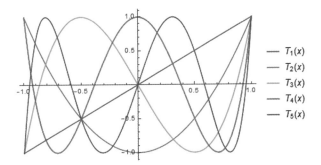

Figure 4. The first five Chebyshev polynomials.

The following trigonometric identities are also of interest:

$$T_n(x) = \begin{cases} \cos(n \cos^{-1}(x)) & \text{for } x \in [-1, 1], \\ \cosh(n \cosh^{-1}(x)) & \text{for } x \geq 1, \\ (-1)^n \cosh(n \cosh^{-1}(-x)) & \text{for } x \leq -1. \end{cases}$$

These identities allow us to deduce some useful results about the extrema and roots of $T_n(x)$ on the interval $[-1, 1]$. Specifically (see Exercise 18.5.26), the extrema of $T_n(x)$ on the interval $-1 \leq x \leq 1$ are given by

$$x_k = \cos\left(\frac{k}{n}\pi\right)$$

for $k = 0, \ldots, n$, and the roots for $T_n(x)$ are

$$x_r = \cos\left(\frac{2r - 1}{2n}\pi\right)$$

for $r = 0, \ldots, n$.

These are important as our analysis of the interpolation error showed us that minimizing error is related to finding x_i that make $\prod_{i=0}^{n}(x - x_i)$ as small as possible. Note, crucially, that this term itself is a polynomial in x of degree $n + 1$. Moreover, it's monic as its leading coefficient is 1. *Thus the problem of finding an ideal way to interpolate a target function is related to finding roots for a monic polynomial of degree $n + 1$ with some type of minimality over a given interval.*

Not surprisingly, the Chebyshev polynomials provide a solution and resolve many of the issues. Before stating that result we first introduce some notation. By $||f||_\infty$ we mean the largest value f attains on our compact set $[-1, 1]$. We read $||f||_\infty$ as the L^∞ **norm** of f. (Recall a norm $||\cdot||$ assigns a positive number to every element in our space save the zero element, which is assigned zero. Furthermore, for any real numbers α, β and elements f, g we have $||\alpha f + \beta g|| \leq |\alpha| \, ||f|| + |\beta| \, ||g||$.)

Theorem 18.3.1 (Minimality of the Monic Chebyshev Polynomials). *If $q(x)$ is any monic polynomial of degree $n + 1$ on $[-1, 1]$, then*

$$\left\| \frac{1}{2^n} T_{n+1} \right\|_\infty \leq \|q\|_\infty \, .$$

In other words, among all monic polynomials of degree $n + 1$, none have a smaller L^∞ norm than $T_{n+1}(x)/2^n$.

Proof. The proof proceeds by contradiction. Suppose there exists a monic polynomial $q(x)$ of degree $n + 1$ such that $\|q\|_\infty < \left\| \frac{1}{2^n} T_{n+1} \right\|_\infty$. As T_{n+1}'s extrema are all $+1$ or -1, since we have assumed q has smaller L^∞ norm we know

$$r(x) \; := \; \frac{1}{2^n} T_{n+1}(x) - q(x)$$

is positive when $x = x_k$ is a maximum of T_{n+1} and negative when $x = x_k$ is a minimum of T_{n+1}. Thus the function $r(x)$, which is at most a degree n polynomial since the x^{n+1} terms cancel in the subtraction, has sign changes whenever x passes from a maximum to a minimum of T_{n+1}. As the extrema of T_{n+1} are located at $x_k = \cos(k\pi/n)$, a little inspection shows that $r(x)$ has at least $n + 1$ sign changes. By the Intermediate Value Theorem, $r(x)$ has at least $n + 1$ zeros. The only way a degree at most n polynomial can have $n + 1$ zeros is if it identically vanishes. This contradicts the assumption of a $q(x)$ with smaller norm, completing the proof. $\square$

Armed with this result, we are now in a position to say something definitive about the ideal set (or at least a very good set) of nodes for polynomial interpolation. Since $\frac{1}{2^n} T_{n+1}$ is monic, degree $n+1$, and of minimal norm on a given interval, we set

$$\prod_{i=0}^{n} (x - x_i) \; = \; \frac{1}{2^n} T_{n+1}(x)$$

by choosing the x-values for our interpolation nodes to be the roots of the Chebyshev polynomial on the interval $[-1, 1]$; i.e.,

$$x_i \; = \; \cos\left(\frac{2i - 1}{2n} \pi \right)$$

for $i \in \{0, \ldots, n\}$. Through a simple transformation, we can of course find the nodes on an arbitrary interval (see Exercise 18.5.28).

With the bulk of the hard theoretical work behind us, we can now reap the benefits. Figure 5 shows Runge's function with Lagrange interpolation, except this time the x_i values of the interpolation nodes are set to be the roots of the Chebyshev polynomial. Notice the greatly improved convergence.

18.4. Splines

We saw in the previous section Chebyshev polynomials allow us to find $n + 1$ nodes for a target function $f(x)$ that avoids the Runge phenomena. If we insist on interpolating these $n + 1$ nodes with one degree n polynomial, then this is our best option (if we wish to minimize our product). However, we saw that the bad behavior of Runge's function occurs with large n, and thus there could be an advantage with

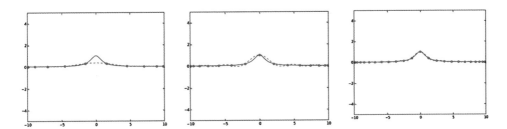

Figure 5. Runge's function with 10, 20 and 100 Chebyshev nodes.

trying to decrease n. The freedom we have for approximating $f(x)$ through these nodes suggests that we try to interpolate with polynomials of lesser degree. Briefly, the idea is to approximate f with smaller degree polynomials on various segments, and then connect them together. The text below is meant to be a brief introduction; the interested reader should consult the literature (see for example [**ANW**]).

18.4.1. Many Polynomials, Fewer Degrees. A **spline function** is defined piecewise by polynomials such that the function has a high degree of smoothness (i.e., many derivatives exist) at places where the polynomials meet, called **knots**. The most naive choice is to use first degree linear polynomials (see Figure 6). This is effectively drawing lines between the nodes, and our intuitions from calculus tell us that this should give a good approximation for nearby points, but of course we don't expect smoothness at the intersections.

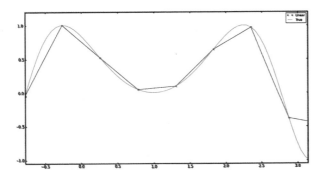

Figure 6. Linear spline interpolation for $\sin(x-1)^2$.

Though we could try quadratic splines, it turns out cubic splines are more useful. In particulary, it will be important to have a twice differentiable polynomial making up our spline. Figure 7 shows cubic spline interpolation for the function $\sin((x-1)^2)$ for eight knots.

As we can see, relatively tight fitting occurs through interpolating on the spline through these eight points. Of course the real test should be whether or not spline interpolation avoids Runge's phenomenon as we originally desired. Figure 8 shows that this is indeed possible, and thus motivates delving deeper into the theory.

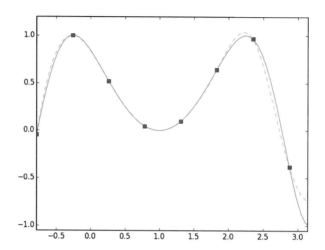

Figure 7.   Cubic spline interpolation for $\sin(x-1)^2$.

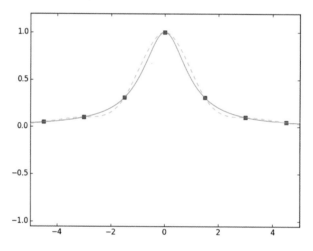

Figure 8.   Cubic spline interpolation for the Runge phenomenon.

18.4.2. Theory of Splines. A function p is called a **spline of degree k on** $[a, b]$ if $a = x_0 < x_1 < \cdots < x_n = b$ and

$$
p(x) \;=\; \begin{cases} q_0(x) \text{ for } (x) \in [a, x_1], \\ q_1(x) \text{ for } (x) \in [x_1, x_2] \\ \;\vdots \\ q_{n-1}(x) \text{ for } (x) \in [a, x_1]. \end{cases}
$$

For $n + 1$ knots (x_i, y_i), $i \in \{0, \ldots, n\}$, we interpolate between all of the pairs of nodes (x_{i-1}, y_{i-1}) and (x_i, y_i) with polynomials $y = q_i(x)$, $i \in \{0, \ldots, n\}$, such that

$$q_i'(x_i) = q_{i+1}'(x_i),$$
$$q_i''(x_i) = q_{i+1}''(x_i)$$

for $1 \leq u \leq n - 1$.

The reason that we require the first and second derivatives to be equal at knots has to do with smoothness, and in turn, curvature. Note that if we allow complete freedom for $q_i(x)$ to interpolate between two knots, there is no guarantee that it will fit the target function $f(x)$, regardless of the number and location of knots we use for our interpolation problem.

As an added requirement, we want to minimize the curvature over the spline. Recall that the curvature for a function $f(x)$ is given by

$$C(x) = \frac{f''(x)}{(1 + f'(x))^{3/2}}.$$

By setting the first and second derivative of the splining polynomials equal to one another at the knots, it effectively minimizes $C(x)$ over the entire spline function, resulting in a close approximation to our target function $f(x)$.

18.4.3. Deriving a Cubic Spline Interpolation. We end by sketching the derivation of a cubic spline. We'll work with the **natural cubic splines**. These are cubic splines under the following "natural" conditions: $p''(x_0) = p''(x_n) = 0$.

Let $\Delta x_i = x_{i+1} - x_i$ and $\Delta y_i = y_{i+1} - y_i$; these represent the change in x and the change in y. We consider the general cubic

$$q_i(x) = a_i + b_i(x - x_i) + c_i(x - x_i)^2 + d_i(x - x_i)^3,$$

where we have chosen to write it in terms of the displacement from the starting position x_i.

We have two initial interpolation conditions. First, $q_i(x_i) = p(x_i) = y_i$, which implies $a_i = y_i$.

Secondly, $q_i(x_{i+1}) = p(x_{i+1}) = y_{i+1}$, which then yields the equation

$$y_i + b_i \Delta x_i + c_i \Delta x_i^2 + d_i \Delta x_i^3 = y_{i+1},$$

which is equivalent to

$$b_i + c_i \Delta x_i + d_i \Delta x_i^2 = \frac{\Delta y_i}{\Delta x_i}.$$

Since cubic polynomials are twice differentiable, we may take the first and second derivatives of the above equation, which must also satisfy the conditions $q_i'(x_i) = q_{i+1}'(x_i)$ and $q_i''(x_i) = q_{i+1}''(x_i)$. The resulting equation can be written entirely in terms of the c_i's in the following form:

$$\Delta x_{i-1} c_{i-1} + 2(\Delta x_{i-1} + \Delta x_i) c_i + \Delta x_i c_{i+1} = 3 \left(\frac{\Delta x_i}{\Delta y_i} - \frac{\Delta x_{i+1}}{\Delta y_{i+1}} \right).$$

Having solved for c_i we can continue and solve for b_i and d_i and hence obtain the entire expression for the spline $p(x)$. Also, note that selecting this equation to solve has some distinct advantages when we consider the overall goal of efficient computation. Obtaining c values for the splining polynomials, much like our original polynomial interpolation problem, involves solving a system of $n + 1$ linear equations. However, in this case the system of equations is much more structured than the general problem we had in the case of the Vandermonde matrix. For notational convenience, let

$$u_i = 3\left(\frac{\Delta x_i}{\Delta y_i} - \frac{\Delta x_{i+1}}{\Delta y_{i+1}}\right).$$

Then we can express the system of equations as a matrix equation:

$$\begin{pmatrix} 2c_0 & c_1 & 0 & \cdots & 0 & 0 & 0 \\ c_0 & 2c_1 & c_2 & \cdots & 0 & 0 & 0 \\ 0 & c_1 & 2c_2 & c_3 & \cdots & 0 & 0 \\ \vdots & \ddots & \ddots & \ddots & \ddots & \vdots & 0 \\ \vdots & \vdots & \ddots & \ddots & \ddots & \ddots & \vdots \\ 0 & 0 & \cdots & c_{n-3} & 2c_{n-2} & c_{n-1} & 0 \\ 0 & 0 & \cdots & 0 & c_{n-2} & 2c_{n-1} & c_n \\ 0 & 0 & \cdots & 0 & 0 & c_{n-1} & 2c_n \end{pmatrix} \begin{pmatrix} c_0 \\ c_1 \\ c_2 \\ \vdots \\ c_{n-2} \\ c_{n-1} \\ c_n \end{pmatrix} = \begin{pmatrix} u_0 \\ u_1 \\ u_2 \\ \vdots \\ u_{n-2} \\ u_{n-1} \\ u_n \end{pmatrix}.$$

Hence, solving for the c_i's involves finding the solution to a tridiagonal matrix; these are highly structured sparse matrices. Unsurprisingly, this added structure in the matrix makes solving the system of linear equations significantly easier than in the general case. If our goal is to get computationally cheap approximations, then we should prefer splines over polynomial interpolations, precisely because obtaining a spline representation of the target function only requires that we deal with a tridiagonal matrix rather than a general Vandermonde matrix. In the worst case, the Vandermonde could take on the order of n^3 operations to solve (though we have seen that there are fast ways to multiply matrices, such as Strassen's algorithm, so perhaps there is hope that here too the exponent could be lowered), while solving a tridiagonal system takes on the order of n operations. Thus, for large n, spline interpolation provides large advantages over polynomial interpolation.

18.5. Exercises

Lagrange Interpolation

Exercise 18.5.1. *Prove that the degree of $V(x_0, \ldots, x_n)$ is $n(n+1)/2$.*

Exercise 18.5.2. *Prove that the constant C in the proof of the determinant of the Vandermonde matrix is 1.*

Exercise 18.5.3. *Prove the Lagrange polynomials are interpolating.*

Exercise 18.5.4. *Plot the Lagrange polynomials for $n \in \{1, 2, 3, 4\}$ and $(x_i, y_i) = (i, F_i)$; here F_i is the ith Fibonacci number (so 1, 1, 2, 3).*

Exercise 18.5.5. *Consider an army with 10 generals. One wants a security system such that any three of them can determine the code to launch nuclear missiles, but no two of them can. It's possible to devise such a system by using a quadratic polynomial $ax^2 + bx + c$; to launch the missiles, input (a, b, c). We can't just tell each general one of a, b, or c (as then it is possible that some subset of three generals* *won't know a, b, and c); however, if a general knows two of (a, b, c), then a set of two generals can launch the missiles! What information should be given to the generals so that any three can find (a, b, c) but no two can? What about the general situation with N generals and any M can launch (but no set of $M - 1$ can)?*

Interpolation Error

Exercise 18.5.6. *Prove* **Rolle's Theorem**: *If f is a continuous function on $[a, b]$ with $f(a) = f(b) = 0$, then there exists a $c \in (a, b)$ such that $f'(c) = 0$.*

Exercise 18.5.7. *Generalize Rolle's Theorem (see the previous exercise): If f is a continuous function on $[a, b]$ with $N + 1$ distinct zeros, then there exists a $c \in (a, b)$ such that $f^{(N)}(c) = 0$.*

Exercise 18.5.8. *Generalize the Mean Value Theorem: If f and g are differentiable functions on $[a, b]$, prove that there is a point $c \in (a, b)$ such that $[f(b) - f(a)]g'(c) = [g(b) - g(a)]f'(c)$.*

Exercise 18.5.9. *Consider*

$$x_0 = 0 \; < \; x_1 \; < \; \cdots \; < \; x_{n-1} \; < \; x_n = 1,$$

and consider, as a function of $x \in (0, 1)$, the error product

$$\prod_{i=0}^{n} (x - x_i).$$

Determine the maximum value of the error product in the case when the points are evenly spaced in $[0, 1]$.

Exercise 18.5.10. *Prove* (18.1).

Exercise 18.5.11. *Show $\prod_{i=0}^{n}(s - j) \leq n!/4$.*

The following problem describes a good way to approximate areas under curves. Are you surprised that the error is of order h^4 and not h^3? For more information, see https://en.wikipedia.org/wiki/Simpson%27s_rule.

Exercise 18.5.12 (Simpson's rule). *Let f be a continuous function on $[a, b]$, and for n a positive integer set $h = (b - a)/n$. Prove*

$$\int_a^b f(x)dx \;\approx\; \frac{h}{3} \sum_{k=1}^{n/2} [f(x_{2k-2}) + 4f(x_{2k-1}) + f(x_{2k})].$$

If f is four times continuously differentiable, show the error between the integral and the sum is bounded by

$$\frac{(b - a)}{180} \max_{a \leq x \leq b} |f^{(iv)}(x)|h^4,$$

where $f^{(iv)}$ is the fourth derivative of f.

The next problems introduce **Monte Carlo integration**, another good way of integrating. For more on the history and theory see [**Met, MU**]; see Figure 9 for an illustration.

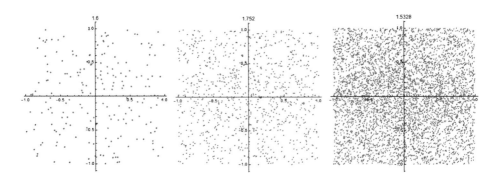

Figure 9. Monte Carlo simulations for area inside the ellipse $x^2 + 4y^2 \le 1$.
Left: 100 points, area estimate 1.6. Center: 500 points, area estimate 1.752.
Right: 2500 points, area estimate 1.5328.

Exercise 18.5.13 (Monte Carlo integration). *Let A be a nice region (say a piece-wise continuous boundary) contained in the unit square $[0,1] \times [0,1]$. Choose N points uniformly at random inside the unit square. We estimate the area of A by the fraction of the N points that are inside A. More formally, for $i \in \{1, \ldots, N\}$ let $X_i = 1$ if the ith point is in A and 0 otherwise, and set $Y_N = (X_1 + \cdots + X_N)/N$. Prove $\mathbb{E}[Y_N] = \text{area}(A)$.*

Exercise 18.5.14. *Generalize the previous problem for an n-dimensional region $\mathcal{R}$ contained in an n-dimensional box $[a_1, b_1] \times \cdots \times [a_n, b_n]$.*

Exercise 18.5.15. *Write a program to do Monte Carlo integration and estimate the area of the ellipse $(x/a)^2 + (y/b)^2 \le 1$ for various a and b (Figure 9 gives an example of such output). Conjecture a formula for the area of the ellipse and prove your conjecture. Hint: what should your formula reduce to if $a = b = r$?*

Chebyshev Polynomials

Exercise 18.5.16. *Write down the first six Chebyshev polynomials.*

Exercise 18.5.17. *Prove the trigonometric identities satisfied by the Chebyshev polynomials.*

Exercise 18.5.18. *Prove the claims about the extrema and roots of the Chebyshev polynomials.*

Exercise 18.5.19. *The Vehsybehc polynomials are defined similarly to the Cheby-shev polynomials, but now $V_0(x) = x$, $V_1(x) = 1$, and $V_{n+1}(x) = 2xV_n(x) - V_{n-1}(x)$. What properties of the Chebyshev polynomials are inherited by these?*

Exercise 18.5.20. *Prove $T_{n+1}(x)/2^n$ is a monic polynomial.*

Exercise 18.5.21. *Consider the unsigned Chebyshev polynomials, which are given by $U_0(x) = 1$, $U_1(x) = x$, and $U_{n+1}(x) = 2xU_n(x) + U_{n-1}(x)$ (thus the negative sign is replaced with a positive sign in the recurrence). What properties do these share with the Chebyshev polynomials?*

Exercise 18.5.22. *Prove that if there is a c and an n such that $T_n(c) = T_{n+1}(c) = 0$, then $T_m(c) = 0$ for all $m \geq n$.*

Exercise 18.5.23. *Let $\mathcal{C}$ be the set of all c such that $T_n(c) = T_{n+1}(c) = 0$ for some n (which may depend on c). Determine $\mathcal{C}$.*

Exercise 18.5.24. *Classify all n for which $T_n(x)$ is an odd function on $[-1, 1]$.*

Exercise 18.5.25. *Prove the following generating function identity for the Chebyshev polynomials:*

$$\sum_{n=0}^{\infty} \frac{T_n(x)t^n}{n!} = e^{tx} \cosh\left(t\sqrt{x^2 - 1}\right).$$

Exercise 18.5.26. *Prove the extrema of $T_n(x)$ are $x_k = \cos(k\pi/n)$ and the roots are $x_r = \cos((2r - 1)\pi/2n)$ for $k, r \in \{0, \ldots, n\}$.*

Exercise 18.5.27. *Prove $\|\cdot\|_{\infty}$ is a norm.*

Exercise 18.5.28. *Generalize the Chebyshev nodes to polynomial interpolation on $[a, b]$; in particular, show that we should now take*

$$x_i = \frac{a + b}{2} + \frac{(b - a)\cos\left(\frac{2i-1}{2n}\pi\right)}{2}$$

for $i \in \{0, \ldots, n\}$.

Splines

Exercise 18.5.29. *Find two cubic splines for the data set $(0, 0), (1, 2), (2, 5)$ by finding cubic polynomials from the first to the second, and from the second to the third, data points, using the freedom we have in choosing these polynomials to have as much differentiability as possible at $x = 1$.*

Exercise 18.5.30. *Redo the previous problem but now using a quartic for the points $(0, 0), (1, 2), (2, 5)$, another quartic for the points $(2, 5), (3, 1), (4, -6)$, using the freedom to have as much differentiability as possible at $x = 2$.*

Exercise 18.5.31. *Explore the Runge by Chebyshev polynomials and cubic splines. Figure out how to make this an apples and apples comparison; i.e., you want to have the same computational cost for each method. Discuss a good way to compare the error between each prediction and the actual value (you could do a random sampling of points or do an L^p norm). Determine which approach is "better".*

Exercise 18.5.32. *It was claimed that setting the first and second derivative of the splining polynomials equal to one another at the knots effectively minimizes the curvature over the entire spline function. Justify this statement.*

Exercise 18.5.33. *Prove that the cubic spline interpolate minimizes the energy integral $\int_a^b (d^2 f/dx^2)^2 dx$.*

Exercise 18.5.34. *Prove that it takes on the order of n operations to solve an $n \times n$ tridiagonal matrix equation.*

Chapter 19

The Four Color Problem

In this and the next chapter, we discuss how computers eventually gave mathematicians the power to prove two long outstanding theorems — the Four Color Theorem (on coloring regions of maps so that adjacent areas have different colors) and the Kepler Conjecture (on the optimal way to stack spheres). Both of these problems were articulated in eras long before the possibility of computational solutions were conceivable. The Four Color Theorem, often considered the first to have a significant part of the proof by intensive computation on computers, was introduced in 1852, and Kepler's Conjecture was first stated in 1611, making it older than Fermat's Last Theorem (it also resisted solution longer).

As this is a first course, we do not have time to go into the details of the proofs of these important problems. This is a shame, as the Kepler problem clearly belongs in a course such as this; linear programming played an important role in the proof. Our goal instead is to work on simpler cases that highlight many of the methods and ideas (such as a proof of the Five Color Theorem) which will prepare you to read some of the many excellent surveys.

Additionally, these problems provide an outstanding opportunity to talk about the nature of proof in the 21st century. There's a rich interplay between theory and computation in each solution. Many centuries' worth of work was required to make each problem computationally tractable in the first place. However, all of that progress still couldn't reduce the problem to something that could be feasibly checked by hand. Hence, in the era of the modern computer, algorithms were developed to do the checks that humans could not; furthermore, significant work was needed to make the algorithms efficient enough to run in a reasonable amount of time on available resources. Perhaps unsurprisingly, many mathematicians were at first hesitant to accept these computerized checkers, holding out hope that there would be a more standard theoretical argument which could be processed by an individual. This has not happened; while the computations have been simplified and better arguments have pruned the number of cases computers must examine, there are now solutions in the literature which rely on large scale computation. This has led to a healthy discussion in the mathematics community as to what should be considered a proof. Thus, based on the historical importance of these problems, as well as the techniques which arise in their analysis, we will spend a substantial amount of time in this and the next chapter on these problems.

19.1. A Brief History

The main focus of this course has been developing methods to quickly find optimal solutions to problems. This often involved using the incredible processing power offered by modern computers. It is unsurprising then, especially given the topic of this chapter, that computing power plays a large role in the eventual resolution of the Four Color Problem. While many mathematicians attempted to solve the problem before modern computing became available, all of whom were unsuccessful, their failures and near-successes were instrumental in paving the way for the final successful resolution.

The problem is easily stated: *The regions of any map can be colored with only four colors, in such a way that any two adjacent regions have different colors* (a more formal definition is given in the next section; we give an illustration in Figure 1). The problem was first proposed in 1852, when the South African mathematician Francis Guthrie suddenly realized that he could color a map of the counties of England with only four different colors. He wondered if this four-colorability was the case for *any* map, and so he posed the question to his former advisor Augustus De Morgan (of De Morgan's Laws) that year. De Morgan had no good answer to Guthrie's question, but thought it was a neat puzzle, and so kept note of it for several years while asking every mathematician he knew if they could find a solution to the problem.

When the conjecture was finally released to the public in 1878 (as a publication by the Royal Society of London), many mathematicians came forward with their attempts at a solution. While all the offered proofs were incorrect, many of the greatest advancements towards solving this mathematical mystery came as a byproduct of these repeated failures. Arthur Cayley, for example, gave some of the first successful formalizations of the problem, allowing for inductive proofs on certain types of maps. In 1879, Sir Alfred Kempe introduced many of the theoretical

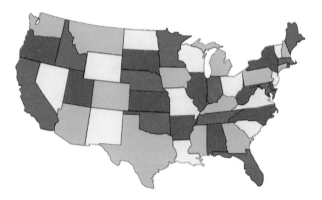

Figure 1. A four coloring of the continental United States (image from [**Mar**]).

sign posts that guide modern proofs, including the reduction of possible counterexamples to subgraphs with a central kernel of fewer than 6 faces. Work in the first half of the 20th century saw that proofs for the four-colorability of finite maps lend strength to the conjecture. In 1922 the four-colorability was shown to hold for any map with at most 25 regions; this numbered had increased by 1975 to 90 regions. At the same time, Heesch refined the concept of region reducibility and was the first to forward an algorithmic method for searching through various cases, called the discharge procedure. However, in a somewhat cruel twist, he was unable to obtain the computing power required to complete the proof in a general case. That result ultimately came in 1976 when Kenneth Appel and Wolfgan Haken announced that they had finally proven the Four Color Theorem.

This announcement was met with wide acclaim and was highly visible in the press at the time; however, there was also a healthy amount of skepticism cast on the result from the professional mathematical community. This was the first proof that was majorly aided by a computer program. Moreover, it appeared that the human verifiable portion of the proof was completely infeasible to do by hand. The algorithm used by Appel and Haken checked 1,936 configurations and took over 2 months to run on a mainframe computer. The enormity of these calculations and the realization that human confirmation of their validity was totally impractical, simply added to the skepticism.

The first to articulate these objections was Thomas Tymoczko, who in his 1979 publication "The Four-Color Problem and Its Philosophical Significance" [**Ty**], stated the following:

> No mathematician has seen a proof of the 4CT, nor has any seen a proof that it has a proof. Moreover, it is very unlikely that any mathematician will ever see a proof of the 4CT.

Tymoczko, among many others, believed that drawn-out computer-aided proofs were not, in a sense, "real" mathematical proofs, because they incorporated too many logical steps that were not *feasibly* verifiable by human beings. He argued that, instead of relying on axiomatic and logical deduction, mathematicians were being asked to blindly *trust* in an unreliable empirical process that was potentially riddled with errors from the computer program, as well as defects in the hardware and run-time environment. Many of the more conservative mathematicians of the time were reluctant to call anything involving a computer a "mathematical proof". They claimed that computer-aided proofs should merely be thought of as "calculations", rather than proofs. On top of that, some people complained that Appel and Haken's proof was nothing more than an irreproducible experiment (due to the amount of time and money required to run the program on a powerful enough computer), which meant that other researchers wouldn't be able to easily re-run the program on their own hardware and check the proof's validity. Accepting this result, some said, would require a degree of faith in the legitimacy of the algorithmic results which flies in the face of rigorous logical arguments that normally legitimize mathematical proofs.

Subsequent proofs, such as the work by Robertson et al. in 1995, have cut down significantly on the number of checks required to complete a valid proof of the Four Color Theorem, and have reduced the complexity of the computer algorithm needed to perform the case analysis. The basic computational aspect is still present, however, and the argument of proof by checking a number of configurations remains. As time has passed, this computer dependence has become less and less problematic for the professional community, most likely because computational methods have implicated themselves so comprehensively into modern mathematics.

The Four Color Theorem, then, is not just interesting for the simplicity of its formulation, but also because it marks a tipping point in the history of mathematics. Theory could now be used to bring problems into the realm of computational tractability, and an appropriate algorithm might accomplish the rest. While adopting some technical simplifications afforded by more recent results, in the next few sections we will outline the mathematical theory behind some of the initial attempts at the proof, and then move on to discuss the computer-aided work done by Appel and Haken in 1976 and the subsequent improvements made by others following their breakthrough.

19.2. Preliminaries

The proof of the Four Color Theorem benefited greatly from Kempe's failed proof given in 1879 [**Kem**]. Though a mistake in the argument was discovered a decade later, it did lead to a proof of the Five Color Theorem and created a structure for the eventual successful attack. The reason for this is that, with some technical elaboration and additional work, the number of cases that must be checked became manageable. Hence, we begin with Kempe's proof, discuss the error and why we obtain the result with five but not four colors, and then move on to the modern version.

19.2.1. The Formalization. The theorem itself is simple enough to state.

Theorem 19.2.1 (The Four Color Theorem). *The regions of any simple planar map can be colored with only four colors in such a way that any two adjacent regions have different colors.*

The complexity of proving the theorem becomes apparent as we start giving formal definitions of each of these terms. A **planar map** is a set of pairwise disjoint subsets of the plane, called regions. A **simple map** is one whose regions are connected open sets. Two regions of a map are **adjacent** if their respective closures have a common point that is not a corner of the map (i.e., their boundaries touch at more than a point). Finally, a point is a **corner** of a map if and only if it belongs to the closures of at least three regions.

As we saw in Figure 1, this map of the continental United States can indeed be four-colored. However, these definitions leave us in somewhat of a bind. Dealing with subsets and regions of the plane means that we inevitably must appeal to the vocabulary of topology and the mathematics of curves. As a gauge of the difficulty that this presents, we can consider the significant work that went into proving the **Jordan Curve Theorem**. A **Jordan curve** in the plane $\mathbb{R}^2$ is the image C of an injective continuous map ϕ of a circle S^1 into the plane, $\phi : S^1 \to \mathbb{R}^2$ (recall a function is injective if distinct inputs go to distinct outputs).

Theorem 19.2.2 (The Jordan Curve Theorem). *Let C be a Jordan curve in the plane $\mathbb{R}^2$. Then its complement, $\mathbb{R}^2 \setminus C$, consists of exactly two connected components. One of these components is bounded (the interior) and the other is unbounded (the exterior), and the curve C is the boundary of each component.*

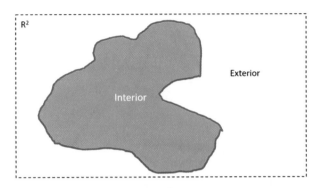

Figure 2.   Illustration of the Jordan Curve Theorem.

Even though this result seems obvious (see Figure 2 for an example), the known proofs are often technical and involved. Even more troubling for our purposes, a result about various regions in $\mathbb{R}^2$ seems to require this result as a fundamental starting point, and anything further appears destined to be phrased in terms of similar analytic and topological language. Kempe's first great innovation was to move the problem out of the realm of the continuous and into a finite, discrete formulation that allowed him to use some fundamental counting arguments. To do this we simply overlay a graph onto the map in question, and examine the colorability of nodes instead of regions (see Figure 3).

Theorem 19.2.3 (Dual of the Four Color Theorem). *The nodes of any planar graph can be colored with only four colors, in such a way that no two adjacent nodes have the same color.*

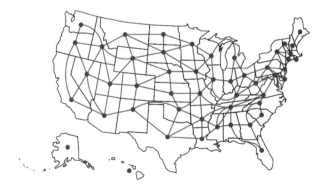

Figure 3.   Associated graph to the United States (image from [**Mar**]).

The second simplification that Kempe made was to only consider maximal planar graphs. A graph is **planar** if it can be drawn on the plane in such a way that its edges intersect only at their endpoints (i.e., it can be drawn with no crossing edges), and a planar graph is **maximal** if adding any additional edges violates the condition of planarity. See Figure 4 for an example.

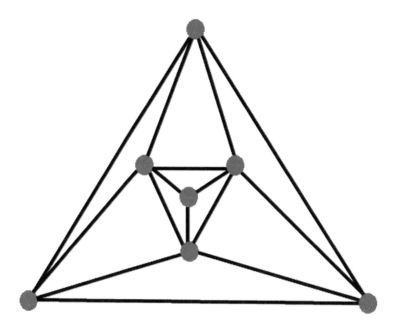

Figure 4. A maximal planar graph.

It's sufficient to prove the dual of the Four Color Theorem for maximal planar graphs because every planar graph can be constructed from a maximal planar graph by simply removing edges. In other words, removing edges doesn't change the colorability of the graph and can only add degrees of freedom with respect to how each node is colored.

With the problem reduced, Kempe's argument attempted to show the impossibility of a counterexample (i.e., a map that cannot be four-colored). In fact, we can add the further simplification that we only search for minimal counterexamples. If there is a graph that is not four-colorable, then there is a smallest such graph.

Even though we haven't said anything explicit yet about colorability, simply rephrasing the problem in graph-theoretical terms is major progress. Where before we seemed to be faced with a topological problem and forced into analysis, we now only have to deal with the properties of maximal planar graphs and candidates for minimal counterexamples to colorability.

19.2.2. Euler's Formula. Now that we have converted our problem to one on graphs, we can apply a useful result from Euler. Note that the number of faces also includes the "external" face going off to infinity.

Theorem 19.2.4 (Euler's Formula). *For a finite, connected, planar graph $v - e + f = 2$, where v is the number of vertices, e is the number edges, and f is the number of faces.*

The most common proof proceeds by induction on either the faces, edges, or vertices. Another interesting method attributed to Von Staudt is called Interdigitating Trees, which constructs the dual of a graph and then appeals to the Jordan Curve Theorem. However, for the sake of variety, and partly to foreshadow some of the computational techniques covered in the following sections, we give a proof due to Thurston [**RSST**] (see also [**Le**]).

Proof by Discharge. Rotate the graph in the plane such that no edge is horizontal (see Exercise 19.6.6), and denote the uppermost vertex U and the lowermost vertex L. Now, place a unit positive charge at each vertex and in the center of each face, and a unit negative change in the middle of each edge. Our goal is to show that there is a total charge of 2.

To do this, we displace each charge of edges and vertices horizontally and to the right so that they accumulate in an adjacent face; see Figure 5.

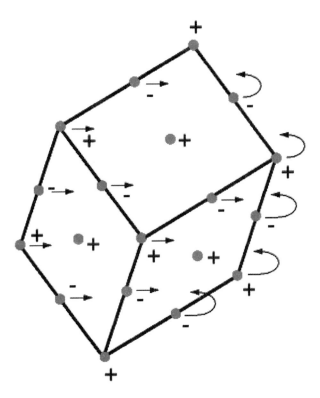

Figure 5. Illustration of Proof by Discharge (image from [**EPPS**]).

Hence, each face receives the net charge from an open interval along its boundary, made up of edges and vertices. These must alternate, and since the first and last are edges, there is a surplus of one negative charge; therefore, once we incorporate the positive charge originally assigned to each face, the total charge in each face is zero. All that is left is +2, coming from one at L and one at U, for a net charge of 1. □

Returning to maximal planar graphs, Euler's formula allows us to prove a result about the average degree of vertices in such a graph. This will then allow us to further narrow our search when analyzing candidates for minimal counterexamples to four-colorability. The reason this is particularly useful for our purposes is that, from this result, it follows that in any planar graph there *must* be a vertex of degree 5 or less. In particular, any minimal counterexample to four-colorability must contain one of the following configurations shown in Figure 6.

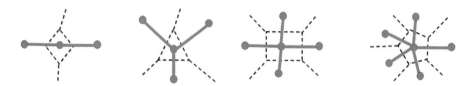

Figure 6. Configurations that must occur in a minimal counterexample.

Lemma 19.2.5 (Average Vertex Degree Less Than 6). *For a maximal planar graph, the average degree of a vertex in the graph is* $6 - \frac{12}{v}$.

Proof. In a maximal planar graph each face is bounded by 3 edges (see Exercise 19.6.9) and each edge is part of the boundary of exactly two faces. Hence, $f = 2e/3$. Plugging this into Euler's formula yields $e = 3v - 6$. Each edge is connected to two vertices, and hence the total degree of all vertices in the graph is $2e = 3v - 6$. From this it follows that the average degree per node is $6 - \frac{12}{v} < 6$.   □

19.2.3. Kempe's Inductive Argument. Kempe's argument proceeds by induction on the number of vertices in the graph. Suppose that a map with N vertices is colorable. We want to show that a map with $N + 1$ vertices is colorable. Now, from Lemma 19.2.5 we know that in an $N + 1$ colorable graph, there must be a vertex with degree 5 or less. This gives us the needed traction to leverage the inductive argument. The argument proceeds as follows:

(1) Remove a vertex of degree 5 or less.

(2) Color the resulting graph via induction.

(3) Reintroduce the removed vertex.

(4) Color the new vertex so as to maintain colorability of the graph.

If this process is successful, then we will have shown that all the candidates for minimal counterexamples to four-colorability yield no genuine counterexamples.

It turns out that the process for degree 2 and 3 nodes is quite simple, given that reintroduction into this graph always leaves a degree of freedom from which to choose a color; see Figure 7.

The issue in the case with degree 4, and the ultimate downfall of the argument in the case with degree 5, is that we may need to alter the global coloration of the graph when we attempt to reintroduce the vertex; see Figure 8. While this obstruction prevents us from proving the Four Color Theorem at this moment, it does allow us to prove the Five Color Theorem (see Exercise 19.6.10).

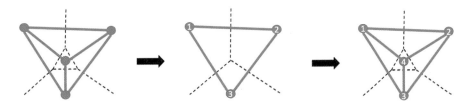

Figure 7. Kempe's inductive argument for a vertex of degree 3.

Figure 8. Issues with Kempe's argument for a vertex of degree 4.

To free a color for the reintroduced vertex, Kempe developed the notion of a color chain, which in his honor is now called a **Kempe chain [RSST]**.

Definition 19.2.6 (C_1C_2 Kempe Chain). *Let G be a planar graph whose vertices have been legally four-colored, and suppose v is a member of the set of vertices of G. Suppose that v is colored C_1. The C_1C_2 Kempe chain containing v is the maximal connected component of G that contains v and contains only vertices colored C_1 or C_2.*

The idea in introducing Kempe chains is to free up a color for the reintroduced node by constructing a chain based on one of the adjacent vertices, and essentially swapping color values throughout the chain; this will free up a viable coloring for the reintroduced node.

Definition 19.2.7 (C_1C_2 Kempe Chain Switch). *Let K be a C_1C_2 Kempe chain. A C_1C_2 Kempe chain switch interchanges the values of C_1 and C_2 in K.*

For the coloring of the reintroduced vertex of degree four construct a chain out of an adjacent vertex v colored C_1 and a vertex adjacent to v which is not also adjacent to the reintroduced vertex. Now, construct the Kempe chain containing v. We have two cases to deal with [**Kem**].

Case 1: K **does not contain any vertices adjacent to the reintroduced vertex.**

In this case, we simply perform a color switch on the constructed chain. From the definition of a Kempe chain, this remains a valid coloring. See Figure 9.

Case 2: K **contains a vertex adjacent to the reintroduced vertex.**

We call this a cyclic Kempe chain, and it results in an invalid coloring because a color swap still results in two adjacent vertices of the same color. However, the

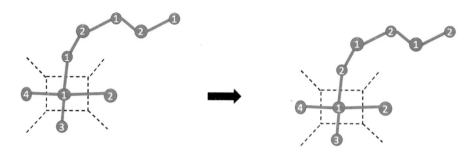

Figure 9. Coloring for Case 1 of a degree 4 vertex.

planarity of the graph saves us here. The greater subgraph bounded by the cyclic Kempe chain must include three vertices adjacent to the reintroduced vertex — two included in the Kempe chain and a third in the interior of the subgraph (see Exercise 19.6.11). This Kempe chain then serves as a barrier guaranteeing that a Kempe chain constructed out of the last vertex adjacent to the reintroduced vertex is not cyclic and thus can be color-swapped. This results in a valid coloring [**Ste**]; see Figure 10.

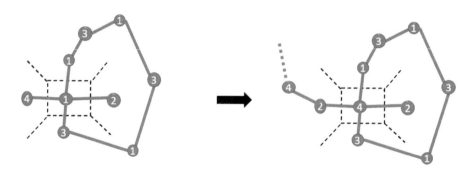

Figure 10. Coloring for Case 2 of a degree 4 vertex.

The problem for Kempe arises in the 5 degree case. Here, we can have multiple cyclic Kempe chains, and this results in cases where minimal counterexamples can persist; see Figure 11. The mistake in Kempe's proof wasn't noticed until a decade after its original publication. What's more, further progress wasn't made until Birkhoff's observations in 1913. However, we shouldn't be discouraged by Kempe's failure here. Indeed, he gives us yet another simplification of the problem. To prove the Four Color Theorem, we must now only make an argument for the graph with a node degree 5. Furthermore, much of the main thrust of the argument still holds for the final proof by Appel and Haken, with some added technical sophistication.

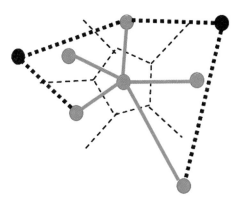

Figure 11. A degree 5 vertex with no feasible Kempe chain.

19.3. Birkhoff and the Modern Proof

Birkhoff's [**Bi**] great contribution was to move away from simply considering the degree of vertices. Instead we consider rings (i.e., partial faces) around a subgraph of the entire map we desire to color. These are called **configurations**.

The argument for such configurations is similar to Kempe's approach. Remove the entire configuration, assume colorability of the smaller map by induction, and reintroduce the configuration and determine a valid coloring. If such colorability is possible upon a reintroduction of the configuration, we say that the configuration is **reducible** [**AH2**].

Birkhoff, applying the same chaining methods used in Kempe's original attempt, easily showed that configurations with rings of 2, 3, and 4 faces are reducible. Eventually, he was able to show that every configuration with 5 faces is reducible (see for example Figure 12) except for the one composed of pentagons, a case analogous with the gap in Kempe's original proof.

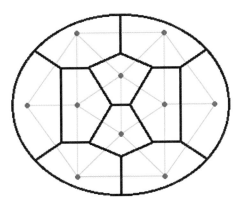

Figure 12. Birkhoff's Diamond, a famous reducible configuration (image from user Snorri95 at Wikimedia Commons; licensed under the Creative Commons Attribution-Share Alike 3.0 Unported (https://creativecommons.org/licenses/by-sa/3.0/deed.en) license).

In the same paper, with an even more thorough application of Kempe's method, Birkhoff proved that all possible counterexamples to the Four Color Theorem are a particular type of configuration. We offer some definitions used in the paper and then state the lemma.

A configuration is said to be **internally 6-connected** if it has no short circuits and a minimum degree of 5. A **circuit** C is a closed, non-intersecting walk on the edges of a graph. Let $E(C)$ be the edges of the circuit C in a graph G. A **short circuit** of G is a C with $|E(C)| \leq 5$ such that each of the two open regions bounded by C contain at least one vertex of G if $|E(C)| \leq 4$ and at least two vertices if $|E(C)| = 5$.

Lemma 19.3.1 (Birkhoff's Lemma [**Bi**]). *A minimal counterexample is an internally 6-connected triangulation.*

Birkhoff's lemma trimmed the number of cases to mere millions. This was still an enormous number that was not feasible to do by hand, but was increasingly close to the realm of possibility with advances in computing power coming in the 1970s. However, before handing the process over to computer-aided methods, we need one last formal piece. Namely, we can't simply show that candidates for minimal counterexamples contain reducible configurations, we must show that *every* such candidate for a minimal counterexample must contain a configuration which is reducible. Hence, we introduce the notion of an unavoidable configuration [**AH2**]: a set of configurations is called **unavoidable** if a member occurs in every internally 6-connected triangulation.

The goal then becomes finding an *unavoidable* set of *reducible* configurations. The theorem will follow from the existence of such a set. There are an enormous number of cases to check in order to confirm the existence of such a set, but with some clever computer programming, it turned out to be computationally tractable with the technology of 1976.

19.4. Appel-Haken Proof

In 1976, nearly a century after the first publication of the conjecture in 1878, the American mathematician Kenneth Appel and his German colleague Wolfgang Haken announced their solution to the Four Color Theorem at the University of Illinois at Urbana-Champaign. Their proof was, essentially, a proof by *computational case analysis*, which borrowed heavily from the theoretical findings of Kempe, Heesch, and Birkoff in each of their previous failed attempts.

In order to prove the existence of an unavoidable set of reducible configurations, one must first find such a set of unavoidable configurations. Remember that it's sufficient to only consider maximal planar graphs, or *triangulations*. Additionally, Birkoff showed that any candidate counterexample to the Four Color Theorem must be an *internally 6-connected* triangulation. So, in order to prove that every internally 6-connected triangulation must contain an unavoidable set of configurations, Appel and Haken developed a method called the **discharge procedure**, which works as follows.

19.4.1. The Discharge Procedure. Let T be an internally 6-connected triangulation. Assign to every vertex v a "charge" of value $6 - \deg(v)$. It follows from Euler's formula and Theorem 19.2.5 that the sum of the charges of all vertices in T is 12 (in particular, that it's *positive*). The process of discharge consists of "redistributing" charge throughout T via a series of rules that move charge between vertices such that the presence of positive charge on a vertex, after redistribution, demonstrates the unavoidability of a given configuration. Since the total charge must remain positive after discharging, some vertex must have positive charge and therefore some unavoidable configuration must appear somewhere in T.

To see how this process works in practice, consider the following example. Let's show that the following set of vertex configurations is unavoidable: a degree-2 vertex (D_2), a degree-3 vertex (D_3), a degree-4 vertex (D_4), a degree-5 vertex connected to another degree-5 vertex ($D_{5,5}$), and a degree-5 vertex connected to a degree-6 vertex ($D_{5,6}$).

In order to prove this, we assume, for the sake of contradiction, that there exists a triangulation T that does not contain any of the above configurations. We know that every minimal counterexample to the Four Color Theorem must contain at least one vertex of degree 5 (denoted D_5). By assumption, a D_5 cannot be adjacent to a D_2, D_3, or D_4, since there are none, or to another D_5 or a D_6 (as that would create a $D_{5,5}$ or a $D_{5,6}$ configuration). Thus, each D_5 is only adjacent to vertices of degree 7 or more. Now we assign a charge to every vertex, based on its degree. To a vertex of degree k, we assign the charge $6 - k$, so that each D_5 receives a charge of 1, each D_6 receives a charge of 0, each D_7 receives a charge of -1, each D_8 receives a charge of -2, and so on.

Using Euler's formula that $(v - e + f = 2)$ we see that the total charge of T is 12. To "discharge" the graph, we take all the positive charges (i.e., the D_5 charges) and spread them equally to all adjacent vertices, in a way that no charge is added or removed (i.e., the total charge remains 12); see Figure 13. Each neighbor of a D_5 receives a charge of $1/5$, since we have not created or removed any charge on the graph, and each D_5 now has a charge of 0.

Consider the D_7's. Since they have initial charge -1, each D_7 needs at least 6 neighboring D_5's to receive enough charge to have a positive charge after discharge. However, this would require that two of the D_5's are adjacent, which is not allowed by our assumption. Thus, after discharge, each D_7 must have a negative charge. Now consider the D_8's, each of which had initial charge -2. This would require 11 neighboring D_5's in order to receive a positive charge after discharge, but since each D_8 only has 8 neighbors, this is impossible. Lastly, consider each D_6. They all had a charge of 0 before discharge, and cannot have any neighboring D_5's, per our assumption, so their charges can only be negative after discharge.

Thus, our assumption is false. In order for the total charge to remain positive, T must contain at least one of the following configurations: D_2, D_3, D_4, $D_{5,5}$, or $D_{5,6}$.

The details on the size of each charge and how to redistribute them can vary widely in order to prove that certain sets must be unavoidable. This is just one example of a fairly simple unavoidable set, and how one proves that it is, in fact, unavoidable. In Appel and Haken's final proof, the number of discharge rules

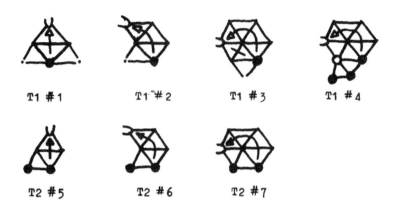

T1 #1   T1 #2   T1 #3   T1 #4

T2 #5   T2 #6   T2 #7

Figure 13. A small subset of the discharge rules discovered by Appel and Haken in their original publication. Image from [**AH2**], copyright 1977, Board of Trustees, University of Illinois. Reprinted with the permission of the *Illinois Journal of Mathematics*, published by the University of Illinois at Urbana-Champaign.

used totaled **487**, requiring the investigation by hand of nearly **10,000** possible triangulations with positive charge. The result of this investigation led to the discovery of **1,936** unavoidable configurations in any maximal planar graph, which constituted the unavoidable set that they needed for their proof (this number was later reduced to 1,482 configurations) [**Ste**].

19.4.2. Reducibility. After painstakingly producing a set of 1,936 unavoidable configurations by hand, Appel and Haken needed a method of checking that each of these configurations was *reducible* in order to conclude the proof. To do this would require checking the reducibility of approximately **1,000,000,000** candidate configuration colorings. This clearly cannot be done by hand in a reasonable amount of time, so Appel and Haken enlisted the aid of modern computing by writing a computer program to run on a mainframe computer (the most powerful type of computer at the time) and enumerate and check candidate colorings until each unavoidable configuration was proven to be reducible.

The program they developed was written with the goal of efficiency in mind. At the time, the cost of renting a mainframe computer, as well as the likelihood of machine failure on a long-standing program, was very high. For this reason, it was in their best interests to write code that would execute as quickly as possible. This is why they chose IBM 370 assembly as their programming language. This code, and its output, was documented in approximately 400 pages of microfiche, appended to the end of Appel and Haken's proof text. The program took approximately 2 months to run on a mainframe computer. The computers that Appel and Haken had access to were all of the IBM variety: a 360-75 at Urbana-Champaign, a 370-158 at the University's Chicago Circle Campus, and later a 370-168 of the University of Illinois administrative data processing unit. The total data saturation at any given time during the program's execution was on the order of 2,000,000 bytes of data (which was a lot for that time period) [**AHK**].

In order to cut down significantly on computing time, Appel and Haken additionally incorporated dynamic programming into their code design. This was done in the sense that, once a coloration was proved "good" (and thus the configuration reducible), its goodness was immediately available for use in the testing of other colorations. This technique, commonly referred to as **memoization**, takes advantage of large amounts of available memory in order to greatly reduce computational time complexity.

For example, when trying to compute the nth Fibonacci number, a naive approach might be to implement the algorithm using just pure recursion. This leads to a very clean, but incredibly inefficient (exponential-time) solution (see Figure 14).

function fib(n)
    function fib(n)
        if n ¡= 1 return n
        return fib(n-1) + fib(n-2)

Figure 14. A recursive implementation to compute the Fibonacci numbers (from [**Wiki**]).

The dynamic programming approach, on the other hand, uses a data structure (often a data table, array, or a hash map) to store previously solved solutions to smaller instances of the problem; in this case, the $(n-1)$th Fibonacci number (see Figure 15).

var m := map(0 → 0, 1 → 1)
    function fib(n)
        if key n is not in map m
            m[n] := fib(n-1) + fib(n-2)
        return m[n]

Figure 15. A dynamic implementation of nth Fibonacci number (from [**Dav**]).

This dynamic approach takes the exponential-time recursive solution and reduces it to quadratic time. Appel and Haken were able to exploit similar savings, and after 2 months of program execution time and nearly 1000 pages of proof text, tiny hand-drawn diagrams, and microfiche documentation, they were able to announce a proof of the Four Color Theorem. Appel famously celebrated this achievement by writing the words, *"Modulo careful checking, it appears that four colors suffice"* on the blackboard of the mathematics department of the University of Illinois [**Le**]; it has also been honored with a postmark (see Figure 16).

19.5. Computational Improvements

The initial Appel and Haken proof was met with a fair bit of skepticism. With such a simple and elegant problem statement like that of the Four Color Theorem, mathematicians expected an equally simplistic proof to solve it. Instead, what they got was thousands of lines of undecipherable IBM 370 assembly code and page

FOUR COLORS
SUFFICE

Figure 16. Postmarks used by Department of Mathematics University of Illinois at Urbana-Champaign: http://www.math.illinois.edu/History/postmarks.pdf.

upon page of small hand-drawn configuration diagrams. On top of that, many rational skeptics knew that computer programming was fundamentally buggy and error-prone, and that the computers themselves that ran the code were prone to hardware malfunctions and other sporadic system failures and bugs. This criticism was only reinforced when several small errors (both in the assembly code and the manual case analysis) were detected, leaving the Appel-Haken proof in the realm of deep uncertainty.

19.5.1. Robertson, Sanders, Seymour, and Thomas. In 1995 Robertson, Sanders, Seymour, and Thomas decided that they wanted to finally feel convinced that the Four Color Theorem proof was correct. After reading through the 1976 proof by Appel and Haken, they isolated two main concerns:

(1) A major portion of the Appel-Haken proof could not be verified by hand, since it relied heavily on the correctness and proper execution of a computer program.

(2) The remaining portion of the proof, which was supposedly checkable by hand, was incredibly arduous and complex, and as far as anyone knew, no one had checked it to its entirety.

To address these concerns, they set out to design a streamlined and more-concise version of the Appel-Haken proof; one that simplified the initial case analysis, and reimplemented the complex computer code used to solve it.

The structure of their proof, as a whole, was very similar to that of Appel and Haken. The goal of the proof was to first find an unavoidable set of configurations, one of which must occur in any internally 6-connected triangulation, and then algorithmically color each of those configurations to prove their reducibility. By systematically considering variants in the original case analysis and taking advantage of certain symmetries in the list of possible triangulations, they reduced the initial set of unavoidable configurations down from 1,478, by Appel and Haken (which they had already reduced from the original 1,936), to only 633. Similarly, in proving the unavoidability of such a set of configurations, their simplified version of the discharge procedure included only 32 different discharge rules, compared to Appel and Haken's 487 (see [**RSST**]).

All of their code was written in the C programming language, as opposed to assembly language, making it much easier to read and verify its logical correctness. The code was able to handle both the unavoidability portion of the proof (which was tediously done by hand in the Appel-Haken proof) and the reducibility portion, which ultimately turned the 1000-page diagram-filled proof by Appel and Haken into a clean and concise 35-page paper.

Moreover, thanks to Moore's Law and the nearly 20 years that had passed since the Appel-Haken proof, the average speed of computers had increased massively. Now, with a greatly reduced case analysis and much faster and more reliable computers, the proof was able to execute in approximately 3 hours on the average PC, as opposed to months on a mainframe computer. This took care of the irreproducibility issue of the Appel-Haken proof, and allowed skeptics of the program to download the source code themselves, on their own machines, and reaffirm the correctness of the proof for their own peace of mind.

As an added bonus to their proof, an algorithm was provided that could four-color any planar graph with n vertices in $O(n^2)$ time. At first glance, it might become evident that an intuitive quadratic algorithm exists, as follows. Given a planar graph G on n vertices (without loss of generality, we can assume that G is a triangulation, because if not we can create a triangulation in linear time by adding edges) we find an unavoidable configuration in G, replace G by a smaller graph G', and recursively repeat this step (this is using the inductive argument described earlier). It appears then, that we can find a reducible configuration of G in linear time, and we can similarly construct a four-coloring of G from a four-coloring of G' in linear time. Since the recursive step is applied at most n times, the overall time complexity would seemingly be quadratic.

The problem here is that Birkoff's lemma only guarantees the appearance of an unavoidable configuration if G is an internally six-connected triangulation. Even if we had started with a highly connected graph, the connectivity would continually drop during each recursive step. Therefore, the algorithm works by either quickly finding an unavoidable configuration appearing in G, in which case we proceed as suggested above, or by finding a set X of vertices of G that *isn't* internally six-connected. In that case we apply recursion to two carefully selected subgraphs of G, and obtain a four-coloring of G by connecting the four-colorings of the two subgraphs. Further details of this procedure can be found in [**RSST**].

19.5.2. The Coq System. Even after the elegant 1995 revision of the Appel and Haken proof, which cleared up most doubts concerning the truth of the Four Color Theorem, Georges Gonthier of Microsoft Research Cambridge Labs still felt that a computer program was too difficult to relate precisely to the formal statement of a mathematical theorem. This is, in a sense, the final step in the clarification effort of the Appel-Haken proof, and works by evicting the two weak links of the proof: the manual case analysis of configurations, and the manual confirmation that computer programs are able to correctly verify the classification of each configuration.

For at least thirty years after Appel and Haken's initial proof, computer scientists had been working out a solution to the problem of **formal program proofs**. The idea was to write code that could perform a formal proof of correctness, by describing not only what the machine should do, but also *why* it should be doing it. The validity of the proof is an objective fact that could be checked empircally by other programs, whose own validity could be verified by being executed on many inputs. However, difficulty still lay in the fact that formal proofs weren't easily produced, even with languages extensive enough to express all of mathematics.

Nevertheless, in 2005 Gonthier announced a proof of the Four Color Theorem which utilized the **Coq system**, a formal logical proof system developed by Thierry Coquand and Gerard Huet in 1984 (with the help of about 40 other researchers). This system, along with a *formal proof script*, effectively covered both the computational *and* mathematical parts of the Four Color Theorem proof. The script could be run through Coq, which could then systematically verify the correctness of *every* aspect of the proof script. Now, although the proof's correctness still relied on the proper functionality of various hardware and software components in the system (e.g., the processor and the operating system), none of these components were directly tied to the proof of theorem itself. All the components came fresh "out-of-the-box", designed to fulfill a more general purpose, and could easily be tested on completion of various other tasks (the idea being that these components could be tested much more extensively than the human mind could be). Also, as an added feature, the Coq system could optionally output a *proof witness* (a detailed log of the chain of logical steps used in completing the proof), which could, in principle, be checked independently by anyone who still felt unconvinced. Therefore, with this new system, the "faith" associated with the validity of the Four Color Theorem proof was redirected toward the computer components and the system itself. Once that was agreed to be correct, the correctness of whole proof would follow.

The Coq system was developed to implement high-level mathematical language called *Gallina* that was derived from the formal mathematical language called the *Calculus of Inductive Constructions* (a derivative of the calculus of constructions), which cleverly integrated higher-order logic into a richly typed functional programming language. The general capabilities of Coq, according to its documentation, are as follows:

(1) Efficient definition/evaluation of functions and predicates.

(2) Clear statement of mathematical theorems and software specifications.

(3) The ability to interactively develop formal proofs of such mathematical theorems.

(4) The ability to systematically and mechanically verify these proofs.

(5) Translating certified programs to structured programming languages like Haskell, OCaml, or Scheme.

As a platform, Coq was designed to be able to formalize any mathematical concept, and Gonthier leveraged this capability in designing his proof. The theory behind his proof was the same as the theory used for both the Appel-Haken and Robertson-Sanders-Seymour-Thomas proofs: *find an unavoidable set of configurations in any internally 6-connected triangulation, and systematically show that each of those unavoidable configurations is reducible.* Using his own rigorously defined data types and mathematical procedures, Gonthier came up with the simple formulation of the theorem shown in Figure 17.

Theorem four_color_hypermap :
    forall g : hypermap, planar_bridgeless g − >
    four_colorable g.

Figure 17. The formal statement of the Four Color Theorem proof [**G**], written in the Coq formal language.

This deceptively simple statement hides a number of fairly technical definitions of hypermaps, cycle-counting, planarity, and other items, and while such a clear formalism spares skeptics from having to wade through the complete proof, Gonthier still provided all the necessary definitions of these terms to thoroughly convince the reader. The above statement alone is based on about 40 lines of elementary topology and 100 lines of axiomatizing real numbers. But, once this is all put together, and compiled into OCaml (the programming language used to design Coq), the proof itself took only one and a half hours to execute in 2005 [**G**].

What all this work ultimately shows us is that you *can* build programs that are *as reliable*, if not more reliable, than mathematical proofs. Gonthier demonstrated that it's possible to build rigorously self-certifying programs/proofs, and that *proof by computation* is a feasible approach to mathematics. On top of that, these computer proof "assistants", like the Coq system, could now be used by mathematicians as "logical calculators", to explore the logical structure of a proof, and uncover some of the hidden logical details that might have been swept under the rug by researchers who previously worked on the problem.

Gonthier's work represented the final nail in the coffin of the Four Color Theorem. After more than a century's worth of philosophical deliberation and mathematical/computational work, the notoriously difficult and deceivingly simple problem was finally laid to rest.

19.6. Exercises

Preliminaries

Exercise 19.6.1. *Does there exist a non-planar graph which requires more than four colors? If yes give an example.*

Exercise 19.6.2. *Consider a planar graph with n vertices. Prove that there is some function f such that at most $f(n)$ colors are needed, and give upper and lower bounds on f. Note the Four Color Theorem states that $f(n) \leq 4$; what results can you obtain elementarily? Can you show $f(n)$ is at most $n!$ or at most n^2?*

Exercise 19.6.3. *A Jordan curve is an injective map; show that the Jordan Curve Theorem can fail for curves which are not injective, and give such an example.*

Exercise 19.6.4. *Check Euler's formula for a regular n-gon, as well as a regular n-gon with an additional vertex added at the center and all perimeter vertices connected to the center.*

Exercise 19.6.5. *Check Euler's formula for a rectangular donut region; take squares centered at $(0,0)$ with side lengths 2 and 4 and additionally connect each vertex of the outside square with the unique vertex on the inner square which is on the path* *from it to the center (thus we connect the two upper left vertices, the two upper right vertices, and so on).*

Exercise 19.6.6. *Prove that if we have a planar graph with n vertices, it's possible to rotate the graph so that no edge is parallel to the x-axis. Is it possible to rotate so that none of the edges are parallel to the x-axis or y-axis? What if we are given a finite set of lines and we wish to rotate so that no edge is parallel to any line in our finite set; can that be done?*

Exercise 19.6.7. *Prove Euler's formula by induction on the number of vertices.*

Exercise 19.6.8. *Try to generalize Euler's formula. What if we have a three-dimensional solid?*

Exercise 19.6.9. *Prove that in a maximal planar graph each face is bounded by 3 edges.*

Exercise 19.6.10. *Use Kempe's argument to prove the Five Color Theorem.*

Exercise 19.6.11. *Prove the claim from Case 2 of the degree 4 case in Kempe's argument: The greater subgraph bounded by the cyclic Kempe chain must include three vertices adjacent to the reintroduced vertex.*

Birkhoff and the Modern Proof

Exercise 19.6.12. *Show that a configuration with rings of 2 faces is reducible.*

Exercise 19.6.13. *Show that a configuration with rings of 2 or 3 faces is reducible.*

Exercise 19.6.14. *Show the diamond configuration in Figure 12 is reducible.*

Chapter 20

The Kepler Conjecture

20.1. Introduction

At the turn of the twentieth century, the German mathematician David Hilbert published a list of 23 problems that he believed were the most important unsolved problems in mathematics. In this famous list of problems, Hilbert included a problem regarding "sphere packing". As part of his 18th problem, he wrote:

> I point out the following question, related to the preceding one, and important to number theory and perhaps sometimes useful in physics and chemistry: How can one arrange most densely in space an infinite number of equal solids of given form, e.g., spheres with given radii or regular tetrahedra with given edges (or in prescribed positions), that is, how can one so fit them together that the ratio of the filled to the unfilled space may be as great as possible?

This problem has its roots in a number of everyday applications. A grocer, for example, may want to stack oranges in a way that he or she can fit as many as possible, given a base of fixed surface area. A more antiquated example of sphere packing comes from the times of Johannes Kepler himself: stacking cannonballs in warships! This problem was stated formally for the first time, concerning the case of equal spheres, by Kepler in his 1611 paper *On the six-cornered snowflake* [**Kep**].

Kepler started studying various ways of packing or arranging spheres as a result of his correspondence with the English mathematician and astronomer Thomas Harriot in 1606. Harriot was a friend and assistant of Sir Walter Raleigh, who had set Harriot the problem of determining how best to stack cannonballs on the decks of his ships; Harriot published a study of various stacking patterns in 1591.

Harriot came across a packing that he called the **cannonball packing** (see Figure 1). Essentially, the scheme involved starting with a base of spheres such that each sphere touched four others, and then stacking each successive layer so that the spheres fit as deep as possible in indentations between the spheres in the

Figure 1. Thomas Harriot's Cannonball Packing Scheme. (From *Good Old-Fashioned Challenging Puzzles* (page 28), H. E. Dudeney, © Summersdale Publishers Ltd. 2007.)

preceding layer. It's also not hard to check that such an arrangement produces a packing density of about 74.05%. This arrangement, of course, is a very natural way of packing spheres and, perhaps unsurprisingly, is the same scheme used to stack spherical produce in supermarkets around the world. The packing scheme also pops up everywhere in nature, including in the packing of molecules into crystals. Chemists use the term **face-centered cubic** or **FCC** to describe this packing arrangement when talking about crystals.

While walking outside on a snowy day, Kepler noticed that all snowflakes fall with six corners before clumping into larger flakes. In his paper *On the six-cornered snowflake* he considered why snowflakes always fall this way rather than with five or seven corners. He proposed that this is because six corners form a hexagon, which is one of three regular shapes that perfectly tile the plane. He noted that cells in beehives share this six-cornered plan. He observed their three-dimensional arrangement, nestling between each other with rhombic walls. Using the rhombi from the bees, Kepler constructed the rhombic dodecahedron (see Figure 2), a shape able to entirely fill a three-dimensional space, similar to hexagons in the plane.

Figure 2. A rhombic dodecahedron (from Wikimedia Commons: Permission is granted to copy, distribute and/or modify this document under the terms of the GNU Free Documentation License, Version 1.2 or any later version published by the Free Software Foundation; and licensed under the Creative Commons Attribution-Share Alike 3.0 Unported (https://creativecommons.org/licenses/by-sa/3.0/deed.en) license).

Taking a space comprised entirely of rhombic dodecahedra and filling each shape with a spherical nucleus gives the FCC. On the basis of these observations, Kepler conjectured that no arrangement of equally sized spheres filling space has a greater average density than that of the face-centered cubic packing, and that the packing density of this arrangement is approximately $\pi/\sqrt{18}$ (or about 74.048%). This conjecture came to be known as **Kepler's conjecture**.

While the conjecture seemed intuitively clear, no proof was found for centuries despite an incredible amount of effort. Hilbert himself said "it seems 'obvious' that Kepler's conjecture is correct. Why is the gulf so large between intuition and proof?" Thomas Hales, the mathematician responsible for finally proving the Kepler conjecture, cited it as an example of how geometry taunts and defies us [**H1**]. "What about stacking tin cans," he said, "can anyone doubt that parallel rows of upright cans give the best arrangement? Could some disordered heap of cans waste less space? We say certainly not, but the proof escapes us. What is the shape of the cluster of three, four, or five soap bubbles of equal volume that minimizes total surface area? We blow bubbles and soon discover the answer but cannot prove it. Or what about the bee's honeycomb? The three-dimensional design of the honeycomb used by the bee is not the most efficient possible. What is the most efficient design?" He went on to say that the Kepler conjecture and many other results in mathematics might have been proved centuries ago if only our mathematical toolbag had matched our intuition in power.

The first significant step in proving Kepler's conjecture was taken by Carl Friedrich Gauss in 1831, over two hundred years after Kepler first formulated the problem. He proved that the Kepler conjecture is true if the spheres have to be arranged in a **regular lattice**. A regular lattice, in the case of equal spheres, can be thought of as an arrangement of spheres in $\mathbb{R}^3$ where the coordinates of the sphere centers form a subgroup of $\mathbb{R}^3$ that is isomorphic to $\mathbb{Z}^3$.

The proof takes only a few lines and requires no calculations. Gauss's proof was thought to be important at the time because it implied that any packing arrangement that disproved the Kepler conjecture would have to be an irregular one. This motivated some mathematicians to look at ways of proving the conjecture by eliminating all possible irregular arrangements. This, however, turned out to be an extremely difficult task, and this was one of the things that made the Kepler conjecture so hard to prove. In fact, there are irregular arrangements that are denser than the face-centered cubic packing over a small enough volume, but any attempt to extend these arrangements to fill a larger volume always reduces their density.

A typical strategy in mathematics is to gain insight into a hard problem by first tackling a simplified version. Norwegian mathematician Alex Thue did exactly this in 1890 when he solved the two-dimensional analog of Kepler's problem [**Th**]. The two-dimensional version of Kepler's conjecture asks for the densest packing of unit disks in the plane. If we inscribe a disk in each hexagon in the regular hexagonal tiling of the plane, the density of the packing is $\pi/\sqrt{12}$, or about 90.69%. His approach used Voronoi cells, a concept that we will visit in detail later in the chapter.

After the publication of Thue's proof, mathematicians tried several different approaches to proving the Kepler conjecture. Despite increased efforts by the mathematical community, it took another 100 years for the Kepler conjecture to finally be proved. Thomas Hales and his graduate student Samuel Ferguson gave a proof in 1998, following an approach described by the Hungarian mathematician László Fejes Tóth in 1953, which they used in conjunction with their own insights and an incredible amount of help from computers.

Kepler's conjecture, in principle, concerns the packing of an infinite number of spheres. Fejes Tóth, however, was able to show that the problem of determining the maximum density of all arrangements (both regular and irregular) could in fact be reduced to a finite (but very large) number of calculations [**To**]. This implied that a proof by exhaustion was, in principle, possible. Thomas Hales was inspired to attempt proving the conjecture based on the realization that a fast enough computer could turn Tóth's theoretical result into a practical approach to the problem.

The rest of the chapter is devoted towards describing the kind of difficulties that one encounters when attempting to prove Kepler's conjecture. We also explore how Hales and Ferguson were able to combine brilliant mathematical insights along with the powers of modern computing in their 1998 proof of Kepler's conjecture. See [**CS**] for an excellent and thorough introduction to sphere packing, and [**Coh**] for a summary of some recent breakthroughs.

20.2. Sphere Packing

The Kepler conjecture says that the face-centered cubic is the most efficient arrangement of equal spheres. Before we describe the difficulties encountered in proving Kepler's conjecture, it's worth some time exploring what packings are optimal and what exactly the FCC is packing.

A **packing of spheres** (or a **sphere packing**) is an arrangement of non-overlapping spheres of unit radius in Euclidean space. Each sphere is determined by its center, so the sphere packing is equivalently a collection of points separated by distances of at least 2. Our aim is to maximize the ratio of the volume of spheres in the space to the volume of the space.

Conjecture 20.2.1 (Kepler Conjecture). *Any packing Ω of unit spheres in $\mathbb{R}^3$ has* upper packing density

$$\bar{\rho}(\Omega) \ \leq \ \frac{\pi}{3\sqrt{2}} \ \approx \ 0.740480.$$

In particular, the face-centered cubic packing achieves this optimal density.

We'll consider in detail what we mean by upper packing density later; for now we concentrate on understanding the face-centered cubic packing, depicted in Figure 3.

Interestingly, there is another packing that is as efficient as the FCC, the **Hexagonal Closed Packing** or **HCP** (Figure 4). This is another natural packing and, just like the FCC, it's found in certain crystals. It turns out that HCP and FCC are closely related, and thus it's worth looking at both when talking about the Kepler conjecture.

Figure 3. Aerial view of the face-centered cubic packing (FCC). The spheres are placed in hexagonal patterns in rows, with the centers of one row directly above the holes of the previous.

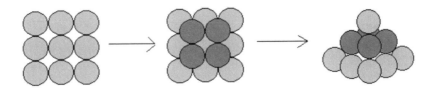

Figure 4. Hexagonal Close Packing (HCP) (image from [**H2**]).

One can imagine FCC as a pyramid with a triangular base and HCP as a pyramid with square base where each layer is placed such that the spheres rest in the indentations created by the preceding layer. The face-centered cubic packing can be thought of in the following manner. Think of space tiled with a lattice of cubes of edge length $\sqrt{8}$. Now, arrange sphere centers in and on the cube as follows: put a sphere center on each of the 8 corners of the cube, and add another six centers at the center of each of the 6 faces of the cube, as shown in Figure 5.

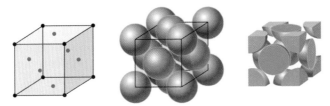

Face-centered cubic structure

Figure 5. Placement of Spheres in FCC Packing (image from [**KB**]; and licensed under the Creative Commons Attribution 4.0 International (CC BY 4.0) (https://creativecommons.org/licenses/by/4.0/) license).

The portions of the spheres placed at the corners contribute an eighth of a sphere. Each sphere center contributes half a sphere. Therefore, we arrive at a total of 4 spheres fitting inside of one cube. Now, we can calculate the optimal density from this arrangement. If we are looking at the unit sphere, the FCC arrangement has a cube of side length $\sqrt{8}$. The density of a packing is the ratio of the volumes of the spheres in the space to the volume of the space. Therefore, the density of the FCC packing is

$$\frac{4 \times 4\frac{\pi}{3}}{\sqrt{8}^3} = \frac{\pi}{\sqrt{18}}.$$

This density, computed locally over a cube, is equal to the optimal density predicted by the Kepler conjecture! Since FCC packing is uniform over the entire space, it's enough to find the density within one cube, and then expand that to the entire space [**H2**].

Just like the face-centered cubic packing, each sphere in the HCP packing touches 12 other spheres. This arrangement utilizes a hexagonal form, as sphere centers are arranged in a hexagonal pattern. A similar analysis shows that the HCP pattern also attains the density of the Kepler bound and hence it's also optimal. Now that we have a basic idea of what two of the most natural optimal packing arrangements look like, we can move on to discussing the problems that make the Kepler conjecture so hard to prove.

20.3. Challenges in Proving the Kepler Conjecture

The following section is influenced heavily by Jeffrey Lagarias' chapter *Bounds for Local Density of Sphere Packing and the Kepler Conjecture* in [**Lag3**].

A first difficulty is defining a rigorous notion of density for sphere packing. Up until now we have talked about density in mostly vague terms. The conjecture concerns the density of a packing arrangement that fills all of three-dimensional Euclidian space. At the same time, we have not yet mentioned how one can go about defining density of a packing. This problem of defining density for sphere packings is resolved using limiting notions of packing a large region and letting its diameter increase to infinity. As natural as it seems, this notion was not clarified until the 20th century. One can then prove a result asserting that a packing exists that attains a maximal limiting density.

A second difficulty is that these notions of limiting density are very crude in the sense that one can always remove spheres from an arbitrarily large finite region without changing the limiting density. Therefore we wish to impose further local restrictions on our notion of a "densest" packing. We require from our packing that it contain no large "holes". Hales and Ferguson formulated this restriction by defining a saturated packing of spheres to be one into which no new sphere can be inserted, i.e., there is no "hole" of diameter 2 or larger in the packing. Another related problem is that one might be able to increase density locally by removing a finite collection of spheres in a region and repacking that region to squeeze in one more sphere. This sort of condition is more difficult to analyze, but it already motivates the study of "local" conditions specifying density of a packing [**BBC**].

A third difficulty is that there exist uncountably many "optimally dense" packings that are, strictly speaking, different from each other. Here we consider packings as essentially different if they are not congruent under a Euclidean motion of space [**Lag3**]. Consider the packing that starts with a planar layer of hexagonally close packed spheres. That is, there is a planar slice through this layer that intersects all the sphere centers, giving a circle packing of the plane, and this packing is the optimal two-dimensional hexagonal circle packing. Now one can fit a second identical layer of spheres on top of this layer, so that the spheres fit as deep as possible in indentations between the spheres in the first layer. If we mark one sphere as the center of the first layer, there are two possible ways to do this: each such packing occupies 3 of the 6 holes formed by a hexagon, and there are two choices. One can repeat this in the layer below the first layer, and continue adding layers in this fashion, making a choice at each layer; see [**Lag3**] for an illustration. All such packings have the same limiting density. If the stacked layers are viewed vertically from above, then spheres in all the layers can be seen to line up in three possible positions of the hexagonal lattice, which we can label A, B, and C.

Therefore, the choices at each layer of the packing can be labeled using one of the letters from the set $\{A, B, C\}$, with no two consecutive letters the same, to uniquely label a packing with a doubly infinite string of such letters. It is easy to show that two packings are essentially the same (up to a Euclidean motion) if their strings of letters can be lined up in a way that they agree with each other. Furthermore, it's also not hard to see that the repeating pattern $ABCABCABC\ldots$ corresponds to the FCC packing while the HCP packing corresponds to a repeating pattern $ABABAB\ldots$ (or $ACACAC\ldots$ or $BCBCBC\ldots$). The FCC packing, as we said before, was found by Kepler, while the HCP packing was first described by English civil engineer William Barlow in 1883 [**Ba**]. The repeating patterns and our two corresponding packing schemes can be seen in Figure 6.

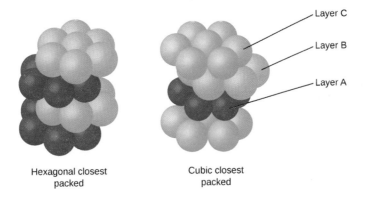

Figure 6. Repeating layers in FCC and HCP (image from [**KB**]; and licensed under the Creative Commons Attribution 4.0 International (CC BY 4.0) (https://creativecommons.org/licenses/by/4.0/) license).

Before we move on to other difficulties, we need to discuss the concept of Voronoi cells. The collection of "optimally dense" packings that we just described are actually locally optimal in a much stronger sense.

A **Voronoi cell** around a given sphere center is the set of all points in space closer to that sphere center than to any other sphere center. In the case of the packing arrangements using hexagonal packing layers that we just described, all Voronoi cells have one of two shapes, each having 12 faces and 14 vertices. Each of these shapes is a type of dodecahedron (see [**Lag3**] for an illustration).

In the FCC packing, all Voronoi cells are of type 1. In the HCP packing, however, all Voronoi cells are of type 2. In all of the other remaining packings that were described, Voronoi cells of both type occur. Two surprising results hold.

Firstly, each packing can be uniquely described by the different arrangements of type 1 and 2 Voronoi cells. Furthermore, for each of these Voronoi cells, the ratio of the volume of the sphere to the total volume of the cell containing it is exactly $\pi/\sqrt{18}$! Therefore, in the case of these 'special' packings, optimality holds locally as well as globally.

Another difficulty is posed by the fact that the optimization problem is essentially infinite dimensional. There are an infinite number of spheres to pack and therefore there are an infinite number of coordinates that need to be determined. The approaches that are aimed at dealing with this issue of infinite-dimensionality seek to prove stronger results which only involve finite-dimensional optimizations that encode local conditions. This is called the **local density inequality approach**. The underlying idea is to assign, by some recipe, to each sphere in a sphere packing a (weighted) sum of the covered and uncovered volume near that sphere center. This recipe is said to be "local" because the weighted sum for a given sphere center is completely determined by the locations of all spheres in the sphere packing nearby, within a fixed distance C of the given sphere center. When the recipe quantities are added up over all spheres in a given sphere packing, it should count all volume with weight 1. As a result, an upper bound on the weighted area will give an upper bound on global sphere packing density. We call such a local density inequality "optimal" if it produces the Kepler upper bound for density of sphere packing, i.e., $\pi/\sqrt{18}$.

A follow-up issue is that of designing candidate local density inequalities that may be "optimal". Due to the work of Fejes Tóth, Hales knew that if an optimal local inequality does exist, then it can in principle be verified by a finite computation, i.e., it comprises a finite-dimensional non-linear optimization problem over a large number of variables, specifying the possible locations of sphere centers in a ball of radius C around a given sphere center. This was where computers came to the fore under the Hales-Ferguson program. Designing local density inequalities that exist and make the computations finite was also one of the most important parts of the Hales-Ferguson proof. Their "formulation" paper, in which they describe how to correctly partition the space and how to "score" the local density measure, ran almost 50 pages!

Finally, the hardest challenge to overcome for "optimal" local density inequalities is the absolutely massive computational size of the problem. Each local density inequality, although finite, is an enormously complicated non-linear optimization problem. The size of the problem makes it computationally unfeasible. The problems run over 150 variables on average, and without any additional structure they are too large to solve. To reduce the needed calculations, the Hales-Ferguson formulation of the Kepler conjecture allowed the local density inequalities to be complicated and inelegant.

20.4. Local Density Inequalities

The purpose of this section is to give a sense of the techniques and notions used by Hales and Ferguson (see [**H4**]), and how they went about designing the required local density inequalities. We follow Lagarias' chapter *Bounds for Local Density of Sphere Packings and the Kepler Conjecture* from [**Lag3**].

20.4.1. Density Definitions. We begin by using local density inequalities to partition and then score the space of each packing. As many of the arguments apply in n-dimensions just as well as in 3, we argue as generally as possible as long as possible. Set

$$B_n := B_n(0;1) = \{x \in \mathbb{R}^n : ||x|| \le 1\};$$

this is the **unit n-sphere** centered at the origin. For a packing, let Ω be the set of all the sphere centers in it; note each center has a distance of at least two from all other vertices in the set. By $\Omega + B_n$ we mean the region covered by unit spheres where their centers are the points in Ω. If S is a subset of $\mathbb{R}^n$, we study the amount of an S covered by the packing $\Omega + B_n$, which gives rise to the **density** $\rho(S)$ given by

$$\rho(S) := \frac{\text{vol}(S \cap (\Omega + B_n))}{\text{vol}(S)}.$$

For a $T > 0$, the **upper density** is the supremum of the density of a packing Ω over all n-cubes with side-length T, defined by

$$\overline{\rho}(\Omega, T) := \sup_{x \in \mathbb{R}^n} \rho([0,T]^n + x).$$

Taking the limit as T goes to infinity gives the **upper packing density** of a packing Ω:

$$\overline{\rho}(\Omega) := \limsup_{T \to \infty} \overline{\rho}(\Omega, T).$$

Finally, the **sphere packing density** is the supremum of the upper packing densities across all configurations of sphere centers:

$$\delta(B_n) := \sup_{\Omega} \overline{\rho}(\Omega).$$

To bound sphere packing densities it suffices to study the saturated packings described earlier; remember a sphere packing Ω is **saturated** if no new spheres can be added to it (i.e., there are no holes in it). Through the rest of this chapter we assume that all packings are saturated unless stated otherwise.

20.4.2. Partitions and Scoring. Since Hales' proof involves taking an infinite problem and making it smaller and finite, we turn our focus to partitions, rules that divide space into finite pieces that we can individually analyze. We begin by clarifying what we mean by a partition.

For a saturated sphere packing $\Omega \in R^n$, an **admissible partition rule** is a collection of closed sets,

$$\mathcal{P}(\Omega) := \{R_\alpha = R_\alpha(\Omega)\},$$

with the following three properties:

1. *Partition*: Each R_α is a finite union of bounded convex polyhedra. The sets R_α cover $\mathbb{R}^n$ and have pairwise-disjoint interiors. This guarantees that when the partition is applied no volume of space is part of two different regions.

2. *Locality*: There exists some positive constant C such that each region R_α has a diameter less than or equal to C, and each R_α is completely determined by the set of sphere centers $w \in \Omega$ with w a distance at most C from R_α. This means that while there exist infinitely many sphere centers outside of R_α, we

are only concerned with those whose centers are within distance C. This is one way that we begin to step down from infinity.

3. *Translation Invariance*: For any R_α that is centered at a point x, if we move its center to y, then it should still be part of the same partition. This means that a region should be treated the same no matter where it appears in space.

Now that we have a way of partitioning the space with admissible partitions, the next step is to create a function that determines the density of an admissible partition. Such a function is called an **admissible scoring function** and assigns to each R_α in a partition of Ω and each sphere center v in Ω a real weight, defined as

$$\sigma(R_\alpha, v).$$

For an admissible scoring function we will need the following:

1. *Weighted Density Average*: There are positive constants A and B, independent of Ω, such that for each set R_α:

$$\sum_{v \in \Omega} \sigma(R_\alpha, v) \ = \ (A\rho(R_\alpha) - B)\mathrm{vol}(R_\alpha),$$

where the value

$$\rho(R_\alpha)\mathrm{vol}(R_\alpha) \ = \ \mathrm{vol}(R_\alpha \cap (\Omega + B_n))$$

is a rearrangement of the general density equation from earlier. The weighted density average measures the volume covered in R_α by the sphere packing Ω with unit spheres.

2. *Locality*: Again, as with partitions, we only want to consider how spheres within a certain radius affect the density, not all spheres in the entire packing. Therefore we determine another absolute constant C such that each value $\sigma(R_\alpha, v) = 0$ if the distance between v and R_α is at most C. This means that a sphere center will not contribute to the sum of the overall score if it is sufficiently far away (distance C) from a region.

3. *Translation Invariance*: Also similar to partitions, if there is a repetition of a region R_α in space, we want it to have the same score in both places. A region's score should not be a function of its absolute position in space.

One thing to notice is that the definition for an admissible scoring function allows for negative weights. This is intentional because the configuration of spheres in one region of a packing may detract from the overall average density, while the spheres of another region may increase it, and we want the score to reflect this.

The locality of any sphere center is not locked into a single R_α. A given sphere center v will be within a distance C of multiple R_α's, so to accurately determine the overall score average we need to define another function, the **vertex decomposition star**, or D-star, denoted $\mathcal{D}(v)$, that considers all sets R_α in a partition rule such that the score of the region relative to that center is not equal to zero:

$$\mathrm{Score}(\mathcal{D}(v)) \ := \ \sum_{R_\alpha \in \mathcal{D}(v)} \sigma(R_\alpha, v).$$

So, the score assigned to a $\mathcal{D}(v)$ at some v in Ω is the sum of all scores for that vertex v for all regions associated to $\mathcal{D}(v)$. Combining our partition and scoring rules, we can get a value for the density of a sphere packing Ω, defined by θ as a function of A and B:

$$\theta := \theta_{\mathcal{P},\sigma}(A, B) := \sup_{\Omega \text{ saturated}} \left(\sup_{v \in \Omega} \operatorname{Score}(\mathcal{D}(v)) \right).$$

This function states that the maximum density of sphere packings Ω for a given partition $\mathcal{P}$ is the supremum over all saturated sphere packings of the supremum of all vertex D-star scores over all sphere centers in Ω. When choosing A and B we can be sure that $\theta < K_n A$, where K_n is the volume of the unit n-sphere. It follows that a maximum sphere density satisfies

$$\delta(B) \leq \frac{K_n B}{K_n A - \theta}.$$

A packing is optimal if these two values are equal. In Hales and Ferguson's proof they use $\delta(B_3) = \pi/\sqrt{18}$ (or about 74.048%), the densities of the FCC and HCP. An optimal density inequality is one achieving this value. For choosing A and B, Hales and Ferguson's score constant ratio is

$$\frac{B}{A} = \delta_{\text{opt}} = \frac{-3\pi + 12 \arccos{(1/\sqrt{3})}}{\sqrt{8}} \approx 0.720903,$$

which results from the decompositions they choose for partitioning. Thus, the problem becomes searching over all possible configurations of sphere centers and determining valid partitions. Each sphere center has 3 degrees of freedom, and the number of sphere centers involved in determining each vertex D-star is around 50, so the search space has roughly 150 dimensions! This quick sketch of the proof captures the massive computational size of the problem.

20.5. Computer-Aided Proof

Building on Fejer Tóth's ideas, Hales and Ferguson found a way of breaking up an infinite-dimensional optimization problem into approximately 150 dimensional "pieces". The final result, however, still depended on a strategy of proof by exhaustion. All of the work done by Hales and Ferguson would have been of little to no use had computers not been around. Throughout the proof, computers were used in various significant ways. Some of the most important ways in which computers helped prove the Kepler conjecture are the discussed briefly here.

- *Proof of inequalities via interval arithmetic:* A key element of the Hales-Ferguson proof was the use of computers to do **interval arithmetic** [**H3**]. The goal was to reduce very large computations to relatively small calculations and to ensure that there are no errors. Computers were used to prove various inequalities in a small number of variables using interval arithmetic. The computers produced upper and lower bounds for a desired value, the score function in our case, in significantly less time than it would have taken to compute an exact result. This significantly reduced the computational time needed to prove all the inequalities needed for the proof. Explicitly, the goal is to show that any packing other than FCC or HCP has a lower density. All we need to do is for each packing to compute an interval containing its true density; if the density of the FCC and HCP is above that interval, we know this packing cannot be optimal. If, however, it's in the interval, then we must redo the calculation in greater detail, hopefully reducing the size of the interval sufficiently to show that the density of FCC and HCP is larger.

- *Combinatorics and Numerical Optimization:* A computer program classified all the planar maps relevant to the Kepler conjecture. If done manually, this task would have been intractable because of the very large number of maps. The exploration of the problem was also substantially aided by non-linear optimization and various symbolic mathematical packages.

- Linear Programming Bounds: This was perhaps the part of the proof where the help of computers was of greatest importance, and the main reason we have included a discussion of this topic. Hales and Ferguson reduced the problem to a finite number of finite-dimensional problems. In practice, however, these problems were still out of the reach of humans because they involved about 150 variables and because of the very large number of individual problems. Thankfully, modern computers were able to save the day yet again, just like they did with the Four Color Theorem. Many of the non-linear optimization problems for the scores of decomposition stars are replaced by linear problems that dominate the original score. They then used computers to solve Linear Programming problems. A typical one has between 100 and 200 variables and 1000 and 2000 constraints. Nearly 100,000 such problems enter into the proof [**Lag3**]. When linear programming methods do not give sufficiently good bounds, they have been combined with branch and bound methods from global optimization. In addition to all of the computations, optimizations, and combinatorics, computers were also employed by Hales and Ferguson to organize their output. The organization of the few gigabytes of code and data that enter into the proof was in itself a non-trivial undertaking.

- Theorem Proving Software: When Hales announced his proof for the Kepler conjecture, people were very skeptical because of the nature of his proof. There was concern regarding the reliability of a proof involving heavy use of computers and large amounts of computations. A panel of 12 referees worked on reviewing the paper for over 4 years; they reported that the panel was "99% certain" of the correctness of the proof, but they could not certify the correctness of all of the computer calculations. Interestingly, this panel of referees was headed by Gábor Fejes Tóth, the son of László Fejes Tóth. This reception inspired Hales to find a "formal" proof that could be verified using a computer. A proof is said to be **formal** if it is a finite sequence of sentences,

```
theorem sqrt2_not_rational:
  "sqrt (real 2) ∉ ℚ"
proof
  let ?x = "sqrt (real 2)"
  assume "?x ∈ ℚ"
  then obtain m n :: nat where
    sqrt_rat: "¦?x¦ = real m / real n" and lowest_terms: "coprime m n"
    by (rule Rats_abs_nat_div_natE)
  hence "real (m^2) = ?x^2 * real (n^2)" by (auto simp add: power2_eq_square)
  hence eq: "m^2 = 2 * n^2" using of_nat_eq_iff power2_eq_square by fastforce
  hence "2 dvd m^2" by simp
  hence "2 dvd m" by simp
  have "2 dvd n" proof-
    from ⟨2 dvd m⟩ obtain k where "m = 2 * k" ..
    with eq have "2 * n^2 = 2^2 * k^2" by simp
    hence "2 dvd n^2" by simp
    thus "2 dvd n" by simp
  qed
  with ⟨2 dvd m⟩ have "2 dvd gcd m n" by (rule gcd_greatest)
  with lowest_terms have "2 dvd 1" by simp
  thus False using odd_one by blast
qed
```

Figure 7. A 'formal' proof by contradiction for the irrationality of $\sqrt{2}$ (from https://en.wikipedia.org/wiki/Isabelle_(proof_assistant); licensed under the Creative Commons Attribution-Share Alike 3.0 Unported (https://creativecommons.org/licenses/by-sa/3.0/deed.en) license).

each of which is an axiom or follows from the preceding sentences in the sequence by a rule of inference; the last sentence in the sequence is a theorem of a formal system.

In January of 2003, Hales announced that he was starting a collaborative project to produce a complete formal proof of the Kepler conjecture. He wanted to remove any remaining uncertainty about the validity of his proof by creating a formal proof that can be checked by an automated proof checking software such as HOL Light or Isabelle (see Figure 7 for an example proof). This project was called **Flyspeck** - the F, P, and K standing for Formal Proof of Kepler. Though Hales initially estimated that producing a complete formal proof would take around two decades of work, surprisingly the project was announced completed on August 10, 2014.

20.6. Exercises

Introduction

Exercise 20.6.1. *Consider a sphere of radius 1, centered in a cube of side length 2. What fraction of the cube's volume is occupied by the sphere?*

Exercise 20.6.2. *Generalize the previous problem and consider a d-dimensional sphere of radius 1 inside a hypercube of side length 2. What is the hyper-volume of the sphere, and what fraction of the hypercube is taken up by it? What happens to this ratio as $n \to \infty$?*

Exercise 20.6.3. *Redo the previous problem, but now calculate the ratio of the surface hyper-area of the d-dimensional sphere of radius 1 to that of the d-dimensional hypercube of side length 2.*

Exercise 20.6.4. *Compute the density of the face-centered cubic packing.*

Exercise 20.6.5. *Three regular n-gons perfectly tile the plane: the equilateral triangle, a square, and a regular hexagon. Consider one of each shape, where the dimensions are chosen so that each has unit area. Clearly the packing density of all three is 100%; compute how the perimeter of the packing grows in each case (remembering to include both internal and external walls). Does your answer depend on how you choose a region of area tending to infinity?*

Exercise 20.6.6. *Prove Gauss' result, namely that if the spheres must be arranged in a regular lattice, then the FCC packing has the greatest density.*

Sphere Packing

Exercise 20.6.7. *In 3-dimensions the FCC and the HCP have the same density, and thus there is not a unique optimal packing in three dimensions. Prove or disprove: there is a unique optimal packing in 2-dimensions.*

The **kissing number** is the maximum number of disjoint unit spheres that can simultaneously touch a fixed unit sphere.

Exercise 20.6.8. *Prove the kissing number is 2 for one-dimensional spheres (which are just line segments) and 6 for two-dimensional spheres (which are just circles).*

Exercise 20.6.9. *Newton thought the kissing number in 3-dimensions was 12, while Gregory thought it was 13. While Newton is right, this result resisted proof for centuries too. Prove elementarily that this is a finite number, and get the best bound you can.*

Exercise 20.6.10. *Generalize the previous problem and show that there is a non-decreasing function $f(d)$ such that the kissing number in d-dimensions is at most $f(d)$. Bound the growth rate of f.*

Local Density Inequalities

Exercise 20.6.11. *Let $\Omega = \{(3i, 3j, 3\ell) : i, j, \ell \in \mathbb{Z}\}$. Determine $\overline{\rho}(\Omega)$.*

Exercise 20.6.12. *Let $\Omega = \{a_1 + a_2\sqrt{3}, b_1 + b_2\sqrt{3}, 2c : a_1, a_2, b_1, b_2, c \in \mathbb{Z}\}$. Determine $\overline{\rho}(\Omega)$.*

Exercise 20.6.13. *Find the score for the Ω's from the previous two problems.*

Bibliography

[ANW]   J. H. Ahlberg, E. N. Nilson and J. L. Walsh, *The Theory of Splines and Their Applications*, Mathematics in Science and Engineering **38** (1967), Academic Press, New York.

[AZ]   M. Aigner and G. M. Ziegler, *Proofs from THE BOOK* (fourth edition), Springer-Verlag, Berlin, 2010.

[AGP]   W. R. Alford, A. Granville, and C. Pomerance, *There are infinitely many Carmichael numbers*, Ann. Math. **139** (1994), 703–722.

[AH1]   K. Appel and W. Haken, *The solution of the four-color-map problem*, Sci. Amer. **237** (1977), 108–121.

[AH2]   K. Appel and W. Haken, *Every planar map is four colorable. Part I: Discharging*, Illinois Journal of Mathematics **21** (1977), no. 3, 429–490.

[AHK]   K. Appel, W. Haken, and J. Koch, *Every planar map is four colorable. Part II: Reducibility*, Illinois Journal of Mathematics **21** (1977), no. 3, 491–567.

[Ar]   J. S. Arora, *Introduction to Optimum Design, 4th Edition*, Academic Press, Inc., 2016.

[BFMT-B] O. Barrett, F. W. K. Firk, S. J. Miller and C. Turnage-Butterbaugh, *From Quantum Systems to L-Functions: Pair Correlation Statistics and Beyond*, in Open Problems in Mathematics (editors John Nash Jr. and Michael Th. Rassias), Springer-Verlag, 2016. Available online at https://arxiv.org/pdf/1505.07481.pdf.

[Ba]   W. Barlow, *Probable nature of the internal symmetry of crystals*, Nature **29** (1883), no. 738, 186–188.

[BC]   A. D. Belegundu and T. R. Chandrupatla, *Optimization Concepts and Applications in Engineering*, Prentice-Hall, 1999.

[B-AC]   I. Ben-Ari and K. Conrad, *Maclaurin's Inequality and a Generalized Bernoulli Inequality*, Math. Mag. **87** (2014), 14–24.

[BBC]   A. Bezdek, K. Bezdek and R. Connelly, *Finite and uniform stability of sphere packings*, Disc. & Comput. Geom. **20** (1998), no. 1, 111–130.

[Bi]   G. D. Birkhoff, *The reducibility of maps*, Amer. J. Math. **70** (1913), 114–128.

[BoDi]   W. Boyce and R. DiPrima, *Elementary differential equations and boundary value problems*, 7th edition, John Wiley & Sons, New York, 2000.

[BP]   S. Butenko and P. M. Pardalos, *Numerical Methods and Optimization: An Introduction*, Chapman & Hall / CRC Numerical Analysis and Scientific Computing Series, 2014.

[Cl]   J. Clausen, *Branch and Bound Algorithms – Principles and Examples*, notes posted for course DM63 - metaheuristics Fall 2005, http://www.imada.sdu.dk/~jbj/heuristikker/ (handout available at http://www.imada.sdu.dk/~jbj/heuristikker/TSPtext.pdf).

[Coh]   H. Cohn, *A Conceptual Breakthrough in Sphere Packing*, Notices of the AMS **64** (2017), no. 2, 102–115.

[CS]   J. Conway and N. J. A. Sloane, *Sphere Packings, Lattices and Groups*, Springer-Verlag, New York, 1999.

[CM]   M. Cozzens and S. J. Miller, *The Mathematics of Encryption: An Elementary Introduction*, AMS Mathematical World Series **29**, Providence, RI, 2013, 332 pages.

[Da1]   G. B. Dantzig, *Linear Programming and Extensions*, Princeton University Press, 1963.

[Da2]   G. Dantzig, *The Diet Problem*, Interfaces **20** (1990), no. 4, The Practice of Mathematical Programming, pp. 43–47. Published by: INFORMS. http://dl.dropboxusercontent.com/u/5317066/1990-dantzig-dietproblem.pdf.

[DaTh]   G. Dantzig and M. N. Thapa, *Linear Programming 1: Introduction*, Springer Series in Operations Research and Financial Engineering, v. 1, 1997.

[Da]   H. Davenport, *Multiplicative Number Theory*, 2nd edition, revised by H. Montgomery, Graduate Texts in Mathematics, Vol. 74, Springer-Verlag, New York, 1980.

[Dav]   S. Davies, *The Four Color Problem*, blog post on "With High Probability", posted on 11 July 2016, accessed on 20 December 2016: https://samidavies.wordpress.com/2016/07/11/the-four-color-problem/.

[Dev]   R. Devaney, *An Introduction to Chaotic Dynamical Systems*, 2nd edition, Westview Press, Cambridge, MA, 2003.

[DP]   M. Devlin and S. Payne, *Discrete bidding games*, Electronic J. Combin. **17** (2010), no. 1, Research Paper 85, 40 pp.

[Edg]   G. Edgar, *Measure, Topology, and Fractal Geometry*, 2nd edition, Springer-Verlag, 1990.

[EPPS]   D. Eppstein, *Twenty Proofs of Euler's Formula: $V - E + F = 2$*, post at "The Geometry Junkyard", accessed 16 December 2016: https://www.ics.uci.edu/~eppstein/junkyard/euler/.

[Fal]   K. Falconer, *Fractal Geometry: Mathematical Foundations and Applications*, 2nd edition, John Wiley & Sons, New York, 2003.

[FL]   R. Feynman and R. Leighton, *Surely you're joking, Mr. Feynman*, W. W. Norton & Company, New York, 1985.

[FH-E]   F. Frick and K. Houston-Edwards, *Achieving Rental Harmony With A Secretive Roommate*, https://arxiv.org/pdf/1702.07325.pdf. See also the video at https://www.youtube.com/watch?v=48oBEvpdYSE.

[Fr]   J. N. Franklin, *Mathematical Methods of Economics: Linear and Nonlinear Programming, Fixed-Point Theorems*, Springer-Verlag, New York, 1980.

[Ga]   J. Gallian, *Contemporary Abstract Algebra*, Brooks Cole, seventh edition, 2009.

[G]   G. Gonthier, *Formal proof – the four-color theorem*, Notices of the AMS **55** (2008), no. 11, 1382–1393.

[GKT]   B. Guenin, J. Könemann and L. Tuncel, *A Gentle Introduction to Optimization*, Cambridge University Press, 2014.

[GS]   D. Gale and L. S. Shapley, *College Admissions and the Stability of Marriage*, American Mathematical Monthly **69** (1962), 9-14. https://www.jstor.org/stable/2312726?seq=1.

[H1]   T. C. Hales, *Cannonballs and honeycombs*, Notices Amer. Math. Soc. **47** (2000), 440–449.

[H2]   T. C. Hales, *Dense Sphere Packings: A Blueprint for Formal Proofs*, London, Cambridge University Press (2012).

[H3]   T. C. Hales, *Some algorithms arising in the proof of the Kepler conjecture* (2012). https://arxiv.org/pdf/math/0205209v1.pdf.

[H4]   T. C. Hales, *A proof of the Kepler conjecture*, Annals of Mathematics, **162** (2005), 1065–1185.

[Hi]   D. Hilbert, *Mathematical Problems*, Bulletin of the American Mathematical Society. **8**, no. 10 (1902), pp. 437–479. Earlier publications (in the original German) appeared in Göttinger Nachrichten, 1900, pp. 253–297, and Archiv der Mathematik und Physik, vol. 1 (1901), pp. 44–63, 213–237.

[JL-B] A. X. Jiang and Kevin Leyton-Brown, *A Tutorial on the Proof of the Existence of Nash Equilibria*, University of British Columbia, accessed 8 December 2016. Available online at http://www.cs.ubc.ca/~jiang/papers/NashReport.pdf.

[Kem]   A. B. Kempe, *On the geographical problem of the four colors*, Amer. J. Math. **2** (1879), 193–200.

[Kep]   J. Kepler, *Strena Seu de Niue Sexangula*, published by Godfrey Tampach at Frankfort on Main, 1611. English Translation in: J. Kepler, "The Six-Cornered Snowflake" (Colin Hardie, Translator), Oxford University Press, Oxford 1966.

[KB]   J. A. Key and D. W. Ball, *Introductory Chemistry – 1st Canadian Edition*, 2011: https://opentextbc.ca/chemistry/chapter/10-6-lattice-structures-in-crystalline-solids.

[Kh]   S. Khan, *Prisoners' Dilemma and Nash Equilibrium*, Khan Academy, accessed 8 December 2016. Available online at https://youtu.be/UkXI-zPcDIM.

[KM]   V. Klee and G. J. Minty, *How good is the simplex algorithm?*, Inequalities III (Proceedings of the Third Symposium on Inequalities held at the University of California, Los Angeles, California, September 1–9, 1969, dedicated to the memory of Theodore S. Motzkin), Academic Press (1972), New York-London, 159–175.

[KonMi] A. Kontorovich and S. J. Miller, *Benford's law, values of L-functions and the $3x + 1$ problem*, Acta Arith. **120** (2005), 269–297.

[KonSi] A. Kontorovich and Ya. G. Sinai, *Structure theorem for (d, g, h)-maps*, Bull. Braz. Math. Soc. (N.S.) 33 (2002), no. 2, 213–224.

[Ku]   S. Kuhn, *Prisoner's Dilemma*, Stanford Encyclopedia of Philosophy, Stanford University, 4 September 1997, accessed 8 December 2016. Available online at https://plato.stanford.edu/entries/prisoner-dilemma/.

[Lag1] J. Lagarias, *The $3x + 1$ problem and its generalizations*. Pages 305–334 in *Organic mathematics (Burnaby, BC, 1995)*, CMS Conf. Proc., vol. 20, AMS, Providence, RI, 1997.

[Lag2]   J. Lagarias, *The Ultimate Challenge: The 3x+1 problem*, American Mathematical Society, Providence, RI 2010.

[Lag3]   J. C. Lagarias, *Kepler Conjecture: The Hales-Ferguson Proof*, Springer-Verlag, New York, 2011.

[LagSo]   J. Lagarias and K. Soundararajan, *Benford's Law for the 3x + 1 function*, J. London Math. Soc. (2) **74** (2006), no. 2, 289–303.

[La]   S. Lang, *Complex Analysis*, Graduate Texts in Mathematics, Vol. 103, Springer-Verlag, New York, 1999.

[Le]   O. Leward, *Four-Color Map Theorem*, Uppsala University (2014). `https://uu.diva-portal.org/smash/get/diva2:749857/FULLTEXT01.pdf`.

[LW]   S. Loepp and W. K. Wootters, *Protecting Information: From classical error correction to quantum cryptography*, Cambridge University Press, 2006.

[Mar]   J. Martin, *The Notorious 4 Color Problem*, presentation at KU Mini College, June 2013: `http://people.ku.edu/~jlmartin/MiniCollege2013/handout.pdf`.

[Ma]   R. M. May, *Simple mathematical models with very complicated dynamics*, Nature **261** (1976), 459–467.

[Me]   A. Messac, *Optimization in Practice with MATLAB*, Cambridge University Press, 2015.

[Met]   N. Metropolis, *The beginning of the Monte Carlo method*, Los Alamos Science, No. 15, Special Issue (1987), 125–130.

[MU]   N. Metropolis and S. Ulam, *The Monte Carlo method*, J. Amer. Statist. Assoc. **44** (1949), 335–341.

[M]   S. J. Miller, *The Probability Lifesaver*, Princeton University Press, to appear.

[MM]   S. J. Miller and D. Montague, *Rational irrationality proofs*, Mathematics Magazine **85** (2012), no. 2, 110–114.

[MS]   S. J. Miller and C. Silva, *If a prime divides a product...*, preprint 2015. `http://arxiv.org/pdf/1012.5866v1`.

[MT-B]   S. J. Miller and R. Takloo-Bighash, *An Invitation to Modern Number Theory*, Princeton University Press, Princeton, NJ, 2006, 503 pages.

[MiPe]   R. Minton and T. J. Pennings, *Do dogs know bifurcations?*, College Math. J. **38** (2007), no. 5, 356–361. `http://www.jstor.org/stable/27646534`.

[Mo]   F. Morgan, *Kepler's Conjecture and Hale's Proof*, A Book Review, Notices of the American Math. Society **52** (2005), no. 1.

[Mu]   R. Murty, *Primes in certain arithmetic progressions*, Journal of the Madras University, (1988), 161–169.

[Na]   J. Nash, *Non-Cooperative Games*, Ph.D. Thesis, Princeton University, May 1950. Available online at `http://rbsc.princeton.edu/sites/default/files/Non-Cooperative_Games_Nash.pdf`.

[vNM]   J. von Neumann and O. Morgenstern, *Theory of Games and Economic Behavior*, Princeton University Press, 1944. Available online at `http://muse.jhu.edu/book/31089`.

[Pe]   T. J. Pennings, *Do dogs know calculus?*, College Math. J. **34** (2003), no. 3, 178–182. `http://www.jstor.org/stable/3595798`.

[Ra]   S. S. Rao, *Engineering Optimization, 4th Edition*, Wiley, 2009.

[RSST]   N. Robertson, D. P. Sanders, P. D. Seymour and R. Thomas, *The four-colour theorem*, J. Combin. Theory Ser. B **70** (1997), 2–44.

[Rud1]   W. Rudin, *Principles of Mathematical Analysis*, 3rd edition, International Series in Pure and Applied Mathematics, McGraw-Hill, New York, 1976.

[Rud2]   W. Rudin, *Real and complex analysis*, 3rd edition, McGraw-Hill, New York, 1986.

[SH]   S. Sen and J. L. Higle, *An Introductory Tutorial on Stochastic Linear Programming Models*, INTERFACES **29** (1999), no. 2, 33–61.

[ShSt]   R. H. Shumway and D. S. Stoffer, *Time Series Analysis and Its Applications*, 4th edition, Springer Texts in Statistics (2009).

[Sil]   J. Silverman, *A friendly introduction to number theory*, Pearson Prentice Hall, 2006.

[Si]   Ya. G. Sinai, *Statistical $(3x+1)$ problem*, Communications in Pure and Applied Mathematics **56** (2003), no. 7, 1016–1028.

[Sp]   W. Spaniel, *Rock-Paper-Scissors*, Game Theory 101, 26 November 2011, accessed 8 December 2016. Available online at https://youtu.be/-1GDMXoMdaY.

[SS]   E. Stein and R. Shakarchi, *Complex Analysis*, Princeton University Press, Princeton, NJ, 2003.

[Ste]   J. Steinberger *An unavoidable set of D-reducible configurations*, Transactions of the American Mathematical Society **362** (2010), no. 12, 6633–6661.

[St]   G. Stigler, *The cost of subsistence*, The Journal of Farm Economics, 1945.

[Su]   F. Su, *Rental Harmony: Sperner's Lemma in Fair Division*, Amer. Math. Monthly **106** (1999), 930–942.

[TaWi]   R. Taylor and A. Wiles, *Ring-theoretic properties of certain Hecke algebras*, Ann. Math. **141** (1995), 553–572.

[Th]   A. Thue, *Über die dichteste Zuzammenstellung von kongruenten Kreisenin der Ebene*, Christiana Vid. Selsk. Skr. **1** (1910), 1–9.

[To]   L. Fejes Tóth, *Lagerungen in der Ebene, auf der Kugel und im Raum, Die Grundlehren der Mathematischen Wissenschaften in Einzeldarstellungen mit besonderer Berúcksichtigung der Anwendungsgebiete*, Band LXV, Berlin, New York: Springer-Verlag Publishing.

[Ty]   T. Tymoczko, *The four-color problem and its philosophical significance*, The Journal of Philosophy **76** (1979), no. 2, 57–83.

[Va]   G. N. Vanderplaats, *Multidiscipline Design Optimization*, VR&D, 2007.

[Wiki]   Wikipedia, *Dynamic Programming*, Wikimedia Foundation, accessed on 20 December 2016, https://en.wikipedia.org/wiki/Dynamic_programming.

[Wi]   A. Wiles, *Modular elliptic curves and Fermat's last theorem*, Ann. Math. **141** (1995), 443–551.

Index

Published Titles in This Series